Lecture Notes in Mathematics 1594

M. Green J. Murre C. Voisin

Algebraic Cycles and Hodge Theory

Lectures given at the 2nd Session of the Centro Internazionale Matematico Estivo (C.I.M.E.) held in Torino, Italy, June 21-29, 1993

Editors: A. Albano, F. Bardelli

Fondazione
C.I.M.E.

Springer-Verlag
Berlin Heidelberg New York
London Paris Tokyo
Hong Kong Barcelona
Budapest

Authors

Mark L. Green
Department of Mathematics
U.C.L.A.
Los Angeles, CA 90024, USA

Jacob P. Murre
Department of Math. and Comp. Science
Leiden University
P. O. Box 9512
2300 RA Leiden, The Netherlands

Claire Voisin
Université de Paris-Sud, Centre d'Orsay
Mathématique, Bâtiment 425
F-91405 Orsay Cedex, France

Editors

Fabio Bardelli
Dipartimento di Matematica
Università di Pisa
Via Buonarroti, 2
I-56127 Pisa, Italy

Alberto Albano
Dipartimento di Matematica
Università di Torino
Via Carlo Alberto, 10
I-10123 Torino, Italy

Mathematics Subject Classification (1991): 14C25, 14C30, 14D07, 19E15, 32J25

ISBN 3-540-58692-X Springer-Verlag Berlin Heidelberg New York

CIP-Data applied for

Printed in Germany

Typesetting: Camera ready by author
SPIN: 10997406 46/3111-54321 - Printed on acid-free paper

FOREWORD

The Second 1993 C.I.M.E. Session "Algebraic Cycles and Hodge Theory" was held at Villa Gualino, Torino, from June 21 to June 29, 1993.

There were three series of main lectures and some seminars: this volume contains the texts of the three series of main lectures and of those seminars most closely related to them, providing results or examples that are directly relevant to some part of these main lectures.

The theory of algebraic cycles is today still one of the most difficult and most beautiful areas of algebraic geometry (and of all mathematics): notable open problems include the Hodge conjecture, the relations among the several equivalence relations between algebraic cycles, the connections with the properties of some cohomology theories.

Our main goal in organizing this C.I.M.E. Session was to gather together some of the leading mathematicians active in this area, to assess the present state of the art and to describe the possible future developments.

Thus the three series of main lectures dealt with:
i) Infinitesimal methods in Hodge Theory, delivered by Mark L. Green (U.C.L.A., USA)
ii) Algebraic cycles and algebraic aspects of cohomology and K-theory, delivered by J.P. Murre (Rijksuniversiteit, Leiden, The Netherlands)
iii) Transcendental methods in the study of algebraic cycles, delivered by Claire Voisin (Université Paris-Sud, Orsay, France)

To complete this rough outline of the volume, it suffices to say a few words about the seminars that have been selected for inclusion in the text.

The first one, by G.P. Pirola, reports on joint work with A. Collino: they compute the infinitesimal invariant of the normal function associated to the cycle C^+-C^- in its Jacobian and derive from this computation a nice refinement of Ceresa's theorem and a Torelli theorem in the spirit of Griffiths in genus three. These results are obtained by applying M. Green's technique of computing the infinitesimal invariant of a normal function and can be regarded therefore as an exemplification and as a striking application of this technique. They are closely related to the first series of main lectures, and in the summer of 1993, new, surprising and important results were obtained from what was proved in the seminar.

The second seminar by Bert Van Geemen ties in closely with the lectures of J.P. Murre and his treatment of the Hodge conjecture: the author restricts his attention to abelian varieties, in particular to those of Weil type, and studies them by means of the Mumford-Tate group, a topic that cannot be left aside in a course like this one. Finally the author points out some relations between theta functions and cycles on some particular abelian fourfolds.

The last seminar, by S. Müller-Stach, deals with height pairings and reveals a connection between mixed Hodge structures (already treated in M. Green's lectures) and Deligne cohomology (see J.P. Murre's lectures) by using the theory of logarithmic currents (see M. Green's lectures). Thus this topic finds here an ideal context in which to be outlined and discussed.

We are very happy to note that the lectures did an outstanding job and that all the participants contributed with interest and enthusiasm to creating a very stimulating atmosphere throughout the session. It is fair to say that its spirit has been captured well by the texts of this volume. We wish to thank the C.I.M.E. Foundation which made all of this possible.

Alberto Albano
Dipartimento di Matematica
Università di Torino, Italy

Fabio Bardelli
Dipartimento di Matematica
Università di Pisa, Italy

TABLE OF CONTENTS

C.I.M.E. NOTES

INFINITESIMAL METHODS IN HODGE THEORY

June 1993

Mark L. Green

M. Green et al.: LNM 1594, A. Albano and F. Bardelli (Eds.), pp. 1–92, 1994.

LECTURE 1

During most of my years as an undergraduate student, I thought that algebra was my favorite subject. However, in my senior year of college, I took a course from Victor Guillemin. This was my first course in geometry, and the main theorem was De Rham's Theorem. This had a lasting effect on my mathematical interests, as the reader can observe. Much as Aristophanes thought that men and women were originally one creature trying (often in vain) to become reunited, so mathematicians often search through life (once again, often in vain) for problems that will bring together the various parts of mathematics that they love. For me, my contact with the area of infinitesimal methods in Hodge theory was one moment when, briefly, this actually happened.

My goal in these lectures is to cover the material necessary to the understanding of the Nori Connectedness Theorem, with stops for other interesting results along the way. Hodge theory, like algebraic geometry as a whole, is rich in having many levels of abstraction at which to approach any given idea. Unfortunately, as one rises to higher levels of abstraction and mathematical power, one tends to get further and further away from the underlying geometry. What I have attempted to do here is to try to make accessible some of these various levels by starting with the most geometric formulation and gradually introducing more abstract formulations. Thus some proofs are given more than once, in hopes that this will clarify how the machinery works.

I would like to thank my fellow lecturers, Jacob Murre and Claire Voisin, for their camaraderie and mathematical inspiration; I feel privileged to have shared a podium with them. I want to express my deep gratitude to Fabio Bardelli, who had the insight to realize that this subject needed a series of expository lectures and who mapped out for the three of us his vision of what should be covered; I could not have wished for a better or wiser organizer of the scientific program. I would also like to thank Alberto Albano for courageously and warm-heartedly stepping in when Fabio became ill, and carrying off the conference in a successful and enjoyable way.

De Rham's Theorem states that every real cohomology class on a smooth manifold M can be represented by a closed $\mathcal{C}^\infty$ differential form ω, and that two closed forms represent the same cohomology class if and only if they differ by an exact form $d\tau$, where τ is a $\mathcal{C}^\infty$ differential form. If we denote by $A^k(M)$ the $\mathcal{C}^\infty$ k-forms on M and $A^\bullet(M), d$ the complex of $\mathcal{C}^\infty$ differential forms with exterior derivative, we denote

$$H^k_{\mathrm{DR}}(M) = H^k(A^\bullet(M)).$$

THEOREM (De Rham's Theorem). *For a smooth manifold M, for all k,*

$$H^k_{\mathrm{DR}}(M) \cong H^k(M, \mathbf{R}).$$

The map is given by

$$\omega \mapsto [\sigma \mapsto \int_\sigma \omega].$$

Once Einstein had discovered general relativity, it was realized that the electromagnetic field ($\vec{E} = (E_1, E_2, E_3)$ the electric field and $\vec{B} = (B_1, B_2, B_3)$ the magnetic field) could be represented on four-dimensional space-time by a 2-form

$$\Omega = \sum_{i=1}^{3} E_i dx_i \wedge dt + B_1 dx_2 \wedge dx_3 + B_2 dx_3 \wedge dx_1 + B_3 dx_1 \wedge dx_2,$$

and that two out of four of Maxwell's equations in free space could be written as

$$d\Omega = 0.$$

The other two equations do not come out as naturally, but if one considers the 2-form

$$*\Omega = E_1 dx_2 \wedge dx_3 + E_2 dx_3 \wedge dx_1 + E_3 dx_1 \wedge dx_2 - \sum_{i=1}^{3} B_i dx_i \wedge dt,$$

then the other two equations are

$$d * \Omega = 0.$$

The relationship between Ω and $*\Omega$ is not invariant under smooth change of coordinates, but it is invariant under changes of coordinates which preserve the Lorentz metric on space-time. It was physical considerations of this kind which led Hodge to discover the Hodge $*$-operator and to formulate the Hodge Theorem.

The most natural mathematical motivation for the Hodge Theorem is to ask whether one can find one "natural" differential form ω representing each cohomology class. It is appealing to find some measure of the "size" of a differential form and then look for the "smallest" element of the set $\{\omega + d\tau\}$ for fixed closed form ω as τ varies over all smooth forms of a given degree. One might define the size first pointwise and then integrate over M. This is done as follows:

We need to remember some standard constructions. If V, W are vector spaces with a positive-definite inner product, then $V \otimes W$ may be given a natural positive-definite inner product so that if e_i and f_j are orthonormal bases for V, W respectively, then $e_i \otimes f_j$ is an orthonormal basis for $V \otimes W$. Secondly, if $W \subseteq V$ and V has a positive-definite inner product, then W inherits one by restriction and V/W by orthogonal projection. If V has a positive definite inner product, then V^* inherits one naturally in such a way that the dual basis of an orthonormal basis is orthonormal.

Let V be an oriented n-dimensional vector space over $\mathbf{R}$ equipped with a positive-definite inner product. Then for any k, $\wedge^k V$ may be given a natural positive-definite inner product by combining these two standard constructions, since $\wedge^k V \subseteq \otimes^k V$. Thus if M is an oriented Riemannian manifold of dimension n and ω is a smooth k-form on M, then for any $p \in M$, we can apply the construction above to $T^*_{p,M}$ with the induced inner product to obtain a length $\|\omega\|_p^2$. If M is an oriented compact Riemannian manifold, then we define

$$\|\omega\|_M^2 = \int_M \|\omega\|_p^2 dV,$$

where dV is the element of volume. This is a positive-definite inner product on the space of smooth k-forms on M.

DEFINITION. *A smooth k-form ω on a compact Riemannian manifold M is* **harmonic** *if $d\omega = 0$ and*

$$\|\omega\|_M \leq \|\omega + d\tau\|_M$$

for all smooth $(k-1)$-forms τ. We denote the set of harmonic k-forms on M by $\mathcal{H}^k(M)$.

There is slightly different way to describe the inner product on forms. The inner product on a vector space V gives a natural map

$$V \otimes V \to \mathbf{R}.$$

This in turn gives a natural isomorphism $V \to V^*$ of V with its dual. If we take $\wedge^k$ of this isomorphism, we obtain an isomorphism $\wedge^k V \to \wedge^k V^*$. Taking volume gives a natural isomorphism $\wedge^n V \cong \mathbf{R}$. Wedge product gives a map

$$\wedge^k V \times \wedge^{n-k} V \to \wedge^n V \cong \mathbf{R},$$

and since this is a non-degenerate pairing, it gives a natural isomorphism $\wedge^k V \cong \wedge^{n-k} V^*$, and now using $\wedge^k$ of the isomorphism induced by the inner product, we can identify the factor on the right with $\wedge^{n-k} V$. Putting all this together, we obtain a natural map

$$*\colon \wedge^k V \to \wedge^{n-k} V,$$

and this is the Hodge $*$-operator. This is defined pointwise and thus extends to a map $*\colon A^k(M) \to A^{n-k}(M)$. The basic facts are:

LEMMA. *(1) For $\alpha, \beta \in \wedge^k V$,*

$$\alpha \wedge *\beta = \beta \wedge *\alpha = (\alpha, \beta)\mathrm{Vol},$$

where $\mathrm{Vol} \in \wedge^n V$ is the element of volume;
(2) If $e_1, \ldots, e_n$ is an oriented orthonormal basis for V, then $*(e_{i_1} \wedge \cdots \wedge e_{i_k}) = \pm e_{j_1} \wedge \cdots \wedge e_{j_{n-k}}$, where $\{j_1, \ldots, j_{n-k}\} = \{1, 2, \ldots, n\} - \{i_1, \ldots, i_k\}$ and the sign

is chosen so that $e_{i_1} \wedge \cdots e_{i_k} \wedge \pm e_{j_1} \wedge \cdots \wedge e_{j_{n-k}} = e_1 \wedge \cdots \wedge e_n$;
(3) For $\alpha \in \wedge^k V$, $*^2\alpha = (-1)^{k(n-k)}\alpha$.

COROLLARY. *For* $\alpha, \beta \in A^k(M)$,

$$(\alpha, \beta)_M = \int_M \alpha \wedge *\beta.$$

We would like to construct an adjoint for $d\colon A^k(M) \to A^{k+1}(M)$.

PROPOSITION. *Let* $d^*\omega = (-1)^{(k+1)(n-k)+1} * d * \omega$ *for all* $\omega \in A^k(M)$. *Then*

$$(d^*\omega, \phi)_M = (\omega, d\phi)_M$$

for all $\omega \in A^k(M), \phi \in A^{k-1}(M)$.

PROOF: By Stokes Theorem. □

DEFINITION. *The* **Laplace operator** $\Delta\colon A^k(M) \to A^k(M)$ *is defined by*

$$\Delta = dd^* + d^*d.$$

PROPOSITION. *For* $\omega \in A^k(M)$, *the following are equivalent:*
(1) ω *is harmonic;*
(2) $d\omega = 0$ *and* $d^*\omega = 0$;
(3) $\Delta\omega = 0$.

PROOF: (1)$\leftarrow$(2): If ω is harmonic, then $d\omega = 0$. Now for any constant ϵ,

$$\begin{aligned} \|\omega + \epsilon d\tau\|_M^2 &= (\omega + \epsilon d\tau, \omega + \epsilon d\tau)_M \\ &= \|\omega\|_M^2 + 2\epsilon(\omega, d\tau)_M + \epsilon^2\|d\tau\|_M^2 \\ &= \|\omega\|_M^2 + 2\epsilon(d^*\omega, \tau)_M + \epsilon^2\|d\tau\|_M^2. \end{aligned}$$

For ϵ small, we see that ω harmonic implies that $(d^*\omega, \tau)_M = 0$ for all $\tau \in A^{k-1}(M)$, and this forces $d^*\omega = 0$.
(2)$\leftarrow$ (1): By the formula above, if $d\omega = 0$ and $d^*\omega = 0$, then

$$\|\omega + d\tau\|_M^2 = \|\omega\|_M^2 + \|d\tau\|_M^2,$$

and thus

$$\|\omega\|_M < \|\omega + d\tau\|_M$$

if $d\tau \neq 0$.
(2)$\rightarrow$ (3) $\Delta\omega = dd^*\omega + d^*d\omega = d(0) + d^*(0) = 0$.
(3) $\rightarrow$(2): We have for any ω that

$$\begin{aligned} (\Delta\omega, \omega)_M &= (dd^*\omega, \omega)_M + (d^*d\omega, \omega)_M \\ &= (d^*\omega, d^*\omega)_M + (d\omega, d\omega)_M \\ &= \|d^*\omega\|_M^2 + \|d\omega\|_M^2. \end{aligned}$$

If $\Delta\omega = 0$, then the left hand side is zero, and hence the right hand side is, which implies $d^*\omega = 0, d\omega = 0$. □

PROPOSTION. *There is a natural injection*

$$\mathcal{H}^k(M) \to H^k_{\mathrm{DR}}(M)$$

sending a harmonic k-form ω to its De Rham cohomology class.

PROOF: The only thing to be proved is that if a harmonic form is exact, then it is 0. If $\omega \in \mathcal{H}^k(M)$ and 0 belongs to the De Rham class of ω, then by minimality $|\omega|_M \leq |0|_M = 0$, so $\omega = 0$. □

Of course, it is not clear that harmonic forms exist, i.e. that there is a form of minimal size in each De Rham class. To see that a sequence of smooth forms in a De Rham class with sizes converging to the infimum of the sizes of forms in that class must converge to a smooth form requires some basic results from the theory of elliptic operators. The final result, which we quote here, is:

THEOREM (The Hodge Theorem). *For M a compact oriented Riemannian manifold, the natural map*

$$\mathcal{H}^k(M) \to H^k_{\mathrm{DR}}(M)$$

is an isomorphism, i.e. every De Rham class is represented by a unique harmonic form.

Beautiful as it is, the Hodge Theorem by itself is not quite enough for the purposes of Hodge theory. One sometimes needs the full package of consequences of elliptic operator theory. The eigenspaces of Δ are finite-dimensional and are spanned by smooth functions, and the eigenvalues are ≥ 0 and march off to infinity. Every L^2 k-form can be expressed as the L^2 limit of sums of eigenforms of Δ. If one takes a k-form, projects it on the space orthogonal to $\mathcal{H}^k(M)$, and then multiplies its projection on the λ-eigenspace of Δ by $\frac{1}{\lambda}$, one obtains the Green's operator. The fact we will need to quote is that:

THEOREM (Existence of Green's Function). *For M a compact oriented Riemannian manifold, there exists a unique operator $G: A^k(M) \to A^k(M)$ such that G commutes with d and d^*, $G(\mathcal{H}^k(M)) = 0$, and*

$$\mathrm{Id} = \pi_{\mathcal{H}} + \Delta G,$$

where $\pi_{\mathcal{H}}$ is the orthogonal projection $A^k(M) \to \mathcal{H}^k(M)$.

If a compact orientable manifold M has a metric with nice differential-geometric properties, it is possible to draw interesting conclusions about the cohomology of M. However, it is really in the case of complex manifolds, especially Kähler manifolds, that the Hodge Theorem pays truly powerful geometric dividends.

As with the Hodge theorem, there is a certain amount of preliminary multilinear algebra that goes into the story. The main difference is how much interesting multilinear algebra goes on, and how subtle some of the results are.

DEFINITION. *Let V be a vector space over* $\mathbf{R}$. *An* **almost-complex structure** *on V is an endomorphism $J \in \mathrm{End}(V)$ such that $J^2 = -\mathrm{Id}$.*

DEFINITION. *Let V, J be a real vector space with almost-complex structure, and $V_{\mathbf{C}} = V \otimes_{\mathbf{R}} \mathbf{C}$, with J extended to $V_{\mathbf{C}}$ in the canonical way. Define $V^{1,0}, V^{0,1}$ respectively as the $+i$ and $-i$ eigenspaces of J on $V_{\mathbf{C}}$.*

PROPOSITION. *Let V, J be a real vector space with almost-complex structure. Then $V_{\mathbf{C}} = V^{1,0} \oplus V^{0,1}$. Further, $\dim_{\mathbf{R}} V = \dim_{\mathbf{C}} V^{1,0} = \dim_{\mathbf{C}} V^{0,1}$.*

PROOF: The eigenvalues of J occur in conjugate pairs, and clearly the direct sum of $V^{1,0}$ and $V^{0,1}$ injects into $V_{\mathbf{C}}$. It thus suffices to prove that $\dim_{\mathbf{C}} V^{1,0} = \dim_{\mathbf{R}} V$. The map $v \mapsto (iv + Jv) \oplus (-iv + Jv)$ takes $V_{\mathbf{C}} \to V^{1,0} \oplus V^{0,1}$ and is injective, which shows that $\dim_{\mathbf{C}} V_{\mathbf{C}} \leq 2\dim_{\mathbf{C}} V^{1,0}$, which is enough. □

DEFINITION. *Let V, J be a real vector space with an almost-complex structure. A positive definite inner product $(\,,)$ is* **hermitian** *if J is an isometry, i.e. $(Jv, Jw) = (v, w)$ for all $v, w \in V$.*

PROPOSITION. *Let V, J be a real vector space with almost-complex structure and hermitian metric $(\,,)$.*
(1) The map $\omega: V \otimes V \to \mathbf{R}$ defined by

$$\omega(v, w) = (Jv, w)$$

is a real alternating form;
(2) If we extend ω to an element of $\wedge^2 V_{\mathbf{C}}^*$, then ω is zero when restricted to $V^{1,0} \otimes V^{1,0}$ and $V^{0,1} \otimes V^{0,1}$;
(3) ω gives a non-degenerate pairing when restricted to $V^{1,0} \otimes V^{0,1}$.
(4) If we extend $(\,,)$ to be complex linear in the first variable and conjugate linear in the second variable, then it is a positive definite Hermitian inner product on $V^{1,0}$.

PROOF: (1) $(Jv, w) = (J^2 v, Jw) = -(v, Jw) = -(Jw, v)$.
(2) For the purposes of proving (2) and (3), extend $(\,,)$ to be complex linear in both entries. If $v, w \in V^{1,0}$, then $(Jv, w) = i(v, w) = (v, Jw) = -(Jv, w)$, so $(v, w) = 0$ and thus $\omega(v, w) = 0$. Similarly for $V^{0,1}$.
(3) If $v \in V^{1,0}$ then for some $w \in V_{\mathbf{C}}$, $(Jv, w) \neq 0$, as otherwise $Jv = 0$ and hence $v = 0$. If $w = w^{1,0} + w^{0,1}$ is the decomposition of w under the direct sum decomposition $V_{\mathbf{C}} = V^{1,0} \oplus V^{0,1}$, then by (2), $(Jv, w) = (Jv, w^{0,1})$, and this proves the pairing is non-degenerate in the first factor. A similar argument works for the second factor.
(4) If $v = a + ib \in V^{1,0}$, where $a, b \in V$, then $(v, \bar{v}) = (a+ib, a-ib) = (a, a) + (b, b)$, from which positive-definiteness is clear. □

DEFINITION. *Let V, J be a real vector space with almost-complex structure and hermitian metric $(,)$. Let ω be as in the preceding Proposition. Then ω is called the* **alternating form associated to** $(,)$.

DEFINITION. *For a complex manifold M, a C^∞* **form of type** (p,q) *is a C^∞ section of the bundle $\wedge^p T^{1,0*} \otimes \wedge^q T^{0,1*}$; we will denote the set of these by $A^{p,q}(M)$.*

DEFINITION. *Let M be a complex manifold with almost-complex structure $J: T_M \to T_M$. A Riemannian metric on the underlying real manifold of M is* **hermitian** *if it is hermitian with respect to J on $T_{M,p}$ for every point $p \in M$. The* **associated (1,1) form** *ω of the hermitian metric is defined by taking ω_p to be the alternating form associated to the metric on $T_{M,p}$ for every $p \in M$.*

COROLLARY. *The associated (1,1)-form of a hermitian metric is a real 2-form on the underlying real manifold of M and has type (1,1).*

DEFINITION. *A hermitian metric on a complex manifold M is said to be a* **Kähler metric** *if the associated $(1,1)$ form ω is closed. In this case, ω is called the* **Kähler form**. *The element of $H^2_{\mathrm{DR}}(M)$ determined by ω is called the* **Kähler class**. *If the Kähler class belongs to the image of $H^2(M, \mathbf{Z})$, the metric is said to be a* **Hodge metric**.

EXAMPLE. *Fubini-Study metric*

On $\mathbf{P}^n$, if we let

$$0 \to S \to V \otimes \mathcal{O}_{\mathbf{P}^n} \to Q \to 0$$

be the tautological sub-bundle sequence, then it is well-known that

$$T_{\mathbf{P}^n} \cong \mathrm{Hom}(S, Q).$$

If we put a Hermitian metric on the complex vector space V, then it induces natural Hermitian metrics on S and Q by restriction and orthogonal projection. This in turn induces a natural metric on $S^* \otimes Q \cong \mathrm{Hom}(S,Q)$. This metric is invariant under the action of the unitary group on V, and this forces the associated (1,1)-form ω to be closed. Since $H^2(\mathbf{P}^n, \mathbf{R})$ is 1-dimensional, adjusting the metric by a constant makes ω integral, and thus $\mathbf{P}^n$ has a Hodge metric, the **Fubini-Study metric**.

PROPOSITION. *A smooth projective variety $M \subseteq \mathbf{P}^N$ has a Hodge metric obtained by restricting the Fubini-Study metric.*

PROOF: It is elementary to see that the restriction of a Hodge metric is Hodge, since the operations of restricting and taking associated (1,1)-form commute. □

We quote the following famous consequence of the Kodaira Embedding Theorem:

THEOREM (Kodaira). *If a compact hermitian complex manifold admits a Hodge metric, then there exists an embedding of M in some $\mathbf{P}^N$ such that the Hodge metric is $\frac{1}{k}$ times the restriction of the Fubini-Study metric, for some positive integer k.*

Using geodesic coordinates, it is easy to see that at every point p of a Riemannian manifold M, there are local coordinates $x_1, \ldots, x_n$ for M centered at p such that

$$\left(\frac{\partial}{\partial x_i}, \frac{\partial}{\partial x_j}\right) = \delta_{ij} + O(\|x\|^2).$$

However, it is not true that at every point p of a hermitian complex manifold, there are local holomorphic coordinates $z_1, \ldots, z_n$ centered at p such that

$$\left(\frac{\partial}{\partial z_i}, \frac{\partial}{\partial z_j}\right) = \delta_{ij} + O(\|z\|^2).$$

The following proposition makes the Kähler condition quite natural (or at least as natural as it is going to get.)

PROPOSITION. *Let M be a complex hermitian manifold. The following are equivalent:*
(1) The metric is Kähler;
(2) At every point p of M, there are local holomorphic coordinates $z_1, \ldots, z_n$ centered at p such that

$$\left(\frac{\partial}{\partial z_i}, \frac{\partial}{\partial z_j}\right) = \delta_{ij} + O(\|z\|^2).$$

PROOF: If $z_1, \ldots, z_n$ are local holomorphic coordinates on a complex manifold M, let

$$h_{ij} = \left(\frac{\partial}{\partial z_i}, \frac{\partial}{\partial z_j}\right).$$

Then

$$\omega = i \sum_{ij} h_{ij} dz_i \wedge d\bar{z}_j.$$

If (2) holds, then all first partials of the h_{ij} vanish at the origin, and hence $d\omega = 0$ there; since the point was arbitrary, $d\omega = 0$ and the metric is Kähler. Conversely, if $d\omega = 0$, and we choose holomorphic coordinates $z_1, \ldots, z_n$ so that

$$h_{ij} = \delta_{ij} + \sum_k a_{ij}^k z_k + \sum_k \bar{a}_{ji}^k \bar{z}_k + O(\|z\|^2),$$

then the change of variables $z_i = w_i + q_i(w, w)$, where the q_i are homogeneous and quadratic in the w's, changes the linear term of h_{ij} by $\frac{\partial q_i}{\partial z_j} + \overline{\frac{\partial q_j}{\partial z_i}}$. Thus a_{ij}^k is changed by $\frac{\partial^2 q_i}{\partial z_j \partial z_k}$. The condition $d\omega = 0$ at p is equivalent to $a_{ij}^k = a_{kj}^i$ for all i, j, k, and thus if we take $q_j = -\sum_{ik} a_{ij}^k$, the coordinates $w_1, \ldots, w_n$ satisfy (2). □

If we now return to the seemingly innocent situation of a vector space V with an almost complex structure J and a hermitian metric $(,)$, we discover that there is a lot of geometry in $\wedge^* V^*$.

Recall that a complex vector space has a natural orientation when veiwed as a real vector space of twice the dimension, because $GL(n, \mathbf{C}) \subseteq GL(2n, \mathbf{R})$ is connected. There is a slight element of convention here, whether one uses the order $x_1, \ldots, x_n, y_1, \ldots, y_n$ or the order $x_1, y_1, x_2, y_2, \ldots, x_n, y_n$ (I will adopt the latter convention here.)

DEFINITION. *Let V, J be a real vector space of dimension $2n$ with an almost-complex structure. The* **natural orientation** *of V is the pullback of the natural orientation of $V^{1,0}$ under the natural isomorphism of real vector spaces $V \to V^{1,0}$ by $v \mapsto iv + Jv$. If $e_1, \ldots, e_n \in V$ are vectors such that $e_1, \ldots, e_n, Je_1, \ldots, Je_n$ are linearly independent, then $e_1, Je_1, e_2, Je_2, \ldots, e_n, Je_n$ is a properly oriented basis for V. Once one has chosen the natural orientation for V, one obtains a natural positive-definite inner product on $\wedge^k V$ and $\wedge^k V^*$ by the construction for an oriented real vector space with a positive-definite inner product; this will also be denoted $(,)$. This may be extended to a positive definite Hermitian inner product on $\wedge^k V_{\mathbf{C}}$ and $\wedge^k V_{\mathbf{C}}^*$, where $V_{\mathbf{C}} = V \otimes \mathbf{C}$ is the complexification of V. One may extend the $*$-operator to $\wedge^k V_{\mathbf{C}}^*$ so as to be conjugate-linear, and then $(\alpha, \beta)\mathrm{Vol} = \alpha \wedge *\beta$.*

PROPOSITION. *For a real vector space of dimension $2n$ with almost-complex structure J and hermitian metric $(,)$, let ω denote the alternating form. Then*

$$\omega^n = n!\mathrm{Vol}.$$

PROOF: Since J is an isometry, by the Gram-Schmidt process, one can choose inductively an orthonormal basis for V of the form $e_1, Je_1, e_2, Je_2, \ldots, e_n, Je_n$. Note that $\omega(e_i, e_j) = (Je_i, e_j) = 0$ for all i, j. Similarly $\omega(Je_i, Je_j) = 0$ for all i, j. Finally, $\omega(e_i, Je_j) = (Je_i, Je_j) = (e_i, e_j) = \delta_{ij}$. The result now follows by a direct calculation from the definition of wedge product. □

DEFINITION (Operators L,Λ, H). *Let V be a real vector space of dimension $2n$ with almost-complex structure J and hermitian metric $(,)$. Let $L: \wedge^* V^* \to \wedge^* V^*$ be given by $L(\alpha) = \omega \wedge \alpha$, where ω is the alternating form associated to the metric. Since the metric of V induced metrics on $\wedge^k V^*$ for all k, the adjoint of L will be denoted Λ, so that*

$$(L(\alpha), \beta) = (\alpha, \Lambda(\beta))$$

for all $\alpha \in \wedge^k V^*, \beta \in \wedge^{k+2} V^*$. Finally, let $H: \wedge^* V^* \to \wedge^* V^*$ be the linear map whose restriction to $\wedge^k V^*$ is $(n-k)\mathrm{Id}_{\wedge^k V^*}$.

PROPOSITION.
(1) $[\Lambda, L] = H$;

(2) $[H, L] = -2L$;
(3) $[H, \Lambda] = 2\Lambda$.

PROOF: (2),(3) are quite easy, since they just say that L raises degree by 2 and Λ drops degree by 2. (1) is more surprising, but it a straightforward computation.□

These relations are exactly those which hold among the standard generators of the Lie algebra $\mathfrak{sl}(2, \mathbf{C})$. It follows that L, Λ, H is a representation of $\mathfrak{sl}(2, \mathbf{C})$ to $\mathrm{End}(\wedge^* V^*)$. Of course, this helps to make many of the results more conceptual for those that know representation theory, but one may also proceed by direct calculation.

DEFINITION (Primitive Forms). *The* **primitive k-forms** *are*

$$P^k = \{\omega \in \wedge^k V^* \mid \Lambda\omega = 0\}.$$

PROPOSITION (Lefschetz decomposition for forms).
(1) $L^{n-k+1}\colon P^k \to \wedge^{2n-k+2} V^*$ *is zero;*
(2) $L^{n-k}\colon P^k \to \wedge^{2n-k} V^*$ *is injective;*
(3) $\wedge^k V^* = \oplus_j L^j P^{k-2j}$.

PROOF: This can be done by a messy calculation. Conceptually, it is nicest to break up $\wedge^* V$ into irreducible representations of $\mathfrak{sl}(2, \mathbf{C})$, and to note that the result holds for each of the irreducible representations of $\mathfrak{sl}(2, \mathbf{C})$; these are just the standard representations on homogeneous polynomials of degree k is 2 variables under change of homogeneous coordinates, which we will denote by W_k. On W_k, $L(p) = yp_x$, $\Lambda(p) = xp_y$, and $H(p) = xp_x - yp_y$. In W_k, the primitive vector is x^k, and since $H(x^{k-i}y^i) = (k-2i)x^{k-i}y^i$, we see that $x^{k-i}y^i$ corresponds to an element of $\wedge^{n+2i-k}V^*$. So $x^k \in P^{n-k}$ and L^k is an isomorphism for this representation. □

The final important bit of multilinear algebra is the following—on P^k, both $*$ and L^{n-k} land in $\wedge^{2n-k}V^*$, and we might hope for them to be multiples of each other, or even more generally, on $L^j P^k$.

PROPOSITION. *If* $\alpha \in L^j P^k$ *and* α *has type* $(p, k-p)$,

$$*\alpha = (-1)^{\frac{k(k+1)}{2}} i^{p-q} \frac{j!}{(n-k-j)!} L^{n-k-2j} \bar{\alpha}.$$

PROOF: This is very messy. The most elegant proof is by Henryk Hecht, inspired by representation theory.

DEFINITION. *On* $\wedge^k V^*$, *define the hermitian inner product* $<,>$ *by*

$$< \alpha, \beta > \mathrm{Vol} = i^{k^2} L^{n-k}(\alpha \wedge \bar{\beta}).$$

The power of i *is necessary to make the form Hermitian.*

COROLLARY. *On* $L^j P^k$, $(-1)^{k+p} <,> = \frac{j!}{(n-k-j)!}(,)$, *and hence is positive-definite.*

PROOF: If $\alpha \in L^j P^k$, then

$$
\begin{aligned}
<\alpha, \alpha> \mathrm{Vol} &= i^{k^2} L^{n-2j-k}(\alpha \wedge \bar{\alpha}) \\
&= (-1)^{\frac{k(k+1)}{2}} i^{k^2-k+2p} \frac{j!}{(n-k-j)!} \alpha \wedge * \alpha \\
&= (-1)^{k+p} \frac{j!}{(n-k-j)!} (\alpha, \alpha) \mathrm{Vol}.
\end{aligned}
$$

□

The operations L, Λ, H are defined pointwise, and hence may be globalized on a Hermitian complex manifold.

DEFINITION. *Let M be a compact Hermitian complex manifold. The maps*

$$
\begin{aligned}
L&: A^k(M) \to A^{k+2}(M); \\
\Lambda&: A^k(M) \to A^{k-2}(M); \\
H&: A^k(M) \to A^k(M)
\end{aligned}
$$

are the global extensions of the pointwise maps already defined on each $T_p(M)$ with the induced Hermitian metric. They satisfy the same commutation relations.

The difficulty, on an arbitrary Hermitian complex manifold, is that neither the decomposition of differential forms

$$A^k(M) = \oplus_{p+q=k} A^{p,q}(M)$$

nor the operations L, Λ, H descend to cohomology. However, at the price of assuming that the metric is Kähler, both of these things work, and we have a situation incredibly rich in geometric structure.

DEFINITION. *On a complex manifold M, the operators*

$$
\begin{aligned}
\partial&: A^{p,q}(M) \to A^{p+1,q}(M), \\
\bar{\partial}&: A^{p,q}(M) \to A^{p,q+1}(M)
\end{aligned}
$$

are uniquely defined by the equation

$$d = \partial + \bar{\partial}.$$

PROPOSITION.

$$
\begin{aligned}
\partial^2 &= 0, \\
\partial\bar{\partial} + \bar{\partial}\partial &= 0, \\
\bar{\partial}^2 &= 0.
\end{aligned}
$$

PROOF: Decompose $d^2 = 0$ by type. □

PROPOSITION. *∂ and $\bar\partial$ have adjoints $\partial^*, \bar\partial^*$ given by*

$$\partial^* = - * \partial *,$$
$$\bar\partial^* = - * \bar\partial * .$$

PROOF: Stokes' Theorem. □

DEFINITION.

$$\Box_\partial = \partial\partial^* + \partial^*\partial,$$
$$\Box_{\bar\partial} = \bar\partial\bar\partial^* + \bar\partial^*\bar\partial.$$

DEFINITION. *The* **p'th Dolbeault complex** *on a complex manifold M is $A^{p,\bullet}(M), \bar\partial$.*

THEOREM (Dolbeault's Theorem). $H^q(\Omega^p_M) \cong H^q_{\bar\partial}(A^{p,\bullet}(M))$.

PROOF: The complex $A^{p,\bullet}(M), \bar\partial$ is a fine resolution of Ω^p_M. □

DEFINITION.

$$\mathcal{H}^{p,q}(M) = \{\omega \in A^{p,q}(M) \mid \Box_{\bar\partial}\omega = 0\}.$$

THEOREM. *On a compact complex Hermitian manifold, the canonical map $\mathcal{H}^{p,q}(M) \to H^q(A^{p,\bullet}(M))$ is an isomorphism.*

PROOF: The same elliptic operator theory used in the proof of the Hodge Theorem. □

THEOREM (Kähler identities). *Let M be a Kähler manifold. Then*
(1) $[\Lambda, \bar\partial] = -i\partial^*$, $[\Lambda, \partial] = i\bar\partial^*$;
(2) $[L, \partial^*] = i\bar\partial$, $[L, \bar\partial^*] = -i\partial$,
(3) $[L, \partial] = 0$, $[L, \bar\partial] = 0$,
(4) $[\Lambda, \partial^*] = 0$, $[\Lambda, \bar\partial^*] = 0$,
(5) $\frac{1}{2}\Delta = \Box_\partial = \Box_{\bar\partial}$,
(6) $[L, \Delta] = 0$, $[\Lambda, \Delta] = 0$, $[H, \Delta] = 0$.

PROOF: Since formulas (1)-(4) depend only on the 1-jet of the coefficients of the Kähler form, and since they may be verified pointwise, the special coordinates guaranteed for a Kähler metric reduce the theorem to verifying the formulas for $\mathbf{C}^n$ with the flat metric. This is now a messy calculation. Using (2) to write $\partial = i[L, \bar\partial^*]$, we obtain the formula

$$\Box_\partial = i(L\bar\partial^*\partial^* - \bar\partial^* L \partial^* + \partial^* L \bar\partial^* - \partial^*\bar\partial^* L).$$

Using (2) to substitute for $\bar\partial$ gives a similar formula for $\Box_{\bar\partial}$, and the two are seen to be equal using the fact $\partial^*, \bar\partial^*$ anti-commute. So $\Box_\partial = \Box_{\bar\partial}$. Expanding $d = \partial + \bar\partial$ and similarly d^*, we see that

$$\Delta = \Box_\partial + \Box_{\bar\partial} + \bar\partial\partial^* + \partial\bar\partial^* + \partial^*\bar\partial + \bar\partial^*\partial.$$

Substituting for ∂ and $\bar\partial$ using (2), the last four terms cancel. This proves (5). Writing out Δ in terms of ∂, ∂^* and using (1)-(4) gives (6). □

DEFINITION. *Let $\pi_{p,q}: A^{p+q}(M) \to A^{p,q}(M)$ denote the canonical projection.*

THEOREM. *Let M be a compact Kähler manifold. Then*

$$\mathcal{H}^k(M) = \oplus_{p+q=k} \mathcal{H}^{p,q}(M).$$

PROOF: Since $\Box_{\bar\partial}$ commutes with $\pi_{p,q}$ and since the various Laplacians coincide (up to a constant) on a Kähler manifold, this is essentially automatic. □

DEFINITION. *It is traditional to denote $H^q(\Omega^p_M)$ by $H^{p,q}(M)$.*

COROLLARY (The Hodge Decomposition).

$$H^k(M, \mathbf{C}) \cong \oplus_{p+q=k} H^{p,q}(M).$$

PROOF: Using Dolbeault's Theorem and the Hodge Theorem, this now follows automatically. □

The Kähler identities have the wonderful consequence that the operators L, Λ, H commute with Δ and thus take harmonic forms to harmonic forms.

DEFINITION (Primitive differential forms).

$$P^k(M) = \{\omega \in A^k(M) \mid \Lambda\omega = 0\};$$
$$P^{p,q}(M) = \{A^{p,q}(M) \mid \Lambda\omega = 0\};$$
$$\mathcal{P}^k(M) = \{\omega \in \mathcal{H}^k(M) \mid \Lambda\omega = 0\}.$$

THEOREM (Lefschetz decomposition for compact Kähler manifolds). *Let M be a compact Kähler manifold. Then*

$$\mathcal{H}^k(M) = \oplus_j L^j \mathcal{P}^{k-2j}(M).$$

DEFINITION. *Let M be a compact Kähler manifold of dimension n with Kähler form ω. Define a hermitian inner product on $A^k(M)$ by*

$$< \alpha, \beta > = i^{k^2} \int_M \omega^{n-k} \wedge \alpha \wedge \bar\beta.$$

Note that the decomposition $A^k(M) = \oplus_{p+q=k} A^{p,q}(M)$ is orthogonal with respect to $<,>$.

REMARK: If ω is integral, then $i^{-k^2} <,>$ takes integral values on the image of integral cohomology.

THEOREM (Hodge Index Theorem). *Let M be a compact Kähler manifold of dimension n. Then $(-1)^q <,>$ is positive definite on $L^j P^{p,q}(M)$, and these spaces are mutually orthogonal.*

PROOF: This is just a global version of a result that we already have for hermitian vector spaces. □

There are yet further ways in which the cohomology of compact Kähler manifolds is special. One is the "principle of two types."

PROPOSITION (Principle of two types). *Let M be a compact Kähler manifold. Let $\alpha \in A^{p,q}(M)$ satisfy $\partial\alpha = 0$, $\bar{\partial}\alpha = 0$ and be either ∂- or $\bar{\partial}$-exact. Then for some $\lambda \in A^{p-1,q-1}(M)$, $\alpha = \partial\bar{\partial}\lambda$.*

PROOF: Surprisingly, one needs to use some of the consequences of elliptic operator theory going beyond the Hodge Theorem, notably the existence of a Green's function.

Say $\alpha = \partial\beta$ for $\beta \in A^{p-1,q}(M)$. If $\gamma \in \mathcal{H}^{p,q}(M)$, then

$$(\alpha, \gamma)_M = (\partial\beta, \gamma)_M = (\beta, \partial^*\gamma)_M = 0,$$

so $\pi_{\mathcal{H}}(\alpha) = 0$. Thus

$$\begin{aligned}\alpha &= \Box_{\bar{\partial}} G\alpha \\ &= \bar{\partial}\bar{\partial}^* G\alpha + \bar{\partial}^* \bar{\partial} G\alpha \\ &= -i\bar{\partial}(\Lambda\partial - \partial\Lambda)G\alpha - i(\Lambda\partial - \partial\Lambda)\bar{\partial}G\alpha \\ &= i\bar{\partial}\partial\Lambda G\alpha \\ &= -i\partial\bar{\partial}\Lambda G\alpha.\end{aligned}$$

□

DEFINITION. *The* **holomorphic De Rham complex** *is the complex $\Omega^\bullet_M$, ∂. The spectral sequence associated to the bi-graded complex*

$$A^{p,q}(M) = A^q(M, \Omega^p), \partial, \bar{\partial}$$

is called the **Hodge-De Rham spectral sequence** *or* **Fróhlicher spectral sequence**; *the total complex of this bi-graded complex is just the (complex-valued) De Rham complex of M, and automatically the Hodge-De Rham spectral sequence has $E_1^{p,q} = H^q(\Omega^p_M)$ and converges to the cohomology of the total complex, hence $H^{p+q}(M, \mathbf{C})$.*

The more abstract way of looking at Hodge theory is due to Grothendieck and Deligne. We begin introducing this point of view with the following result:

THEOREM (Degeneration of the Hodge-De Rham spectral sequence).

Let M be a compact Kähler manifold. Then the Hodge-De Rham spectral sequence degenerates at the E_1-term.

PROOF: The way to compute the differentials $d_1, d_2, d_3, \ldots$ is as follows—pick any $\bar{\partial}$-closed form α representing a Dolbeault class in $H^q(\Omega^p_M)$, and take $\partial\alpha$, which is $\bar{\partial}$-closed and represents a class in $H^q(\Omega^{p+1}_M)$; this is $d_1(\alpha)$. If this class is zero, i.e. if $\partial\alpha = \bar{\partial}\alpha_2$, then $\partial\alpha_2$ is $\bar{\partial}$-closed; this is $d_2(\alpha)$. Continuing inductively, if $d_1(\alpha) = 0, \ldots, d_k(\alpha) = 0$, then $\partial\alpha_{k-1} = \bar{\partial}\alpha_k$. Now $d_{k+1}(\alpha) = \partial\alpha_k$.

At each step, we can pick any α_i we want that satisfy the equations. Start with α harmonic. Then $\partial\alpha = 0$, so we may take $\alpha_2 = 0$. Now continuing inductively, at each stage we may take $\alpha_k = 0$, and thus $d_k(\alpha) = 0$ for all $k \geq 1$. □

It is worth noting that from the perspective of the Hodge-De Rham spectral sequence, we naturally obtain a filtration rather than a direct sum decomposition of $H^k(M, \mathbf{C})$. As we will see later, this is not accidental.

DEFINITION. *Let M be a compact Kähler manifold. The Hodge filtration*

$$0 = F^{k+1}H^k(M,\mathbf{C}) \subseteq F^kH^k(M,\mathbf{C}) \subseteq \cdots$$
$$\subseteq F^1H^k(M,\mathbf{C}) \subseteq F^0H^k(M,\mathbf{C}) = H^k(M,\mathbf{C})$$

is defined by

$$F^pH^k(M,\mathbf{C}) = \oplus_{i\geq p}H^{i,k-i}(X).$$

DEFINITION. *A* **Hodge structure of weight k** *is:*
(1) A real vector space V together with a lattice $\Gamma \subseteq V$;
(2) A decreasing filtration $F^\bullet V_{\mathbf{C}}$ on $V_{\mathbf{C}} = V \otimes \mathbf{C}$ with $F^0V_{\mathbf{C}} = V_{\mathbf{C}}$ and $F^{k+1}V_{\mathbf{C}} = 0$ such that:
(3) $V_{\mathbf{C}} = F^pV_{\mathbf{C}} \oplus \overline{F^{k-p+1}V_{\mathbf{C}}}$ for all $p = 0, \ldots, k$.

PROPOSITION. *On a compact Kähler manifold M, $V = H^k(M,\mathbf{R})$, $\Gamma = H^k(M,\mathbf{Z})$, and F^p the Hodge filtration is a Hodge structure of weight k.*

PROOF: Since

$$F^pH^k(M,\mathbf{C}) = \oplus_{i\geq p}H^{i,k-i}(M)$$

and

$$\overline{F^{k-p+1}H^k(M,\mathbf{C})} = \oplus_{j\geq k-p+1}H^{k-j,j}(M),$$

one easily sees that (3) is satisfied. □

REMARK: For a compact Kähler manifold,

$$H^{p,q}(M) = F^pH^{p+q}(M,\mathbf{C}) \cap \overline{F^qH^{p+q}(M,\mathbf{C})};$$

for a general Hodge structure of weight $p+q$, we may take this as the definition. It is worth noting that the Hodge filtration arises from the Hodge-DeRham spectral sequence, and that while we need a Kähler metric to know that the spectral sequence degenerates, the filtration itself does not depend on the choice of Kähler metric. By the definition just made, neither does the Hodge decomposition.

DEFINITION. *Let $V, F^\bullet, \Gamma$ be a Hodge structure of weight k. We will say that $V', F'^\bullet, \Gamma'$ is a* **sub-Hodge structure** *of V, F, Γ if $V' \subseteq V$ is a linear subspace and*
(1) $\Gamma' = \Gamma \cap V'$;
(2) $F'^pV'_{\mathbf{C}} = F^pV_{\mathbf{C}} \cap V'_{\mathbf{C}}$ for all p.
We say that $V', F'^\bullet, \Gamma'$ is a sub-Hodge structure of weight l is $F'^{l+1}V'_{\mathbf{C}} = 0$. Note that a Hodge structure of lower weight can be a sub-Hodge structure of a Hodge structure of higher weight.

It is worth noting that not every Hodge structure arises from a compact Kähler manifold. As we will see later on, the infinitesimal period relations imply that once there are "enough" non-zero Hodge groups $H^{p,q}$, every family of varieties gives rise to a family of Hodge structures having positive codimension in the space of all Hodge structures. It is an open problem to describe what Hodge structures of weight ≥ 2 occur as Hodge structures of a compact Kähler manifold. It is an open question whether there is a reasonable class of geometric objects giving rise to all (or even an open set of) Hodge structures in an interesting way; Witten conjectures that this is true for the Hodge structures arising from Calabi-Yau manifolds.

Something we will want when it comes time to define Deligne cohomology and mixed Hodge structures is:

DEFINITION. *Let $\Omega_X^{<p}$ denote the complex $\mathcal{O}_X \rightarrow \Omega_X^1 \cdots \rightarrow \Omega_X^{p-1}$. Let $\Omega_X^{\geq p}$ denote the complex $K^\bullet$ with $K^i = \Omega_X^i$ if $i \geq p$ and $K^i = 0$ otherwise.*

PROPOSITION. *For X a compact Kähler manifold,*
(1) $\mathbf{H}^k(\Omega_X^{\geq p}) \cong F^p H^k(X, \mathbf{C})$;
(2) $\mathbf{H}^k(\Omega_X^{<p}) \cong H^k(X, \mathbf{C})/F^p H^k(X, \mathbf{C})$.

PROOF: (1) The hypercohomology spectral sequence has $E_1^{i,j} = H^j(\Omega_X^i)$ for $0 \leq j \leq n$ and $i \geq p$. This spectral sequence degenerates at the E_1 term, since the differentials are just those of the Hodge-DeRham spectral sequence. Thus $\mathbf{H}^k(\Omega_X^{\geq p}) = H^{p,k-p}(X) \oplus \cdots \oplus H^{k,0}(X) = F^p H^k(X, \mathbf{C})$.
(2) This follows by the same kind of argument, or by an exact sequence. □

Note in particular that under the inclusion of complexes $\Omega_X^{\geq p} \rightarrow \Omega_X^\bullet$, we have

$$\operatorname{im}(\mathbf{H}^k(\Omega_X^{\geq p}) \rightarrow \mathbf{H}^k(\Omega_X^\bullet)) = F^p H^k(X, \mathbf{C}).$$

The significance of this result will come later, when it motivates the Hodge filtration on the mixed Hodge structure of a quasi-projective variety.

Bibliographic Notes

The material in this lecture is covered in several places:

A. Weil, *Variétés Kählériennes*
P. Griffiths and J. Harris, *Principles of Algebraic Geometry*
R. Wells, *Differential Analysis on Complex Spaces*

Weil's book is the classic introduction to Kähler manifolds. Some progress has been made even in the foundations of the subject since his book appeared, notably the approach to the Lefschetz decomposition using the representation theory of $\mathfrak{S}L(2, \mathbf{C})$ and the use of the hypercohomology spectral sequence.

Griffiths and Harris cover essentially everything one needs to know to get started in Hodge theory with great insight and intuition, starting with De Rham's Theorem. The reader who has never seen hypercohomology might get his or her feet wet by studying the section in Griffiths and Harris on spectral sequences; I similarly recommend their very down-to-earth treatment of the representation theory of $\mathfrak{S}L(2, \mathbf{C})$.

Wells is an excellent expositor and his book in its later edition contains Hecht's beautiful computation relating L^{n-k} and the Hodge $*$-operator using representation theory.

LECTURE 2

PROPOSITION. *On a compact Kähler manifold M of dimension n, the Hodge $*$-operator induces an isomorphism $*: \mathcal{H}^{p,q}(M) \to \mathcal{H}^{n-p,n-q}(M)$.*

PROOF: Since $*$ commutes with the Laplacian, it takes harmonic forms to harmonic forms, and takes (p,q)-forms to $(n-p, n-q)$-forms. Since $*^2 = \pm\mathrm{id}$, it is an isomorphism. □

COROLLARY. *The map $H^k(M, \mathbf{C}) \otimes H^{2n-k}(M, \mathbf{C}) \to \mathbf{C}$ given by $(\alpha, \beta) \mapsto \int_M \alpha \wedge \beta$ is a perfect pairing, and the duality it induces*

$$H^k(M, \mathbf{C}) \cong H^{2n-k}(M, \mathbf{C})^*$$

induces dualities $H^{p,q}(M) \cong H^{n-p,n-q}(M)^*$.

We recall:

THEOREM (Poincaré Duality). *On a compact oriented manifold M of dimension n, the map*

$$H_k(M, \mathbf{Z}) \to H^{n-k}(M, \mathbf{Z})$$

defined by taking a k-cycle α to the cochain $\beta \mapsto \alpha \cap \beta$, where we choose a representative for α that meets β and $\partial\beta$ transversally, is an isomorphism.

DEFINITION. *If $\sigma \in H_k(M, \mathbf{Z})$, then $\psi_\sigma \in H^{n-k}(M, \mathbf{Z})$ is the image of σ under Poincaré duality, called the* **Poincaré dual class** *of σ. In N is an oriented manifold of dimension $n-p$, and $f: N \to M$ is continuous, then*

$$\psi_N = \psi_{f_*[N]} \in H^p(M, \mathbf{Z})$$

is the **Poincaré dual class** *of N.*

PROPOSITION. *For and $\sigma \in H_k(M, \mathbf{Z})$ and $\phi \in H^k(M, \mathbf{Z})$,*

$$< \phi, \sigma > = < \psi_\sigma \cup \phi, [M] > .$$

PROOF: This is just another way of stating the relation between intersection and cup product. □

We now assume that X is a smooth projective variety of dimension n. A codimension p algebraic subvariety $Z \subseteq X$ can be represented by a desingularization $\hat{Z}$ and a holomorphic map $i: \hat{Z} \to X$. The map

$$\omega \mapsto \int_{\hat{Z}} i^*\omega$$

defines an element

$$\int_{\hat{Z}} i^* \in A^{2n-2p}(X)^*.$$

Since exact forms are killed by Stokes' Theorem, this descends to cohomology to give an element of $H^{2n-2p}(X, \mathbf{C})^*$. Since

$$\int_{\hat{Z}} i^*\omega = 0$$

for all $(2n-2p)$-forms ω except those of type $(n-p, n-p)$, we have that in fact

$$\int_{\hat{Z}} i^* \in H^{n-p,n-p}(X)^*.$$

Using the duality above, we obtain the Poincaré dual class

$$\psi_Z \in H^{p,p}(X).$$

By construction, for all closed $(2n-2p)$-forms ω,

$$\int_{\hat{Z}} i^*\omega = \int_X \psi_Z \wedge \omega.$$

If

$$\omega \in \operatorname{im}(H^{2n-2p}(X,\mathbf{Z}) \to H^{2n-2p}(X,\mathbf{C})),$$

then

$$i^*\omega \in \operatorname{im}(H^{2n-2p}(\hat{Z},\mathbf{Z}) \to H^{2n-2p}(\hat{Z},\mathbf{C})),$$

so $\int_{\hat{Z}} i^*\omega \in \mathbf{Z}$. We thus conclude by Poincaré duality that

$$\psi_Z \in \operatorname{im}(H^{2p}(X,\mathbf{Z}) \to H^{2p}(X,\mathbf{C})) \cap H^{p,p}(X).$$

DEFINITION. *For a compact Kähler manifold M,*

$$H^{p,p}_{\mathbf{Z}}(X) = \operatorname{im}(H^{2p}(M,\mathbf{Z}) \to H^{2p}(M,\mathbf{C})) \cap H^{p,p}(M).$$

These are called the **integral (p,p)-classes** *of M. We also define*

$$\operatorname{Hdg}^p(M) = \{\eta \in H^{2p}(X,\mathbf{Z}) \mid j_*\eta \in H^{p,p}(X)\},$$

where $j_*\colon H^{2p}(X,\mathbf{Z}) \to H^{2p}(X,\mathbf{C})$ is the coefficient map.

Collecting what we have above, we get:

PROPOSITION. *Given a smooth projective variety X of dimension n and a codimension p algebraic subvariety $Z \subseteq X$, the Poincaré dual class ψ_Z lies in* $\operatorname{Hdg}(X)$. *For all closed $(2n-2p)$-forms ω,*

$$\int_{\hat{Z}} i^*\omega = \int_X j_*\psi_Z \wedge \omega.$$

DEFINITION. *ψ_Z is called the* **integral cycle class** *of Z, and $j_*\psi_Z$ the* **cycle class** *of Z.*

DEFINITION. *A codimension-p algebraic cycle on X is a finite formal linear combination of codimension p algebraic subvarieties of X with integral coefficients; $Z^p(X)$ will denote the set of all codimension p algebraic cycles on X. We will denote by $Z^p_h(X)$ the codimension p algebraic cycles homologically equivalent to 0 on X.*

DEFINITION. *We extend the notion of cycle class to cycles by*

$$\psi_{\sum_i n_i Z_i} = \sum_i n_i \psi_{Z_i}.$$

Note that

$$Z_h^p(X) = \{Z \in Z^p(X) \mid \psi_Z = 0\}.$$

An important general class of examples is if $E \to X$ is a rank r analytic vector bundle and $s \in H^0(X, E)$ is a section meeting the zero locus in a reduced variety, and $Z = \{s = 0\}$, then

$$\psi_Z = c_r(E).$$

There are of course similar results relating $c_k(E)$ to the locus where $r - k + 1$ sections are linearly dependent. For a very ample line bundle L, a section of L corresponds to a hyperplane section of $X \cap H$ under the projective embedding associated to L. Then

$$\psi_{X\cap H} = \omega,$$

the Kähler form, and the operator (unfortunately also called L) on differential forms satisfies

$$L = \wedge\psi_{X\cap H}.$$

As an example of how to use this, we mention:

PROPOSITION. *If* $\dim(X) = n$, *the restriction map*

$$r_k: H^k(X, \mathbf{C}) \to H^k(X \cap H, \mathbf{C})$$

has

$$\ker(r_k) = \ker(L).$$

Further,

$$\operatorname{coker}(r_k) = \begin{cases} P^{n-1}(X \cap H) & \text{if } k = n-1, \\ 0 & \text{otherwise.} \end{cases}$$

PROOF: The composition of $H^k(X, \mathbf{C}) \to H^k(X \cap H, \mathbf{C}) \to H^{k+2}(X, \mathbf{C})$ by r_k followed by $\wedge\psi_{X\cap H}$ is L, and therefore $\ker(r_k) \subseteq \ker(L)$. If $\alpha \in H^k(X, \mathbf{C})$ and $L\alpha = 0$, then $\alpha = L^{k-n}\beta$ for some $\beta \in P^{2n-k}(X)$. Using the Lefschetz Theorem, since $k \geq n$ and thus $2n - k \leq n$, we know that $r_k\beta \in P^{2n-k}(X \cap H)$. Thus

$$r_k\alpha = L_H^{n-k} r_{2n-k}\beta = 0,$$

proving that

$$\ker(r_k) = \ker(L).$$

Now by the Lefschetz Theorem we know that r_k is surjective for $k < n - 1$. For $k > n - 1$, every element of $H^k(X \cap H, \mathbf{C})$ is the image under a power of L of an element of some $H^m(X \cap H, \mathbf{C})$ for some $m < n - 1$, and therefore r_k is surjective. The same argument works for $\operatorname{im}(L) \subseteq H^{n-1}(X \cap H, \mathbf{C})$. On the other hand, if $\alpha \in H^{n-1}(X \cap H, \mathbf{C})$ is $r_k(\beta)$, then $L\alpha = r_k L\beta$, and $L\beta \notin \ker(L)$, so $L\alpha \neq 0$. We conclude that the cokernel of r_{n-1} is $P^{k-1}(X \cap H)$. □

The integral cycle class of an algebraic cycle associates to each cycle a natural topological invariant. It is quite wonderful that when this invariant vanishes, there is a natural analytic invariant associated to the cycle, which we will now define. Later, we will discuss a cohomology group in which these two natural invariants coexist. If $Z \in Z_h^p(X)$, then we can write $Z = \partial U$ for some topological $(2n-2p+1)$-chain. Integration over U gives a map

$$\int_U : A^{2n-2p+1}(X) \to \mathbf{C}$$

by

$$\omega \mapsto \int_U \omega.$$

If $\omega = d\phi$, then

$$\int_U \omega = \int_U d\phi = \int_Z \phi.$$

Now if $\omega \in F^{n-p+1}H^{2n-2p+1}(X, \mathbf{C})$, then if ω is exact, we can write $\omega = d\phi$ with $\phi \in \oplus_{m \geq n-p+1} A^{m,2n-2p+1-m}(X)$. Now

$$\int_U \omega = \int_U d\phi = \int_Z \phi = 0.$$

It follows as a consequence that $\int_U$ gives a natural element of

$$F^{n-p+1}H^{2n-2p+1}(X, \mathbf{C})^*.$$

Under the pairing

$$H^q(X, \mathbf{C}) \times H^{2n-q}(X, \mathbf{C}) \to \mathbf{C}$$

given by

$$(\alpha, \beta) \mapsto \int_X \alpha \wedge \beta,$$

we see that

$$\text{annihilator}(F^k H^q(X, \mathbf{C})) = F^{n-k}H^{2n-q}(X, \mathbf{C})$$

(annihilation is easy, and using the fact that for $\alpha \neq 0$, $\int_X \alpha \wedge (*\alpha) > 0$, we see that nothing else is in the annihilator), and thus

$$F^k H^q(X, \mathbf{C})^* \cong H^{2n-q}(X, \mathbf{C})/F^{n-k}H^{2n-q}(X, \mathbf{C}).$$

In particular, $F^{n-p+1}H^{2n-2p+1}(X, \mathbf{C})$ is naturally dual to

$$H^{2p-1}(X, \mathbf{C})/F^p H^{2p-1}(X, \mathbf{C}).$$

If we replace U by $\hat{U}$ so that $\partial\hat{U} = \partial U = Z$, then $\hat{U} - U$ is a topological $(2n-2p+1)$-cycle, and thus represents an element

$$\lambda \in H_{2n-2p+1}(X, \mathbf{Z}).$$

Thus

$$\int_{\hat{U}} - \int_U = \int_\lambda,$$

and the image of $\int_\lambda$ in $H^{2p-1}(X, \mathbf{C})$ is just what is usually defined to be the Poincaré dual class

$$j_*\psi_\lambda \in \text{im}(H^{2p-1}(X, \mathbf{Z})).$$

If we map $\int_U, \int_{\tilde{U}}$ to the quotient

$$J^p(X) = H^{2p-1}(X,\mathbf{C})/(F^p H^{2p-1}(X,\mathbf{C}) + H^{2p-1}(X,\mathbf{Z})),$$

then they map to the same element, which thus depends only on Z, and which we denote as $AJ_X(Z)$.

DEFINITION. *The* **p'th intermediate Jacobian** *or* **p'th Griffiths intermediate Jacobian** *of a compact Kähler manifold X is*

$$J^p(X) = H^{2p-1}(X,\mathbf{C})/(F^p H^{2p-1}(X,\mathbf{C}) + H^{2p-1}(X,\mathbf{Z})).$$

The **Abel-Jacobi map** *of X is*

$$AJ_X : Z^p_h(X) \to J^p(X),$$

defined by $Z \mapsto AJ_X(Z)$.

We note that if $\lambda_1, \ldots, \lambda_{2k}$ is a basis for the free part of $H^{2p-1}(X,\mathbf{Z})$, then if

$$\sum_i r_i \lambda_i \in F^p H^{2p-1}(X,\mathbf{C}),$$

then by conjugation

$$\sum_i r_i \lambda_i = 0,$$

and thus all $r_i = 0$. Thus $H^{2p-1}(X,\mathbf{Z})$ maps to a lattice

$$\Lambda \subseteq H^{2p-1}(X,\mathbf{C})/F^p H^{2p-1}(X,\mathbf{C}).$$

Therefore,

$$J^p(X) \cong \mathbf{C}^k/\Lambda$$

is a complex torus whose complex dimension is half the $(2p-1)$'st Betti number of X.

If X is a smooth projective variety and $p \leq \frac{n}{2}+1$, then the map $\Lambda \times \Lambda \to \mathbf{Z}$ given by

$$(\lambda,\mu) \mapsto \lambda \cup \mu \cup h^{n+2-2p},$$

where h is the hyperplane class in $H^2(X,\mathbf{Z})$ gives a non-degenerate alternating form of Λ, and hence an element of $H^2(J^p(X),\mathbf{Z})$. Furthermore, the condition that if

$$\alpha,\beta \in \oplus_{k\geq p} H^{2p-1-k,k}(X),$$

then

$$\int_X \alpha \wedge \beta \wedge \omega^{n+1-2p} = 0,$$

where ω is the Kähler form, translates into saying that the foregoing class in $H^2(J^p(X),\mathbf{Z})$ is of type $(1,1)$ and therefore represents an analytic line bundle L_J on $J^p(X)$.

Unfortunately, the Hodge Index Theorem says that $c_1(L_J)$ as a Hermitian form has

$$n^+ = \sum_k h^{p-1-2k,p+2k}(X)$$

positive eigenvalues and

$$n^- = \sum_k h^{p-2-2k,p+2k+1}$$

negative eigenvalues. If $n^- > 0$, one cannot hope that L_J has any holomorphic sections. It is nevertheless tempting to try to do something with L_J. If one goes back to the proof of Frobenius theorem on theta functions, it is not hard to see that for $m > 0$,

$$H^q(J^p(X), L_J^m) = 0$$

for $q \neq n^-$. Further,

$$H^{n^-}(J^p(X), L_J^m) = (-1)^{n^-} c_1(L_J^m)^{\dim(J^p(X))}.$$

If we look at the harmonic forms $\mathcal{H}^{n^-}(J^p(X), L_J^m)$, these are all proportional to the wedge of all the $d\bar{z}$'s in the negative eigenspace of $c_1(L_J)$, i.e.

$$\sum_k H^{p-2-2k,p+2k+1}(X).$$

Factoring this out gives a (non-holomorphic) map

$$\Phi: J^p(X) \to \mathbf{P}^N.$$

The image of this map is a projective variety, namely the image of the Weil intermediate Jacobian under the m-th power of its natural polarization. I suspect that the failure to exploit Φ may represent the influence of "holomorphic chauvinism." On the period domain, the direct image sheaves $\mathcal{R}^{n^-}_{\pi *}(L_J)$ give holomorphic vector bundles, which are worth studying even though they will not have holomorphic sections.

These unfortunate complications evaporate if $p = 1$, in which case we are dealing with divisors on X and $J^1(X) \cong \mathrm{Pic}^0(X)$; they evaporate if $p = n$ and we obtain the Albanese variety of X, or in the intermediate cases when

$$H^{2p-1}(X) = H^{p,p-1}(X) \oplus H^{p-1,p}(X),$$

as happens in the celebrated case of the cubic threefold.

There are many alternate ways to describe the intermediate Jacobian and Abel-Jacobi map, of ascending levels of elegance. In order to understand the first of these, it is useful to introduce logarithmic differentials.

DEFINITION. *Let X be a smooth projective variety and D a codimension 1 algebraic subvariety of X. We say that D has* **normal crossings** *if at every point $p \in X$, there exist local holomorphic coordinates $z_1, \ldots, z_n$ for X at p and a $k \geq 0$ such that on a neighborhood U of p, $U \cap D = \{z_1 z_2 \cdots z_k = 0\}$.*

We will call $z_1, \ldots, z_n$ **adapted local coordinates** *for D. We will denote by $D_{[i]} = \{p \in X \mid \mathrm{mult}_p(D) \geq i\}$.*

DEFINITION. *Let X be a smooth projective variety and D a divisor with normal crossings on X. A* **logarithmic p-form** *on X, D is a meromorphic p-form on X such that in terms of adapted local coordinates for D at every point of X, it has a local expression which is a linear combination with holomorphic coefficients of terms which are wedges of p differentials $dz_1/z_1, \ldots, dz_k/z_k, dz_{k+1}, \ldots, dz_n$; we will denote this by $\Omega_X^p(\log(D))$. Exterior derivative makes this into a complex $\Omega_X^\bullet(\log(D))$.*

REMARK: It is equivalent to say that $\omega \in \Omega^p(\log(D))$ and that $\omega, d\omega$ are meromorphic forms with only a simple pole on D. This latter formulation is more obviously invariant.

EXAMPLE.

For X a smooth projective curve and $D = \{p_1, \ldots, p_N\}$ a collection of points, there is a natural residue map

$$\mathrm{Res}\colon \Omega_X^1(\log(D)) \to \mathcal{O}_D \to 0$$

taking $\frac{1}{2\pi i} dz/z$ at p_i to 1_{p_i}. This fits into an exact sequence

$$0 \to \Omega_X^\bullet \to \Omega_X^\bullet(\log(D)) \to \Omega_D^{\bullet-1} \to 0.$$

The connecting map $H^0(\mathcal{O}_D) \to H^1(\Omega_X^1)$ takes 1_{p_i} to the fundamental class of X, which we may think of as ψ_{p_i}. It follows that $Z = \sum_i n_i p_i$ with $\sum_i n_i = 0$ lifts to a logarithmic 1-form

$$\eta_Z \in H^0(\Omega_X^1(\log(D))),$$

uniquely defined up to adding an element of $H^0(\Omega_X^1)$.

Our knowledge of the topology of $X - D$ leads us to note that

$$H^0(\Omega_X^1(\log(D))) \oplus H^1(\mathcal{O}_X) \cong H^1(X, \mathbf{C}),$$

which we may put in a more invariant way as

$$H^1(X - D, \mathbf{C}) \cong \mathbf{H}^1(\Omega_X^\bullet(\log(D))),$$

which fits into an exact sequence

$$0 \to H^1(X, \mathbf{C}) \to H^1(X - D, \mathbf{C}) \to H^0(D, \mathbf{C}) \to H^2(X, \mathbf{C}).$$

There is of course the complexification of the Gysin sequence for the pair X, D (we will discuss this sequence later on in general.) Given any $Z = \sum_i n_i p_i$ with $\sum_i n_i = 0$, it may be written as the boundary of a smooth chain γ on X so that $Z = \partial\gamma$. This defines an element of $H_1(X, D, \mathbf{Z})$. This in turn defines an element $\phi_Z \in H^1(X - D, \mathbf{Z})$ by $\sigma \mapsto \sigma \cap \gamma$. The difference $\eta_Z - \phi_Z$ belongs to the image of $H^1(X, \mathbf{C})$ in $H^1(X - D, \mathbf{C})$, and is determined modulo a class in

$H^1(X,\mathbf{Z})$ for the choice of γ and a class in $F^1H^1(X,\mathbf{C})$ for the choice of η_Z. We thus obtain a canonically determined element

$$u(Z) = [\eta_Z - \phi_Z] \in H^1(X,\mathbf{C})/(F^1H^1(X,\mathbf{C}) + H^1(X,\mathbf{Z})).$$

As one might suspect, $u(Z) = AJ_X(Z)$. To see this, we note that $\eta_Z - \phi_Z$ determines the element $\alpha \in H^1(X,\mathbf{C})$ given by

$$\int_\lambda \alpha = \int_\lambda \eta_Z - \phi_Z$$

for all $\lambda \in H_1(X,\mathbf{Z})$. However,

$$\int_\lambda \phi_Z \in \mathbf{Z},$$

so

$$\int_\lambda \eta_Z = \int_\lambda \alpha (\mathrm{mod}\ \mathbf{Z}).$$

Let X_ϵ denote the complement of a small neighborhood of γ in X. On $X-\gamma$, we may write $\eta_Z = \alpha + df$, where f increases by 1 as we cross γ. Thus for any $\omega \in H^0(X,\Omega^1_X)$,

$$\begin{aligned}\int_\gamma \omega &= \lim_{\epsilon\to 0} \int_{\partial X_\epsilon} \omega f \\ &= \lim_{\epsilon\to 0} \int_{X_\epsilon} \omega \wedge df \\ &= \lim_{\epsilon\to 0} \int_{X_\epsilon} \omega \wedge (\eta_Z - \alpha) \\ &= \int_X \omega \wedge \alpha.\end{aligned}$$

This means that α and $\int_\gamma$ represent the same element of $H^0(\Omega^1_X)^*$, and hence $u(Z) = AJ_X(Z)$.

A consequence of this is that if $AJ_X(Z) = 0$, then we may choose η_Z, ϕ_Z so that $\eta_Z - \phi_Z$ has integral periods on all 1-cycles on $X - D$. If we choose a base point $p \in X - D$ and define

$$f(q) = e^{2\pi i \int_p^q (\eta_Z - \phi_Z)},$$

then this is a well-defined independent of the choice of path joining p to q. It is holomorphic on $X - D$ and

$$df/f = 2\pi i(\eta_Z - \phi_Z),$$

and hence

$$\mathrm{Res}_{p_i}(df/f) = n_i,$$

and hence

$$\mathrm{ord}_{p_i}(f) = n_i,$$

so

$$\operatorname{div}(f) = Z.$$

This proves that if $AJ_X(Z) = 0$, then Z is the divisor of a meromorphic function.

The point is that much of this picture generalizes to an arbitrary divisor with normal crossings, as we will see in a later lecture.

Another way to proceed is to consider for the curve X the complex $\mathbf{Z} \to \mathcal{O}_X$ on X, which we will denote by $D^\bullet$. The hypercohomology $\mathbf{H}^2(D^\bullet)$ fits into an exact sequence

$$0 \to H^1(X, \mathbf{Z}) \to H^1(X, \mathcal{O}_X) \to \mathbf{H}^2(D^\bullet) \to H^2(X, \mathbf{Z}) \to 0.$$

Given a 0-cycle

$$Z = \sum_i n_i p_i$$

on X, we choose an open cover $\mathcal{U}$ of X by open sets U_α so that

$$Z \cap U_\alpha = \operatorname{div}(s_\alpha),$$

and now choose an element

$$(\xi_{\alpha\beta}) \in \mathcal{C}^1(\mathcal{U}, \mathcal{O}_X)$$

given by

$$\xi_{\alpha\beta} = log(s_\alpha / s_\beta),$$

for some branch of log. We note that the coboundary $\delta(\xi)$ is integral valued, hence is the image of some

$$\phi \in \mathcal{C}^2(\mathcal{U}, \mathbf{Z}).$$

Taken together, ξ, ϕ gives a cycle in the total complex which computes $H^2(D^\bullet)$. If we replace s_α by $s_\alpha u_\alpha$, where $u_\alpha \in \mathcal{O}_X^*(U_\alpha)$, then $log(u_\alpha)$ gives an element

$$\eta \in \mathcal{C}^0(\mathcal{U}, \mathcal{O}_X).$$

The coboundary $\delta\eta$ gives an element

$$\psi \in \mathcal{C}^1(\mathcal{U}, \mathbf{Z}),$$

and the new element of the total complex we would get differs from the old one by the coboundary of (η, ψ). We thus obtain an element of $H^2(D^\bullet)$ that depends only on Z. By a simple computation, ϕ represents $\sum_i n_i$ times the fundamental class in $H^2(X, \mathbf{Z})$. Thus, if $\sum_i n_i = 0$, we obtain a class in

$$H^1(X, \mathcal{O}_X)/H^1(X, \mathbf{Z}).$$

By a residue calculation we can see that this is $AJ_X(Z)$.

This approach to defining the Abel-Jacobi map also generalizes, leading to Deligne cohomology.

DEFINITION. *The 2p'th* **Deligne cohomology group** *is*

$$H^{2p}_{\mathcal{D}}(X) = \mathbf{H}^{2p}(\mathbf{Z} \to \Omega_X^{<p}).$$

PROPOSITION. *There is a natural exact sequence*

$$0 \to J^p(X) \to H^{2p}_{\mathcal{D}}(X) \to \mathrm{Hdg}^p(X) \to 0.$$

PROOF: There is an exact sequence of complexes

$$0 \to \Omega_X^{<p}[-1] \to (\mathbf{Z} \to \Omega_X^{<p}) \to \mathbf{Z} \to 0.$$

The long exact sequence for hypercohomology gives

$$H^{2p-1}(X,\mathbf{Z}) \to H^{2p-1}(X,\mathbf{C})/F^p H^{2p-1}(X,\mathbf{C}) \to H^{2p}_{\mathcal{D}}(X) \to$$
$$H^{2p}(X,\mathbf{Z}) \to H^{2p}(X,\mathbf{C})/F^p H^{2p}(X,\mathbf{C}),$$

where we are using a result of Lecture 1 to compute $\mathbf{H}^\bullet(\Omega_X^{<p})$. The desired sequence now follows. □

DEFINITION. *If $Z = \sum_i n_i Z_i$ is an algebraic cycle, the* **support** *of Z will be $|Z| = \cup_i Z_i \subseteq X$.*

Assume now that we have a codimension p algebraic cycle $\mathcal{Z} \subseteq X \times T$, and let p_1, p_2 be the projections of $X \times T$ to X, T respectively. If $m = \dim(T)$, then we have a natural map

$$\rho_{\mathcal{Z}}: Z^m(T) \to Z^p(X)$$

given by

$$W \mapsto p_{1*}(p_2^* W \cap \mathcal{Z}).$$

This gives a map $Z^m_h(T) \to Z^p_h(X)$. The claim is that this map factors through $J^m(T) \to J^p(X)$. To see this, if ω is a closed $(2n-2p+1)$-form on X, and if $W = \partial V$, then $p_2^* W \cap \mathcal{Z} = \partial U$, where $U = p_2^* V \cap \mathcal{Z}$. Now

$$\begin{aligned}\int_{p_{2*}(U)} \omega &= \int_U p_1^*\omega \\ &= \int_{p_2^* V \cap \mathcal{Z}} p_1^*\omega \\ &= \int_{p_1^* V} p_1^*\omega \wedge \psi_{\mathcal{Z}} \\ &= \int_V p_{2*}(p_1^*\omega \wedge \psi_{\mathcal{Z}}).\end{aligned}$$

The map

$$H^{2n-2p+1}(X,\mathbf{C}) \to H^1(T,\mathbf{C})$$

given by $\omega \mapsto p_{2*}(\omega \wedge \psi_{\mathcal{Z}})$ takes $F^{n-p+1}H^{2n-2p+1}(X,\mathbf{C})$ to $F^1H^1(T,\mathbf{C})$, and takes integral classes to integral classes; it thus induces a map

$$\tau_{\mathcal{Z}}: J^m(T) \to J^p(X)$$

such that

$$AJ_X \circ \rho_Z = \tau_Z \circ AJ_T.$$

We note that $J^m(T) = \mathrm{Alb}(T)$, which is an abelian variety if T is projective. If $T = \mathbf{P}^m$, then $\mathrm{Alb}(T) = 0$. We have thus proved:

PROPOSITION. *If $Z^p_{\mathrm{alg}}(X)$ denotes the codimension p cycles algebraically equivalent to 0, then*

$$AJ_X(Z^p_{\mathrm{alg}}(X)) = A \subseteq J^p(X),$$

for some abelian subvariety A of $J^p(X)$. If $Z^p_{\mathrm{rat}}(X)$ denotes the codimension p algebraic cycles on X rationally equivalent to 0, then

$$AJ_X(Z^p_{\mathrm{rat}}(X)) = 0.$$

In particular, AJ_X factors

$$AJ_X : CH^p(X) \to J^p(X).$$

REMARK: A is called the **algebraic part** of $J^p(X)$. A lifts to a sub-Hodge structure of weight 1 in $H^{2p-1}(X, \mathbf{C})$. The generalized Hodge conjecture would imply that A is equal to the maximal abelian subvariety of $J^p(X)$, or equivalently that the lifting of A is equal to the maximal sub-Hodge structure of weight 1 of $H^{2p-1}(X, \mathbf{C})$.

We want an infinitesimal version of the formula above. Let

$$Z_t = p_{1*}(\mathcal{Z} \cap (X \times \{t\})),$$

then if Z_t is smooth, we may choose a lifting $\tilde{\nu} \in A^0(Z_t, \Theta_X|_{Z_t})$ lifting any $v \in T_{T,t}$ to tangent vectors to $\mathcal{Z}$, and projecting down a section $\nu \in H^0(N_{Z_t/X})$. We obtain ν by any $\mathcal{C}^\infty$ lifting of v to $T_{\mathcal{Z}}|_{Z_t}$. If we take V_ϵ to be a small curve originating in t with tangent vector v, and $U_\epsilon = p_2^* V_\epsilon \cap \mathcal{Z}$, then by the formula above,

$$\lim_{\epsilon \to 0}\left(\int_{U_\epsilon} \omega\right) = \lim_{\epsilon \to 0} \int_{p_2^*(V_\epsilon) \cap \mathcal{Z}} \omega = \int_{Z_t} (< \omega, \nu >).$$

Note that if

$$\omega \in F^{n-p+1} H^{2n-2p+1}(X, \mathbf{C}),$$

then this is zero if $\omega \in F^{n-p+2} H^{2n-2p+1}(X, \mathbf{C})$. The derivative of the Abel-Jacobi map, called the **infinitesimal Abel-Jacobi map** $AJ_{X,*}$ is thus a map $T_{T,t} \to H^{p-1,p}(X)$. If we identify $H^{p-1,p}(X) \cong H^{n-p+1,n-p}(X)^*$, then given $\omega \in H^{n-p+1,n-p}(X)$, our calculations so far give the formula

$$< AJ_{X,*}(v), \omega >= \int_Z < \omega, \nu > .$$

The infinitesimal Abel-Jacobi map may be regarded as a map

$$H^0(N_{Z/X}) \to H^{p,p-1}(X),$$

or taking the codifferential of the Abel-Jacobi map we have a map

$$H^{n-p}(\Omega_X^{n-p+1}) \to H^0(N_Z)^* \cong H^{n-p}(N_Z^* \otimes K_Z).$$

The normal sequence

$$0 \to \Theta_Z \to \Theta_X|_Z \to N_{Z/X} \to 0$$

induces a natural map

$$\Omega_X^{n-p+1}|_Z \to N_{Z/X}^* \otimes K_Z,$$

and the formula above for the infinitesimal Abel-Jacobi map is just that the codifferential is the composition of the restriction map

$$H^{n-p}(\Omega_X^{n-p+1}) \to H^{n-p}(\Omega_X^{n-p+1}|_Z)$$

with the induced map on cohomology

$$H^{n-p}(\Omega_X^{n-p+1}|_Z) \to H^{n-p}(N_{Z/X}^* \otimes K_Z).$$

Neither the image nor the kernel of the Abel-Jacobi map are well understood. For $p = 1, n$, the Abel-Jacobi map is surjective, but in the intermediate case it generally is not. This is rather clear, since any algebraic family of cycles must have tangent space parallel to $H^{p-1,p}(X)$, and thus we have at most as image a countable collection of lower-dimensional subsets of $J^p(X)$. One of the outstanding open problems in Hodge theory (at least, outstanding to me) is to understand better the image of the Abel-Jacobi map.

Bibliographic Notes

Poincaré duality may be approached from a number of different points of view—see Greenberg and Harper, *Algebraic Topology: A First Course* or Fomenko, *Differential Geometry and Topology* for geometric approaches, Griffiths and Harris, *Principles of Algebraic Geometry* for an analytic approach over **R** using Hodge theory, and Iversen, *Cohomology of Sheaves* for a highly abstract approach.

The Griffiths intermediate Jacobian was introduced in a remarkable series of papers by Griffiths, with "On the periods of certain rational integrals: I and II," Ann. Math. 460-495, 498-541 an excellent introduction to both the intermediate Jacobian and the Abel-Jacobi map. The Weil intermediate Jacobian is described in Weil, *Variétés Kählériennes.* Griffiths once told me that as a student he was supposed to give a lecture on Weil's work but did not have Weil's book available when he was preparing the talk, and so he tried to reconstruct the intermediate Jacobian by using the criterion (which Weil's Jacobian does not satisfy) that it should vary holomorphically for holomorphic families of varieties, with the result that, as he put it, he "got the signs wrong."

The best reference I know for Deligne cohomology is Esnault and Viehweg, "Deligne-Beilinson cohomology," in Rapoport, Schappacher and Schneider, *Beilinson's Conjectures on Special Values of L-Functions.* Even this very nice reference does not contain everything about Deligne cohomology one needs to know for these notes.

LECTURE 3

We want to assume now that we have an analytic family $\pi: \mathcal{X} \to D$ of compact Kähler manifolds X_t parametrized by the unit disc D. We assume that π is a submersion (i.e. that we have a flat family), and thus all the X_t are smooth. We wish to regard the X_t as deformations of $X = X_0$ and study how the Hodge structure varies with t.

The holomorphic vector field $\frac{\partial}{\partial t}$ on D may be lifted (there are many choices) to a $\mathcal{C}^\infty$ vector field Y on $\mathcal{X}$; we may think of $Y \in A^0(\Theta_{\mathcal{X}}|_X)$ by identifying the real tangent space to $\mathcal{X}$ and $\Theta_{\mathcal{X}}$. We obtain a $\mathcal{C}^\infty$ map $F: X_0 \times D \to \mathcal{X}$ by

$$F(x,t) = \Phi_{tY}(1),$$

where in general $\Phi_Y(t)$ denotes the flow associated to the vector field Y. This has the property that $\pi \circ F = p_2$, where p_2 is the projection $p_2: X \times D \to D$. Note that $F|_{X_0 \times \{t\}} \to X_t$ is a diffeomorphism, which we denote by F_t. The induced map $F_t^*: H^k(X_t, \mathbf{C}) \to H^k(X_0, \mathbf{C})$ is an isomorphism and does not depend on the choice of Y; it preserves the integral lattice.

The exact sequence

$$0 \to \Theta_{X_0} \to \Theta_{\mathcal{X}}|_{X_0} \to \pi^*\Theta_D \to 0$$

shows that if we take $\bar\partial Y \in A^{0,1}(\Theta_{\mathcal{X}}|_{X_0})$, it maps to zero in $\pi^*\Theta_D$, and thus comes from an element $\eta \in A^{0,1}(\Theta_{X_0})$. If we replace Y by a different lifting $\tilde Y$ of $\frac{\partial}{\partial t}$, then $\tilde Y - Y$ is the image of a smooth element $\psi \in A^0(\Theta_X)$, and then $\tilde\eta = \eta + \bar\partial\psi$. Thus η represents a well-defined element of $H^1(X_0, \Theta_{X_0})$, which is the **Kodaira-Spencer class** associated to $\frac{\partial}{\partial t}$.

The first thing to remark is that for any holomorphic family of vector bundles E_t on X_t, the support loci $\{t \mid h^q(X_t, E_t) \geq m\}$ are analytic subvarieties. This, combined with the fact that $\sum_{p+q=k} h^q(\Omega^p_{X_t}) = h^k(X_0, \mathbf{C})$, shows that the numbers $h^{p,q}(X_t) = h^q(X_t, \Omega^p_{X_t})$ are constant. We will now fix a k and denote by $f^p = \dim(F^p H^k(X_t, \mathbf{C}))$, which are constants independent of t. By $f^\bullet$ we will denote $(f^k, f^{k-1}, \ldots, f^0)$.

DEFINITION. *For a complex vector space V, the* **flag variety** $\mathrm{Fl}(f^\bullet, V)$ *will denote the set of nested subvarieties $F^k \subseteq F^{k-1} \subseteq \cdots \subseteq F^0 = V$ with $\dim(F^p) = f^p$ for all p.*

If we use F_t^* to identify $H^k(X_t, \mathbf{C})$ with $H^k(X_0, \mathbf{C})$, then we may regard the Hodge filtration $F^p H^k(X_t, \mathbf{C})$ as giving a varying filtration $F^p(t) H^k(X_0, \mathbf{C}) = F_t^* F^p H^k(X_t, \mathbf{C})$. We thus obtain a $\mathcal{C}^\infty$ map

$$P_{\mathcal{X}}: D \to \mathrm{Fl}(f^\bullet(X_0), H^k(X_0, \mathbf{C}))$$

defined by

$$t \mapsto F^\bullet(t) H^k(X_0, \mathbf{C}).$$

DEFINITION. *The map $P_{\mathcal{X}}$ is called the* **local period map** *of the family $\mathcal{X}$.*

The space $\mathrm{Fl}(f^\bullet, V)$ naturally has the structure of a complex manifold. If we pick locally a holomorphically varying basis $e_1, \ldots, e_N$ for V such that $e_1, \ldots, e_{f^p}$ is a basis for F^p for every p, then $e_i \mapsto de_i$ gives a natural map $F^p \to V \otimes \Omega^1_{\mathrm{Fl}(f^\bullet,V)}$. If we change basis by $\tilde{e}_i = \sum_j a_{ij} e_j$, then we have that

$$d\tilde{e}_i = \sum_j a_{ij} de_j + \sum_j (da_{ij}) e_j.$$

We thus see that the map $\delta^p \colon F^p \to (V/F^p) \otimes \Omega^1_{\mathrm{Fl}(f^\bullet,V)}$ does not depend on the choice of basis. We may rewrite δ^p equivalently as a map

$$\delta^p \colon T_{\mathrm{Fl}(f^\bullet,V)} \to \mathrm{Hom}(F^p, V/F^p).$$

Elaborating a bit on this argument gives that:

PROPOSITION.

$$T_{\mathrm{Fl}(f^\bullet,V),F^\bullet} = \{\gamma^\bullet \in \oplus_p \mathrm{Hom}(F^p, V/F^p) \mid \gamma^p \circ i^p = \pi^p \circ \gamma^{p+1} \text{ for all } p\},$$

where $i^p \colon F^{p+1} \to F^p$ is the natural inclusion and $\pi^p \colon V/F^{p+1} \to V/F^p$ is the natural projection.

In the foregoing proposition, the map is given by $\oplus_p \delta^p$.

Griffiths discovered the following three beautiful facts about the period map:

THEOREM.
(1) The local period map $P_{\mathcal{X}} \colon D \to \mathrm{Fl}(f^\bullet_X, H^k(X, \mathbf{C}))$ is holomorphic;
(2) The derivative of the period map $P_{\mathcal{X}*}$ lands in $\oplus_p \mathrm{Hom}(F^p, F^{p-1}/F^p)$;
(3) By part (2), we have that F^{p+1} goes to zero in F^{p-1}/F^p, and thus we have an induced element of $F^p/F^{p+1} \to F^{p-1}/F^p$, or equivalently $H^{k-p}(\Omega^p_{X_0}) \to H^{k-p+1}(\Omega^{p-1}_{X_0})$. This map is the composition of the cup product map

$$H^{k-p}(\Omega^p_{X_0}) \xrightarrow{\cup \eta} H^{k-p+1}(\Theta_{X_0} \otimes \Omega^p_{X_0}),$$

where η is the Kodaira-Spencer class associated with $\frac{\partial}{\partial t}$, with the map

$$H^{k-p+1}(\Theta_{X_0} \otimes \Omega^p_{X_0}) \to H^{k-p+1}(\Omega^{p-1}_{X_0})$$

induced by the contraction map

$$\Theta_{X_0} \otimes \Omega^p_{X_0} \to \Omega^{p-1}_{X_0}.$$

REMARK: Part (2) is usually called the **infinitesimal period relation** or **Griffiths transversality**.

PROOF: Since the Hodge numbers are constant for the family, the sheaves $\mathcal{R}^{k-i}_{\pi*}\Omega^i_{\mathcal{X}/D}$ are holomorphic vector bundles, and we may choose a local basis of analytic sections. From the exact sequence

$$0 \to \Omega^{p-1}_{\mathcal{X}/D} \otimes \pi^* \Omega^1_D \to \Omega^p_{\mathcal{X}} \to \Omega^p_{\mathcal{X}/D} \to 0,$$

we may lift a local analytic section of $\mathcal{R}^{k-p}_{\pi*}\Omega^p_{\mathcal{X}/D}$ to an element of $A^{p,k-p}(\mathcal{X})$.

If $G\colon \mathcal{X} \to X_0 \times D$ is the inverse of F, then $p_1 \circ G\colon \mathcal{X} \to X_0$ is a retraction, inducing an isomorphism on cohomology. This induces a natural isomorphism $\mathcal{R}^k_{\pi*}\mathbf{C} \cong \mathcal{O}_D \otimes H^k(X_0, \mathbf{C})$, and thus naturally gives $\mathcal{R}^k_{\pi*}\mathbf{C}$ the structure of a flat vector bundle, with associated flat connection $\nabla_{\mathcal{X}/D}$, the **local Gauss-Manin connection.**

If $e(t)$ is a smoothly varying section of $F^pH^k(X_t, \mathbf{C})$, we may find a form $\phi \in \sum_{i\geq p} A^{i,k-i}(\mathcal{X})$ such that $\phi|_{X_t}$ represents $e(t)$ for all t. Now

$$\begin{aligned}\frac{\partial e}{\partial \bar{t}} &= \mathcal{L}_{\frac{\partial}{\partial \bar{t}}}(F^*\phi)\\ &=< dF^*\phi, \frac{\partial}{\partial \bar{t}} > +d < F^*\phi, \frac{\partial}{\partial \bar{t}} >\\ &= F^* < d\phi, \bar{Y} > +F^*d < \phi, \bar{Y} > .\end{aligned}$$

However, $d\phi$ has at least p dz's and thus so does $< d\phi, \bar{Y} >$; similarly, $d < \phi, \bar{Y} >$ has at least p dz's. Thus

$$\frac{\partial e(t)}{\partial \bar{t}}(0) \in F^pH^k(X_0, \mathbf{C}),$$

and therefore goes to zero in $H^k(X_0, \mathbf{C})/F^pH^k(X_0, \mathbf{C})$. It follows that the filtration $F^pH^k(X_t, \mathbf{C})$ varies holomorphically.

Likewise,

$$\frac{\partial e(t)}{\partial t} = F^* < d\phi, Y > +F^*d < \phi, Y >= F^*\mathcal{L}_Y\phi.$$

This has at least $p-1$ dz's, from which we conclude that it lies in $F^{p-1}H^k(X_t, \mathbf{C})$. This proves part (2). Furthermore, the Kodaira-Spencer class associated to $\frac{\partial}{\partial t}$ is $\eta = \bar{\partial}Y|_{X_0}$. We note that $\mathcal{L}_Y d\bar{z}_i = d(Y\bar{z}_i) = 0$ and $\mathcal{L}_Y dz_i = d\alpha_i$, where locally $Y = \sum_i \alpha_i \frac{\partial}{\partial z_i}$. If $\phi = \sum \phi_{I,J} dz_I \wedge d\bar{z}_J$, then the only term of $\mathcal{L}_Y\phi$ that has fewer than p dz's is

$$\sum_{i,I,J} \bar{\partial}\alpha_i \wedge \phi_{I,J} dz_{I-i} \wedge d\bar{z}_J$$

which is the contraction of *phi* with the Kodaira-Spencer class $\bar{\partial}Y$. □

The definitions of period map and Gauss-Manin connection can be globalized. Let $\pi\colon \mathcal{X} \to T$ be a holomorphic family of compact Kähler manifolds, with π a submersion. Then $\mathcal{R}^k_{\pi*}\mathbf{C}$ is a local system on T and $H^k = \mathcal{R}^k_{\pi*}\mathbf{C} \otimes \mathcal{O}_T$ is a flat vector bundle with flat connection $\nabla_{\mathcal{X},T}$ which on each holomorphic disc in T agrees with the local definition. We note that $\Omega^\bullet_{\mathcal{X}/T}$ is a free resolution of $\pi^*\mathcal{O}_T$, and thus

$$\mathbf{R}^k_{\pi*}(\Omega^\bullet_{\mathcal{X}/T}) \cong \mathcal{R}^k_{\pi*}\mathbf{C} \otimes \mathcal{O}_T,$$

and furthermore

$$\mathrm{im}(\mathbf{R}^k_{\pi*}(\Omega^{\geq p}_{\mathcal{X}/T}) \to \mathbf{R}^k_{\pi*}(\Omega^\bullet_{\mathcal{X}/T}))$$

is a holomorphic subbundle $\mathbf{F}^p$ of the flat vector bundle H^k whose fiber at t is $F^pH^k(X_t, c)$; these are the **Hodge bundles** of the family $\pi\colon \mathcal{X} \to T$; the sheaf of holomorphic sections of $\mathbf{F}^p$ will be denoted $\mathcal{F}^p$.

We have the exact sequence

$$0 \to \pi^*\Omega^1_T \to \Omega^1_{\mathcal{X}} \to \Omega^1_{\mathcal{X}/T} \to 0.$$

If we let $\mathcal{S}$ be the kernel of $\Omega^p_{\mathcal{X}} \to \Omega^p_{\mathcal{X}/T}$, then there is a natural map $\mathcal{S} \to \Omega^{p-1}_{\mathcal{X}/T} \otimes \pi^*\Omega^1_T$. We thus obtain maps $\mathcal{R}^q_{\pi*}\Omega^p_{\mathcal{X}/T} \to \mathcal{R}^{q+1}_{\pi*}\mathcal{S} \to \mathcal{R}^{q+1}_{\pi*}\Omega^{p-1}_{\mathcal{X}/T} \otimes \Omega^1_T$. Taking the fiber at 0, we obtain $H^q(\Omega^p_X) \to H^{q+1}(\Omega^{p-1}_X) \otimes T^*_{T,0}$, which is the derivative of the period map. Indeed, we note that $\Omega^\bullet_{\mathcal{X}}$ is filtered by $G^p\Omega^\bullet_{\mathcal{X}} = \mathrm{im}(\pi^*\Omega^p_T \otimes \Omega^{\bullet-p}_{\mathcal{X}} \to \Omega^\bullet_{\mathcal{X}})$. We note that

$$\mathrm{Gr}^p_G\Omega^\bullet_{\mathcal{X}} = \pi^*\Omega^p_T \otimes \Omega^{\bullet-p}_{\mathcal{X}/T}.$$

This filtration gives a spectral sequence which abuts to $\mathbf{R}^k_{\pi*}(\Omega^\bullet_{\mathcal{X}})$ with E_1 term

$$E_1^{p,q} = \Omega^p_T \otimes \mathbf{R}^{p+q}_{\pi*}(\Omega^{\bullet-p}_{\mathcal{X}/T}) \cong \Omega^p_T \otimes \mathcal{R}^q_{\pi*}\mathbf{C}.$$

In this spectral sequence, the first differentials are

$$d_1\colon \Omega^p_T \otimes \mathcal{R}^q_{\pi*}\mathbf{C} \to \Omega^{p+1}_T \otimes \mathcal{R}^q_{\pi*}\mathbf{C},$$

and these are ∇. If we filter the complex $\Omega^{\bullet-p}_{\mathcal{X}/T}$ by

$$B^m\Omega^{k-p}_{\mathcal{X}/T} = \begin{cases} \Omega^{k-p}_{\mathcal{X}/T} & \text{if } k-p \geq m, \\ 0 & \text{otherwise}, \end{cases}$$

then

$$B^m\mathbf{R}^{p+q}(\Omega^{\bullet-p}_{\mathcal{X}/T}) = \mathcal{F}^{m+p}.$$

The fact that d_1 preserves the filtration is the infinitesimal period relation.

The extension class of the exact sequence

$$0 \to \Theta_{\mathcal{X}/T} \to \Theta_{\mathcal{X}} \to \pi^*\Theta_T \to 0$$

is a class

$$e \in H^1(\pi^*\Omega^1_T \otimes \Theta_{\mathcal{X}/T}).$$

By the Leray spectral sequence, e goes to a class

$$\bar{e} \in \Omega^1_T \otimes \mathcal{R}^1_{\pi*}\Theta_{\mathcal{X}/T},$$

and this is equivalent to the Kodaira-Spencer map

$$\Theta_T \to \mathcal{R}^1_{\pi*}\Theta_{\mathcal{X}/T}.$$

From the dual of the exact sequence above, if T has dimension 1, we obtain an exact sequence

$$0 \to \pi^*\Omega^1_T \otimes \Omega^{p-1}_{\mathcal{X}/T} \to \Omega^p_{\mathcal{X}} \to \Omega^p_{\mathcal{X}/T} \to 0$$

whose extension class lies in

$$H^1(\pi^*\Omega^1_T \otimes \operatorname{Hom}(\Omega^p_{\mathcal{X}/T}, \Omega^{p-1}_{\mathcal{X}/T}))$$

which is the image of e under the natural map

$$\Theta_{\mathcal{X}/T} \to \operatorname{Hom}(\Omega^p_{\mathcal{X}/T}, \Omega^{p-1}_{\mathcal{X}/T}).$$

Using the Leray spectral sequence, we see that the connecting map

$$\mathcal{R}^q_{\pi *}\Omega^p_{\mathcal{X}/T} \to \Omega^1_T \otimes \mathcal{R}^{q+1}\Omega^{p-1}_{\mathcal{X}/T}$$

is just cup product with the Kodaira-Spencer map $\Omega^1_T \otimes \mathcal{R}^1_{\pi *}\Theta_{\mathcal{X}/T}$. This is Griffiths' formula for the derivative of the period map.

The flat vector bundle H^k gives a natural representation

$$\rho: \pi_1(T, 0) \to \operatorname{Aut}(H^k(X_0, \mathbf{Z}))$$

called the **monodromy representation** of the family $\mathcal{X} \to T$; $\rho(\gamma)(\phi)$ is defined by parallel translating $\phi \in H^k(X_0, \mathbf{Z})$ around the loop γ. The image of the monodromy representation is restricted by the fact that any topological information is preserved. If $\dim(X) = n$, then for every γ, $\rho(\gamma)$ must preserve the intersection pairing $H^n(X_0, \mathbf{Z}) \times H^n(X_0, \mathbf{Z}) \to H^{2n}(X, \mathbf{Z})$.

DEFINITION. *If there exists an analytic line bundle $\mathcal{L} \to \mathcal{X}$ such that $L_t = \mathcal{L}|_{X_t}$ is ample for all t, we say that $\mathcal{X}, \mathcal{L}$ is a* ***family of polarized varieties.***

For a family of polarized varieties, if $k \leq n$, we have the pairing $<,>$ on $H^k(X, \mathbf{Z})$ defined in the first lecture, and the monodromy representation must preserve this. Let us call Γ the subgroup of $\operatorname{Aut}(H^k(X_0, \mathbf{Z}))$ preserving this pairing. Let

$$D \subseteq \operatorname{Fl}(f^\bullet(X_0), H^k(X_0, \mathbf{C}))$$

be the set of flags corresponding to Hodge structures. Then Γ acts on D, and D/Γ is called the **period domain**. The local period map gives rise to a (global) **period map** $P: T \to D/\Gamma$, well-defined and holomorphic.

The period domain and period map have been extensively studied, although we will not do this here. The period domain has a lot of geometry, although in the general case not enough to allow Hodge theorists to carry out their program of understanding the original variety from its Hodge structure. Luckily, this can be done sometimes, as we will see later on in these lectures.

The most famous example is for X a curve of genus g. The cup product map is

$$H^1(\Theta_X) \to \operatorname{Hom}(H^0(K_X), H^1(\mathcal{O}_X)).$$

Since cup product behaves well under duality, we may use

$$H^1(\mathcal{O}_X) \cong H^0(K_X)^*$$

and

$$H^1(\Theta_X) \cong H^0(K_X^2)^*$$

to get the equivalent map

$$\mu: H^0(K_X) \otimes H^0(K_X) \to H^0(K_X^2)$$

which must be cup product, i.e. multiplication. We thus conclude that the dual of P_* is multiplication, and that the kernel of P_* is the annihilator of the image of μ. If g is a non-hyperelliptic curve of genus ≥ 2, then μ is surjective (Noether's Theorem) and P_* is injective.

An interesting construction is to iterate the derivative of P in the following way: If X has dimension n, then each $\eta \in H^1(\Theta_X)$ gives maps

$$\phi_p(\eta): H^{p,n-p}(X) \to H^{p-1,n-p+1}(X).$$

Thus

$$\phi_1 \circ \phi_2 \circ \cdots \circ \phi_n = \Phi(\eta): H^0(K_X) \to H^n(\mathcal{O}_X).$$

Now $H^n(\mathcal{O}_X) \cong H^0(K_X)^*$, and we may show that $\Phi(\eta) \in S^2H^0(K_X)^*$. This gives us a well-defined map

$$\Phi: S^n H^1(\Theta_X) \to S^2 H^0(K_X)^*,$$

often called the **Yukawa coupling**. If K_X is trivial, this gives a holomorphically varying class (mod scalars) of homogeneous polynomial of degree n on the tangent space to moduli space. For surfaces, where $H^1(\Theta_X) \cong H^1(\Omega_X^1)$, this map is just wedge product, but for 3-folds it gives a rather mysterious cubic form. More generally, $S^2H^0(K_X)$ gives a linear system of homogeneous polynomials of degree n on the tangent space to moduli, which one might call the **Yukawa linear system**. For curves, this is just the image of μ again, but it has not really been studied in other cases. For a smooth hypersurface $\{F = 0\}$ of degree $d \geq n+1$ in $\mathbf{P}^n$, this linear system is just the kernel of the multiplication map

$$S^2 R_{H^d,F} \to R_{H^{2d},F};$$

see Lecture 4 for this notation. One may generalize this by noticing that one also gets a map

$$S^{n-2p}H^1(\Theta_X) \to S^2H^{p,n-p}(X)^*$$

by iterating the differential, and thus a linear system of homogeneous polynomials of degree $n-2p$ on the tangent space to moduli for each $0 \leq p < n/2$; one might call these the **generalized Yukawa linear systems**. They are nested in an obvious way. The maps

$$S^{n-2p}H^1(\Theta_X) \to S^2H^{p,n-p}(X)^*$$

factor through

$$H^{n-2p}(\wedge^{n-2p}\Theta_X) \cong H^{2p}(\Omega_X^{n-2p} \otimes K_X)^*,$$

where the map

$$S^2H^p(\Omega_X^{n-p}) \to H^{2p}(\Omega_X^{n-2p} \otimes K_X)$$

is cup product followed by the canonical map

$$S^2\Omega_X^{n-p} \to \Omega_X^{n-2p} \otimes K_X.$$

It is worth mentioning that one need not stop with the first derivative of the period map. There are a number of interesting ways to look at higher derivatives. For example, for curves, Griffiths and I computed the **second fundamental form** of the period map. Locally, the period map sends an open subset of $\mathcal{M}_g$ to $G(g, H^1(X, \mathbf{C}))$ which sits in $\mathbf{P}(\wedge^g H^1(X, \mathbf{C}))$ under the Plücker embedding. Now given an (local) analytic subvariety $f: M \to \mathbf{P}^k$, Griffiths and Harris define a second fundamental form

$$d^2 f \in H^0(M, S^2\Omega_M^1 \otimes N_{M/\mathbf{P}^k})$$

given by taking the second partials of the position vector and projecting on the normal bundle. Now

$$\Omega^1_{\mathcal{M}_g,X} \cong H^0(K_X^2)$$

and if D is the period domain,

$$\Theta_{D,P(X)} \cong S^2 H^0(K_X)^*,$$

and

$$N^*_{\mathcal{M}_g/D} \cong I_2(\phi_K(X)),$$

where

$$I_2(\phi_K(X)) = \ker(S^2 H^0(K_X) \to H^0(K_X^2)).$$

The second fundamental form of the Grassmannian under the Plücker embedding, restricted to $\mathcal{M}_g$, is the image of

$$\wedge^2 H^0(K_X) \otimes \wedge^2 H^0(K_X) \to S^2 H^0(K_X^2)$$

under the map

$$(\alpha \wedge \beta) \otimes (\gamma \wedge \delta) \mapsto (\alpha\gamma) \otimes (\beta\delta) - (\alpha\delta) \otimes (\beta\gamma).$$

If

$$S^2 H^0(K_X) \to H^0(K_X^2)$$

is onto and $I_*(\phi_K(X))$ is generated by quadrics (i.e. if Enriques-Petri holds), then the quotient of $S^2 H^0(K_X^2)$ by the image of the second fundamental form of the Grassmannian is $H^0(K_X^4)$. The second fundamental form

$$d^2 P: I_2(\phi_K(X)) \to H^0(K_X^4).$$

If we consider

$$H^0(K_X) \otimes H^0(K_X) \cong H^0(X \times X, p_1^* K_X \otimes p_2^* K_X),$$

then this is filtered by

$$H^0(X \times X, \mathcal{I}_\Delta^k \otimes p_1^* K_X \otimes p_2^* K_X).$$

$I_2(\phi_X)$ is just the symmetric part of the first piece of this filtration, and any such element in fact lies in the second such piece. We thus obtain a map

$$I_2(\phi_K(X)) \to H^0(p_1^*K_X \otimes p_2^*K_X \otimes (\mathcal{I}_\Delta^2/\mathcal{I}_\Delta^3)) \cong H^0(K_X^4);$$

up to a non-zero constant, we showed that this is d^2P. This construction is part of a family of constructions called **Wahl maps**.

Another interesting use of higher derivatives of the period map is the work of Yakov Karpishpan.

Bibliographic Notes

An elementary introduction to deformation theory and the Kodaira-Spencer class is in Morrow and Kodaira, *Complex Manifolds*.

The global properties of the period map are discussed in the beautiful lectures of Cornalba and Griffiths at the Arcata algebraic geometry conference in 1973.

The infinitesimal period relations appear in Griffiths, "On the periods of certain rational integrals: I and II," Ann. Math. 460-495, 498-541.

The Yukawa coupling is discussed at length, but not under that name, in Carlson, Green, Griffiths, and Harris, "Infinitesimal variation of Hodge structure (I),"" Comp. Math 50 (1983), 109-205.

The second fundamental form of Griffiths-Harris appears in their paper "Algebraic geometry and local differential geometry," Ann. Sci. Ec. Nor. Sup 12 (1979), 355-432. The computation of the second fundamental form for curves is unpublished work of Green-Griffiths.

LECTURE 4

In this lecture we will discuss the relationship between the Hodge theory of a smooth subvariety X of a smooth projective variety Y. Usually, we will only be able to make this relationship explicit when X is the transverse zero section of a holomorphic vector bundle $E \to Y$ whose rank equals the codimension of X, and then only if we have a lot of vanishing theorems about the cohomology of E. This tends to happen if E is sufficiently ample. The main case will be complete intersection varieties in projective space.

Before we start, we recall:

THEOREM (Bott Vanishing Theorem). *On* $\mathbf{P}^n$,

$$H^q(\Omega^p_{\mathbf{P}^n}(k)) = 0$$

unless

$$\begin{cases} q = 0, n, \text{or} \\ p = q, k = 0. \end{cases}$$

REMARK: The point of the theorem is that these are homogeneous bundles on a homogeneous space. In fact, it computes the non-zero cohomology groups, which turn out to be irreducible representations of the general linear group $GL(n+1, \mathbf{C})$. Of course, $H^p(\Omega^p_{\mathbf{P}^n}) \cong \mathbf{C}$, and the other non-zero groups can be computed from the Euler sequence.

In order to begin, assume that on a compact complex manifold X we have an exact sequence of vector bundles

$$0 \to A \to B \to C \to 0,$$

where the bundles have ranks a, b, c respectively. The obstruction to splitting this exact sequence analytically is the **extension class** $e \in H^1(X, \mathrm{Hom}(C, A))$. It is a useful general fact that there is an exact sequence

$$0 \to S^k A \to S^{k-1} A \otimes B \to \cdots \to \wedge^k B \to \wedge^k C \to 0$$

for all $k \geq 0$, and that the extension class of this sequence, which belongs to

$$H^k(X, \mathrm{Hom}(\wedge^k C, S^k A))$$

can be computed in terms of e. Since this kind of sequence is useful in Hodge theory, I want to give a derivation of this particular one as an example.

Let us now work on $P = \mathbf{P}(A)$, with $p: P \to X$ the natural projection and $0 \to \xi^{-1} \to p^* A$ the tautological sub-bundle. The Euler sequence for the fibers of p is

$$0 \to \Omega^1_{P/X} \to \xi^{-1} \otimes p^* A^* \to \mathcal{O}_P \to 0.$$

Note that $\omega_{P/X} = \det(A^*) \otimes \xi^{-a}$. If we compose $\mathcal{O}_P \to \xi \otimes p^* A \to \xi \otimes p^* B$, we obtain a map $\xi^{-1} \to p^* B$, which we complete to an exact sequence

$$0 \to \xi^{-1} \to p^* B \to Q \to 0.$$

This in turn induces exact sequences

$$0 \to \wedge^{k-1}Q \otimes \xi^{-1} \to \wedge^k p^* B \to \wedge^k Q \to 0.$$

Taking the Koszul complex associated to the map $0 \to \mathcal{O}_P \to \xi \otimes p^* B$, we obtain a long exact sequence

$$0 \to \mathcal{O}_P \to \xi \otimes p^* B \to \xi^2 \otimes \wedge^2 p^* B \to \cdots \to \xi^{b-1} \otimes \wedge^{b-1} p^* B \to \xi^b \otimes \wedge^b p^* B \to 0.$$

Alternatively, using the exact sequences above, we may tensor this with $\omega_{P/X} \otimes \xi^{-k}$ and break off this complex to obtain

$$0 \to \omega_{P/X} \otimes \xi^{-k} \to \omega_{P/X} \otimes \xi^{1-k} \otimes p^* B \to \cdots$$
$$\to \omega_{P/X} \otimes \wedge^k p^* B \to \omega_{P/X} \otimes \wedge^k Q \to 0.$$

If

$$\tilde{e} \in H^1(P, Q^* \otimes \xi^{-1})$$

is the extension class of the exact sequence defining Q, then

$$\wedge^k \tilde{e} \in H^k(P, \wedge^k Q^* \otimes \xi^{-k})$$

is the extension class of this new sequence.

If we now push down this sequence, then we obtain

$$0 \to \mathcal{R}^{a-1}_{p*}(\omega_{P/X} \otimes \xi^{-k}) \to \mathcal{R}^{a-1}_{p*}(\omega_{P/X} \otimes p^* B \otimes \xi^{-k+1}) \to \cdots$$
$$\to \mathcal{R}^{a-1}_{p*}(\omega_{P/X} \otimes \wedge^k p^* B) \to \mathcal{R}^{a-1}_{p*}(\omega_{P/X} \otimes \wedge^k Q) \to 0.$$

Now by Serre duality,

$$\mathcal{R}^{a-1}_{p*}(\omega_{P/X} \otimes \xi^{-m}) \cong \mathcal{R}^0_{p*}(\xi^m)^* \cong S^m A.$$

The long exact sequence above becomes an exact sequence

$$0 \to S^p A \to S^{p-1} A \otimes B \to S^{p-2} A \otimes \wedge^2 B \to \cdots$$
$$\to \wedge^k B \to \mathcal{R}^{a-1}_{p*}(\omega_{P/X} \otimes \wedge^k Q) \to 0.$$

Using the fact that the inclusions of ξ^{-1} in A and B are obtained by composing with $A \to B$, we see that there is a natural exact sequence

$$0 \to \Theta_{P/X} \otimes \xi^{-1} \to Q \to p^* C \to 0.$$

This implies that the kernel of

$$\omega_{P/X} \otimes \wedge^k Q \to \omega_{P/X} \otimes \wedge^k p^* C$$

is filtered by

$$\Omega^{a-1-i}_{P/X} \otimes \xi^{-i} \otimes p^* \wedge^{k-i} C$$

for $1 \leq i \leq a-1$. Since both $\mathcal{R}^{a-1}_{p*}$ and $\mathcal{R}^a_{p*}$ of these vanish by Bott vanishing, we conclude that

$$\mathcal{R}^{a-1}_{p*}(\omega_{P/X} \otimes \wedge^k Q) \to \mathcal{R}^{a-1}_{p*}(p^* \wedge^k C) \cong \wedge^k C$$

is an isomorphism. We thus obtain a natural long exact sequence

$$0 \to S^p A \to S^{p-1} A \otimes B \to S^{p-2} A \otimes \wedge^2 B \to \cdots \to \wedge^k B \to \wedge^k C \to 0.$$

The map on the right is $\wedge^k$ of the original map $B \to C$ and the other maps are contraction with the map $A \to B$.

For $k = 1$, we obtain the original exact sequence back again, and thus $\tilde{e}$ maps to e under the natural map

$$H^1(\mathrm{Hom}(Q, \xi^{-1})) \to$$
$$H^1(\mathrm{Hom}(\mathcal{R}^{a-1}_{p*}(\omega_{P/X} \otimes Q), \mathcal{R}^{a-1}_{p*}(\omega_{P/X} \otimes \xi^{-1}))) \cong H^1(\mathrm{Hom}(C, A)).$$

But then the image of $\wedge^k \tilde{e}$ is $\wedge^k e$ under the natural map

$$H^k(\mathrm{Hom}(\wedge^k Q, \xi^{-k})) \to H^k(\mathrm{Hom}(\mathcal{R}^{a-1}_{p*}(\omega_{P/X} \otimes \wedge^k Q), \mathcal{R}^{a-1}_{p*}(\omega_{P/X} \otimes \xi^{-k})))$$
$$\cong H^k(\mathrm{Hom}(\wedge^k C, S^k A)),$$

and thus the extension class of the exact sequence constructed is

$$\wedge^k e \in H^k(\mathrm{Hom}(\wedge^k C, S^k A)),$$

where, if e is represented by a $C^* \otimes A$-valued $(0,1)$-form, we wedge on the $(0,1)$ and C^* parts and multiply symmetrically on the A part–this gives something symmetric in e.

If we now have an embedding of compact complex manifolds $X \subseteq Y$, then we have the normal bundle sequence

$$0 \to \Theta_X \to \Theta_Y|_X \to N_{X/Y} \to 0,$$

or dually,

$$0 \to N^*_{X/Y} \to \Omega^1_Y|_X \to \Omega^1_X \to 0.$$

This gives us a long exact sequence

$$0 \to S^p N^*_{X/Y} \to \Omega^1_Y|_X \otimes S^{p-1} N^*_{X/Y} \to \cdots \to \Omega^p_Y|_X \to \Omega^p_X \to 0.$$

One may use this to try to compute the Hodge groups of X, provided that one can compute various cohomology groups that come up along the way.

An alternative approach is the following, which I worked out with Stefan Müller-Stach. Given a holomorphic line bundle $L \to Y$ on a compact complex manifold Y, we want to consider the vector bundles F^p_L defined by the exact sequence

$$0 \to \Omega^p_Y \to F^p_L \to \Omega^{p-1}_Y \to 0,$$

having extension class

$$e^p \in H^1(\mathrm{Hom}(\Omega^{p-1}_Y, \Omega^p_Y))$$

defined to be the image of $c_1(L)$ under the map on cohomology induced by

$$\Omega^1_Y \to \mathrm{Hom}(\Omega^{p-1}_Y, \Omega^p_Y)$$

given by

$$\alpha \mapsto (\beta \mapsto \alpha \wedge \beta).$$

If ξ_{ij} are the transition functions of L, then the transition functions of F_L^p are given by

$$\begin{pmatrix} \xi_{ij} \otimes \mathrm{id}_{\Omega_Y^p} & d\xi_{ij} \wedge \mathrm{id}_{\Omega_Y^{p-1}} \\ 0 & \xi_{ij} \otimes \mathrm{id}_{\Omega_Y^{p-1}} \end{pmatrix}.$$

By convention, $F^0 = \mathcal{O}_Y$. There is a natural map of sheaves of sections $d\colon F^p \to F^{p+1}$ given by

$$(\alpha, \beta) \mapsto (d\alpha, \alpha - d\beta).$$

One easily verifies that $d^2 = 0$, and thus we obtain a complex $F_L^\bullet, d$.

We note that $F_L^p = \wedge^p F_L^1$. We note that locally (ds_α, s_α) patch together to give a global section of $\tilde{ds} \in H^0(F_L^1 \otimes L)$; $F_L^1 \otimes L$ is the **first prolongation bundle** of L.

The significance of the complex constructed above is the following:

DEFINITION. *Let $\Omega_{Y,X}^\bullet$ be defined by the exact sequence*

$$0 \to \Omega_{Y,X}^\bullet \to \Omega_Y^\bullet \to \Omega_X^\bullet \to 0,$$

where the map on the right is restriction.

PROPOSITION. *Let $s \in H^0(Y, L)$ define a smooth submanifold $X \subseteq Y$. Then there is an exact sequence of complexes*

$$\cdots F_L^{\bullet-2} \otimes L^{-3} \to F_L^{\bullet-1} \otimes L^{-2} \to F_L^\bullet \otimes L^{-1} \to \Omega_{Y,X}^\bullet \to 0.$$

The maps

$$F_L^p \otimes L^{-k-1} \to F_L^{p+1} \otimes L^{-k}$$

are given by

$$(\alpha, \beta) \mapsto (ds \wedge \alpha, s\alpha - ds \wedge \beta).$$

The maps

$$F_L^p \otimes L^{-1} \to \Omega_Y^p$$

are given by

$$(\alpha, \beta) \mapsto s\alpha - ds \wedge \beta;$$

these go to zero in Ω_X^p.

PROOF: There are a lot of little things to check here. The main point is that the smoothness of X is equivalent to saying locally that $s, \frac{\partial s}{\partial z_1}, \ldots, \frac{\partial s}{\partial z_n}$ have no common zeros, and this in turn is equivalent to saying $\tilde{ds}$ is never 0. This is essentially just the Koszul complex for $\tilde{ds}$, except for checking exactness at $\Omega_{Y,X}^\bullet$, which is easy. □

PROPOSITION. *For $Y = \mathbf{P}^n$ and L a power of the hyperplane bundle H,*

$$F_L^p \cong \wedge^p V \otimes H^{-p},$$

where $V = H^p(\mathbf{P}^n, H)$.

PROOF: The dual of the Euler sequence is

$$0 \to \Omega^1_{\mathbf{P}^n} \to V \otimes H^{-1} \to \mathcal{O}_{\mathbf{P}^n} \to 0$$

and has extension class $c_1(H) = \frac{1}{d}c_1(L)$, hence since the extension class mod non-zero scalars determines the extension,

$$F_L^1 \cong V \otimes H^{-1}.$$

Since

$$F_L^p = \wedge^p F_L^1,$$

the proposition follows. □

DEFINITION. *We will state that a property of a line bundle $L \to Y$ holds for L* **sufficiently ample** *if there is a line bundle $L_0 \to Y$ such that the property holds whenever $L \otimes L_0^{-1}$ is ample.*

DEFINITION. *The* **pseudo-Jacobi ideal** *$J_{E,s}$ for $E \to Y$ a line bundle is the image of the map*

$$H^0(Y, F_L^{1*} \otimes L^{-1} \otimes E) \to H^0(Y, E)$$

induced by the natural map

$$F_L^{1*} \otimes L^{-1} \to \mathcal{O}_Y$$

determined by $s \in H^0(Y, L)$. We will use the notation

$$R_{E,s} = H^0(Y, E)/J_{E,s}.$$

PROPOSITION. *If* $\dim(Y) = n$ *and L is sufficiently ample and $X \in |L|$ is smooth, then*
(1) $H^q(\Omega^p_{Y,X}) = 0$ for $p + q < n$;
(2) $H^{n-p}(\Omega^p_{Y,X}) \cong (H^0(Y, K_Y \otimes L^{p+1})/J_{K_Y \otimes L^{p+1},s})^*$.

PROOF: This may be computed by the resolution for $\Omega^\bullet_{Y,X}$ obtained above. □

PROPOSITION. *If $F \in H^0(\mathbf{P}^n, H^d)$ is a homogeneous polynomial of degree d, then $J_{H^k,F}$ is the ideal in degree k generated by $\frac{\partial F}{\partial z_0}, \ldots, \frac{\partial F}{\partial z_n}$, i.e. the usual* **Jacobi ideal***.*

PROOF: We know that

$$F^1_{H^d} = V \otimes H^{-1},$$

and that

$$0 \to \Omega^1_{\mathbf{P}^n} \to F^1_{H^d} \to \mathcal{O}_{\mathbf{P}^m} \to 0$$

is the Euler sequence. It remains to compute

$$\tilde{dF}: F^{1*}_{H^d} \to H^d.$$

If we use, for example, local coordinates $u_1 = z_1/z_0, \ldots, u_n = z_n/z_0$, then let

$$f(u_1, \ldots, u_n) = F(1, z_1/z_0, \ldots, z_n/z_0).$$

Now

$$\frac{\partial f}{\partial u_i} = \frac{\partial F}{\partial z_i}(1, u_1, \ldots, u_n).$$

If $e_0, \ldots, e_n$ is the basis for V^* dual to $z_1, \ldots, z_n$, then we may locally split the Euler sequence and write $e_i = (0, \frac{\partial}{\partial u_i})$ for $i \geq 1$ and $e_0 = (1, -\frac{1}{d}\sum_{i=1}^n u_i \frac{\partial}{\partial u_i})$. Now

$$\tilde{dF}(e_i) = \frac{\partial F}{\partial z_i}$$

if $i \geq 1$, while

$$\tilde{dF}(e_0) = F - \frac{1}{d}\sum_{i=1}^n u_i \frac{\partial F}{\partial z_i} = \frac{\partial F}{\partial z_0}.$$

We conclude that the map

$$\tilde{dF}: F^{1*}_{H^d} \to H^d$$

is given by

$$(\frac{\partial F}{\partial z_0}, \ldots, \frac{\partial F}{\partial z_n}).$$

The statement about the pseudo-Jacobi ideal follows. □

PROPOSITION. *If $X \subseteq \mathbf{P}^n$ is a smooth hypersurface of degree d given by an irreducible homogeneous polynomial F, then*
(1) $H^{p,q}(\mathbf{P}^n) \cong H^{p,q}(X)$ for $p + q \neq n - 1, 2n$;
(2) There is an exact sequence

$$0 \to H^{n-1-q,q}(\mathbf{P}^n) \to H^{n-1-q,q}(X) \to R_{H^{d-n-1+qd},F} \to 0.$$

PROOF: This follows by using the resolution for $\Omega^\bullet_{\mathbf{P}^n,X}$, Bott vanishing, and the dualities $H^{p,q}(X) \cong H^{n-1-p,n-1-q}(X)^*$. □

REMARK. The representation above is sometimes called the **Poincaré residue representation** of the cohomology of X. The important thing is that it is a purely algebraic way of representing the cohomology of X.

A very down-to-earth way to get the Poincaré residue representation is as follows: on X, we have the exact sequence

$$0 \to \Theta_X \to \Theta_Y|_X \to N_{X/Y} \to 0$$

having extension class $e \in H^1(\Theta_X(-X))$, using the fact that

$$N_{X/Y} \cong \mathcal{O}_X(X).$$

From the exact sequence

$$0 \to \Omega_X^{p-1} \otimes N_{X/Y}^* \to \Omega_Y^p|_X \to \Omega_X^p \to 0,$$

which twisted by $\mathcal{O}_X(mX)$ becomes

$$0 \to \Omega_X^{p-1}((m-1)X) \to \Omega_Y^p(mX)|_X \to \Omega_X^p(mX) \to 0,$$

we obtain a map $H^0(\Omega_X^p(mX)) \to H^1(\Omega^{p-1}((m-1)X))$ which is just cup product with e, followed by contraction. Given suitable vanishing theorems, we have that

$$H^0(\Omega_X^p(mX)) \cong H^m(\Omega_X^{p-m})/\mathrm{im}(H^m(\Omega_Y^{p-m})).$$

In particular, composing with Res, we obtain

$$H^0(K_Y((q+1)X) \to H^q(\Omega_X^{n-1-q})/\mathrm{im}(H^q(\Omega_Y^{n-1-q})).$$

If we note the exact sequence

$$0 \to F_L^{1*} \otimes K_Y(-X) \to \Omega_Y^n \to \Omega_X^{n-1} \to 0,$$

then the kernel of the map is

$$H^0(F_L^{1*} \otimes K_Y(qX)) = J_{K_Y \otimes L^{q+1},s},$$

where $L = \mathcal{O}_Y(X)$ and $X = \mathrm{div}(s)$. We thus obtain

$$H^q(\Omega_X^{n-1-q})/\mathrm{im} H^q(\Omega_Y^{n-1-q}) \cong R_{K_Y \otimes L^{q+1},s}$$

for L sufficiently ample. One advantage of this approach is that since the isomorphisms are obtained from cupping with the extension class of the normal bundle sequence

$$0 \to \Theta_X \to \Theta_Y|_X \to L \to 0,$$

and cupping with this extension class also gives the Kodaira-Spencer map

$$\delta : H^0(X, L) \to H^1(\Theta_X),$$

one sees:

PROPOSITION. *The derivative of the period map*

$$H^{n-1-q,q}(X)/H^{n-1-q,q}(Y) \otimes H^0(X,L) \to H^{n-2-q,q+1}(X)/H^{n-2-q,q+1}(Y)$$

is computed by the multiplication map

$$R_{K_Y \otimes L^{q+1},s} \otimes R_{L,s} \to R_{K_Y \otimes L^{q+2},s}.$$

PROPOSITION. *Let $X \subseteq \mathbf{P}^n$ be a smooth hypersurface of degree d given by a homogeneous polynomial F, $n \geq 2$. Then*
(1) The Kodaira-Spencer map $H^0(\mathbf{P}^n, H^d)/(F) \to H^1(\Theta_X)$ is surjective unless $n = 2, d \geq 4$ or $n = 3, d = 4$;
(2) The Kodaira-Spencer map induces an isomorphism $R_{H^d,F} \cong H^1(\Theta_X)$ unless $n = 2, d = 3$;

(3) The tangent space at F to hypersurfaces of degree d modulo projective equivalence is $R_{H^d,F}$.

PROOF: From the exact sequence

$$0 \to \Theta_X \to \Theta_{\mathbf{P}^n}|X \to N_{X/\mathbf{P}^n} \to 0$$

and identifying $N_{X/\mathbf{P}^n} \cong H^d$, we have an exact sequence

$$H^0(\Theta_{\mathbf{P}^n}|X) \to H^0(\mathbf{P}^n, H^d) \to H^1(\Theta_X) \to H^1(\Theta_{\mathbf{P}^n}|X).$$

The result now follows by the Euler sequence for $\Theta_{\mathbf{P}^n}$, the restriction sequence to X, and the Bott Vanishing Theorem. The one other ingredient is that the composition of the maps

$$V^* \otimes H \to \Theta_{\mathbf{P}^n}$$

and

$$\Theta_{\mathbf{P}^n} \to N_{X/\mathbf{P}^n} \cong H^d$$

is the map $\tilde{dF}$ given by

$$e_i \mapsto \frac{\partial F}{\partial z_i}.$$

This shows (1) and (2).

The tangent space at the identity to $PGL(n+1)$ is

$$H^0(\Theta_{\mathbf{P}^n}) \cong (V^* \otimes V)/\mathbf{C}I.$$

The infinitesimal action of this on the point F is

$$\sum_i A_i \frac{\partial}{\partial z_i} \mapsto \sum_i A_i \frac{\partial F}{\partial z_i},$$

where

$$A_i \in V = H^0(\mathbf{P}^n, H).$$

Part (3) follows. □

THEOREM (Macaulay's Theorem). *Let $X \subseteq \mathbf{P}^n$ be a smooth hypersurface of degree d given by a homogeneous polynomial F. Let $e = (n+1)(d-2)$. Then*
(1) $R_{H^k,F} = 0$ for $k > e$,
(2) $R_{H^e,F} \cong \mathbf{C}$;
(3) The multiplication map $R_{H^k,F} \otimes R_{H^{e-k},F} \to R_{H^e,F} \cong \mathbf{C}$ is non-degenerate on each factor, provided $k, e-k \geq 0$.

PROOF: The partials of F give a map

$$V^* \otimes \mathcal{O}_{\mathbf{P}^n} \to H^{d-1} \to 0.$$

From this we can build a Koszul complex

$$0 \to \wedge^{n+1} V^* \otimes H^{-(n+1)(d-1)} \to \dots \to V^* \otimes H^{-(d-1)} \to \mathcal{O}_{\mathbf{P}^n} \to 0.$$

The hypercohomology of this complex is zero, since the complex is exact. On the other hand, we may compute the hypercohomology another way so that its E_1 term is the cohomology groups of each piece of the complex. The upshot is that we obtain an isomorphism

$$R_{H^k,F} \cong \ker(H^n(\wedge^{n+1}V^* \otimes H^{k-(n+1)(d-1)}) \to H^n(\wedge^n V^* \otimes H^{k-n(d-1)})),$$

and the latter by Serre Duality is $R^*_{H^{e-k},F}$. Since

$$R_{H^m,F} = 0$$

if $m < 0$ and

$$R_{H^0,F} \cong \mathbf{C},$$

we obtain (1),(2). The extension class of the Koszul complex lies in

$$H^n(\wedge^{n+1}V^* \otimes H^{-(n+1)(d-1)}) \cong H^0(H^e)^*,$$

and thus must be the canonical projection

$$H^0(H^e) \to R_{H^e,F} \cong \mathbf{C}.$$

It immediately follows that the duality

$$R_{H^k,F} \cong R^*_{H^{e-k},F}$$

is given by multiplication, proving (3). □

Exactly analogously, one has:

THEOREM. *Let Y be a smooth projective variety of dimension n and L a sufficiently ample line bundle. Let $X = \mathrm{div}(s) \in |L|$ be a smooth hypersurface. Then*

(1) $R_{K_Y^2 \otimes L^{n+1},s} \cong \mathbf{C}$;

(2) For any fixed bundle A, for L sufficiently ample, the multiplication map

$$R_{L^k \otimes A,s} \otimes R_{K_Y^2 \otimes L^{n+1-k} \otimes A^*,s} \to R_{K_Y^2 \otimes L^{n+1},s} \cong \mathbf{C}$$

is non-degenerate in each factor for $1 \le k \le n$.

THEOREM (Infinitesimal Torelli for Hypersurfaces). *Let $X \subseteq \mathbf{P}^n$ be a smooth hypersurface of degree d. The derivative of the period map from hypersurfaces modulo projective equivalence to the period domain is injective, with the exception $n = 3, d = 3$.*

PROOF: We have computed that

$$T_F(\text{moduli of hypersurfaces of degree } d/\text{projective equivalence}) = R_{H^d,F}.$$

The derivative of the period map

$$P_*: R_{H^d,F} \to \oplus_q \mathrm{Hom}(H^{n-1-q,q}(X), H^{n-2-q,q+1}(X))$$
$$\cong \oplus_q \mathrm{Hom}(R_{H^{d-n-1+qd},F}, R_{H^{d-n-1+(q+1)d},F})$$

is given by multiplication. We thus need to know that for some $q \geq 0$, the multiplication map

$$R_{H^d,F} \otimes R_{H^{d-n-1+qd},F} \rightarrow R_{H^{d-n-1+(q+1)d},F}$$

is non-degenerate in the first factor. By Macaulay's Theorem, it is equivalent to know that the multiplication map

$$R_{H^{d-n-1+qd},F} \otimes R_{H^{(d-1)(n+1)-(q+2)d},F} \rightarrow R_{H^{(n+1)(d-2)-d},F}$$

is surjective for some q. Since multiplication on polynomials is surjective, it is enough to find a q so that both powers of H on the left are ≥ 0. This is equivalent to the inequalities

$$n - 1 - \frac{n+1}{d} \geq q \geq \frac{n+1}{d} - 1.$$

The case $d = 2$ is automatic, since all smooth quadrics are projectively equivelent. If $d \geq 3$, it is easy to see that a q can be found unless $d = 3, n = 3$. □

Once again, this generalizes to arbitrary $X \subseteq Y$ for L sufficiently ample.

DEFINITION. *Let $L \rightarrow Y$ be a line bundle. The* **automorphisms** *of Y* **preserving** *L are* $\mathrm{Aut}(Y, L) = \{f: Y \rightarrow Y \mid f \text{ holomorphic and } f^*L \cong L\}$.

PROPOSITION. *The Zariski tangent space at the identity of* $\mathrm{Aut}(Y, L)$ *is*

$$T_I(\mathrm{Aut}(Y, L)) = \mathrm{im}(H^0(F_L^{1*} \rightarrow H^0(\Theta_Y)).$$

PROOF: We know

$$T_I(\mathrm{Aut}(Y)) = H^0(\Theta_Y).$$

The derivative of the map

$$\mathrm{Aut}(Y) \rightarrow \mathrm{Pic}(Y)$$

given by $f \mapsto f^*L$ is

$$\cup c_1(L): H^0(\Theta_Y) \rightarrow H^1(\mathcal{O}_Y).$$

Since F_L^{1*} sits in an exact sequence

$$0 \rightarrow \mathcal{O}_Y \rightarrow F_L^{1*} \rightarrow \Theta_Y \rightarrow 0$$

with extension class $c_1(L)$, the result follows. □

COROLLARY. *The Zariski tangent space at X to* $|L|/\mathrm{Aut}(Y, L)$ *is* $R_{L,s}$.

THEOREM (Infinitesimal Torelli for Hypersurfaces of High Degree). *Let Y be a smooth projective variety of dimension $n \geq 2$ and L a sufficiently ample line bundle. Let $X = \mathrm{div}(s) \in |L|$ be a smooth hypersurface. Then the derivative of the period map from* $|L|/\mathrm{Aut}(Y, L)$ *to the period domain is injective.*

PROOF: The periods of $\mathrm{im}(H^{n-1-q,q}(Y))$ do not change, so the derivative of the period map is described by

$$p_*: R_{L,s} \rightarrow \oplus_q \mathrm{Hom}(R_{K_Y \otimes L^{q+1},s}), R_{K_Y \otimes L^{q+2},s}),$$

where the map is given by multiplication. By the generalization of Macaulay's Theorem, it is enough to show that for some q the map

$$R_{K_Y\otimes L^{q+1},s}\otimes R_{K_Y\otimes L^{n-1-q},s}\to R_{K_Y^2\otimes L^n,s}$$

is surjective. A fortiori, it would be enough to show that

$$H^0(Y,K_Y\otimes L^{q+1})\otimes H^0(Y,K_Y\otimes L^{n-1-q})\to H^0(Y,K_Y^2\otimes L^n)$$

is surjective. If $\Delta\subseteq Y\times Y$ is the diagonal, and $p_1,p_2\colon Y\times Y\to Y$ are the projections, then it is enough to show

$$H^1(Y\times Y,\mathcal{I}_\Delta\otimes p_1^*(K_Y\otimes L^{q+1})\otimes p_2^*(K_Y\otimes L^{n-1-q}))=0.$$

For L sufficiently ample on Y, $p_1^*L\otimes p_2^*L$ is sufficiently ample on $Y\times Y$, so this cohomology group vanishes provided $q+1>0$ and $n-1-q>0$. This is true for $q=0$ as soon as $n\geq 2$. □

We now want to compute the Poincaré dual class of a cycle from its geometry, either in the case of projective hypersurfaces or hypersurfaces of high degree. Assume we have the following situation: X,Y as above and $Z\subseteq X$ a smooth codimension p cycle. Assume $\dim(Y)=2p+1$. The normal bundle sequence for the triple Z,X,Y is

$$0\to N_{Z/X}\to N_{Z/Y}\to N_{X/Y}|_Z\to 0$$

has extension class

$$e\in H^1(N_{X/Y}^*\otimes N_{Z/X}).$$

Since $N_{X/Y}\cong L$, we may regard a $(0,1)$-form representing e as locally an $(n-1)\times p$ matrix with L-valued coefficients. The maximal minors of this matrix constitute a natural element we will write

$$\wedge^p e\in H^p(L^{-p}\otimes\wedge^p N_{Z/X}).$$

Since

$$\wedge^p N_{Z/X}\cong K_Z\otimes K_X^{-1},$$

and $\dim(Z)=p$, we see that

$$\wedge^p e\in H^0(K_X\otimes L^p)^*.$$

PROPOSITION. *Assume $\psi_X(Z)\in H^{p,p}(X)$ annihilates the restriction of $H^{p,p}(Y)$. The image of $\wedge^p e$ in $H^0(K_Y\otimes L^{p+1})^*$ in fact lies in $R^*_{K_Y\otimes L^{p+1},s}$, which we may identify with the part of $H^{p,p}(X)$ annihilating the image of $H^{p,p}(Y)$, and represents $\psi_X(Z)$.*

PROOF: We note that if we take $\wedge^p$ of the natural map $\Theta_X|_Z\to N_{Z/X}$ and tensor with $K_X|_Z$, we obtain the restriction map $\Omega_X^p\cong K_X\otimes\wedge^p\Theta_X\to K_X\otimes\wedge^p N_{Z/X}\cong K_Z$. If $\tilde{e}$ is the extension class of

$$0\to\Theta_X\to\Theta_Y|_X\to L\to 0,$$

then we have a commutative diagram

$$\begin{array}{ccc} H^0(X, K_X \otimes L^p) & \longrightarrow & H^0(Z, K_X \otimes L^p|_Z) \\ \downarrow & & \downarrow \\ H^p(\Omega_X^p) & \longrightarrow & H^p(K_Z) \end{array}$$

where the vertical arrows are $\wedge^p \tilde{e}$ and $\wedge^p e$, and the horizontal arrows are restriction. Dualizing this diagram, we obtain

$$\begin{array}{ccc} H^0(\mathcal{O}_Z) & \longrightarrow & H^p(\Omega_X^p) \\ \downarrow & & \downarrow \\ H^p(Z, \wedge^p N_{Z/X} \otimes L^{-p}) & \longrightarrow & H^{2p}(X, L^{-p}) \end{array}$$

where the top horizontal map is ψ_Z and the vertical maps are $\wedge^p \tilde{e}$ and $\wedge^p e$. However, in the long exact sequence

$$0 \to L^{-p} \to L^{-(p-1)} \otimes \Omega_Y^1|_X \to \cdots \to \Omega_Y^p|_X \to \Omega_X^p \to 0$$

constructed at the beginning of this section, the extension class is $\wedge^p \tilde{e}$, and thus the identification of $H^p(\Omega_X^p)/\mathrm{im} H^p(\Omega_Y^p)$ with $R^*_{K_Y \otimes L^{p+1}, s}$ is given by $\wedge^p \tilde{e}$. This, together with the commutative diagram above, gives the result. □

EXAMPLE. *(C. Peskine)*

Let $X = \{F = 0\} \subseteq \mathbf{P}^3$ be a smooth surface of degree d containing a complete intersection $Z_1 = \{G_1 = 0, G_2 = 0\}$ of type a_1, a_2 as a smooth subvariety. Let $Z_2 = X \cap H$ be a hyperplane section, and $Z = dZ_1 - a_1 a_2 Z_2$, so that Z has degree 0. Let $F = A_1 G_1 + A_2 G_2$. Then the ideal (G_1, G_2, A_1, A_2) has codimension 1 in degree $2d - 4$, and is easily seen to contain J_F. It thus defines an element of $R^*_{H^{2d-4}, F}$, and this is ψ_Z.

To see this, we note that A_1, A_2 is a regular sequence on Z, and thus defines a Koszul complex

$$0 \to \mathcal{O}_Z(a_1 + a_2 - 2d) \to \mathcal{O}_Z(a_1 - d) \oplus \mathcal{O}_Z(a_2 - d) \to \mathcal{O}_Z \to 0.$$

Now $K_Z \cong \mathcal{O}_Z(a_1 + a_2 - 4)$, and we have an exact sequence

$$H^0(\mathcal{O}_Z(2d - 4 - a_1) \oplus H^0(\mathcal{O}_Z(2d - 4 - a_2) \to H^0(\mathcal{O}_Z(2d - 4) \to H^1(K_Z) \to 0.$$

Thus A_1, A_2 span a codimension 1 linear subspace on Z in degree $2d - 4$, and since Z is projectively normal, (A_1, A_2, G_1, G_2) spans a codimension 1 ideal in degree $2d - 4$. The normal bundle

$$N_{Z/\mathbf{P}^3} \cong \mathcal{O}_Z(a_1) \oplus \mathcal{O}_Z(a_2),$$

and the map to $N_{X/\mathbf{P}^3}|_Z \cong \mathcal{O}_Z(d)$ is given by (A_1, A_2), and thus the normal bundle sequence

$$0 \to N_{Z/X} \to N_{Z/\mathbf{P}^3} \to N_{X/\mathbf{P}^3}|_Z \to 0$$

is just the Koszul complex above, and hence its extension class is as claimed.

Peskine points out that this example may be generalized to curves given by the vanishing of the maximal minors of a $k \times (k+1)$ matrix on $\mathbf{P}^3$.

Bibliographic Notes

The version of the Bott Vanishing Theorem quoted here is just a pale shadow of the full Borel-Bott-Weil Theorem, which computes (as a representation) the cohomology groups of homogeneous vector bundles on complex homogeneous spaces. Even for projective space, there is a far more general statement involving Young tableaux that algebraic geometers have not exploited sufficiently. A proof of the theorem is in Jantzen, *Representations of Algebraic Groups*; a statement for Young tableaux is in Lascoux, "Syzygies des variétés déterminantales," Adv. Math. 30 (1978), 202-237.

The Poincaré residue construction is discussed in Griffiths paper "On the periods of certain rational integrals:I and II" already cited. It appears in a more abstract form in Deligne, "Théorie de Hodge,II," Publ IHES 49 (1971), 5-58 and "Théorie de Hodge III," Publ IHES (1974), 6-77.

The pseudo-Jacobi ideal appears in my paper "The period map for hypersurface sections of high degree on an arbitrary variety," Comp. Math. 55 (1984), 135-156.

The role of Macaulay's Theorem is interesting. Griffiths realized that using the Poincaré residue representation of the cohomology of a hypersurface, he could reduce the infinitesimal Torelli Theorem in that case to an algebraic computation, which he carried out by hand for the Fermat hypersurface. Mumford then told him about Macaulay's Theorem. Thus one of the first significant infinitesimal computations done in Hodge theory brought in precisely the type of algebraic results that, pushed far enough, lead to the results of Lectures 7 and 8.

LECTURE 5

In this section, we will introduce mixed Hodge structures, a remarkable discovery of Deligne.

It is worth pointing out, since even for quasi-projective varieties

$$H^k(X, \mathbf{C}) \cong \mathbf{H}^k(\Omega_X^\bullet),$$

why we do not go ahead and, for any quasi-projective variety X, put a filtration

$$G^p H^k(X, \mathbf{C}) = \mathrm{im}(\mathbf{H}^k(\Omega_X^{\geq p}) \to \mathbf{H}^k(\Omega_X^\bullet)).$$

This is the **Grothendieck filtration**, and it makes a brief appearance in the proof of Nori's Connectedness Theorem.. The reason it is not the right generalization of the Hodge filtration is shown by the following example: Let

$$X = T - D,$$

where T is an elliptic curve and D is a non-empty set of points. Then $\Omega_X^1 \cong \mathcal{O}_X$, and thus, since X is Stein,

$$\mathbf{H}^1(\Omega_X^{\geq 1}) \cong H^0(\mathcal{O}_X).$$

Now

$$\mathbf{H}^1(\Omega_X^\bullet) = \mathrm{coker}(H^0(\mathcal{O}_X) \xrightarrow{d} H^0(\mathcal{O}_X)),$$

and thus

$$G^1 H^1(X, \mathbf{C}) = H^1(X, \mathbf{C}).$$

However, there is a long exact sequence

$$H^1(T, \mathbf{C}) \to H^1(X, \mathbf{C}) \to H^0(D, \mathbf{C}) \to H^0(T, \mathbf{C}),$$

and we would like to have our filtration induce the usual Hodge filtration on $H^1(T, \mathbf{C})$. Since $F^1 H^1(T, \mathbf{C}) \neq H^1(T, \mathbf{C})$, it is clear that $G^\bullet$ is not the "correct" choice of filtration. The correct choice is Deligne's mixed Hodge structure using logarithmic differential forms.

We begin with a general example. Let $f: Z \to X$ be an analytic map of smooth projective varieties of dimension $n - p, n$ respectively. Consider the complex

$$\Omega_f^\bullet = \Omega_X^\bullet \oplus \Omega_Z^{\bullet - 1}$$

with the differential

$$\partial(\alpha, \beta) = (\partial\alpha, f^*\alpha - \partial\beta).$$

If C denotes singular cochains, let

$$C^\bullet(f, G) = C^\bullet(X, G) \oplus C^{\bullet - 1}(Z, G),$$

with

$$\delta(\alpha, \beta) = (\delta\alpha, f^*\alpha - \delta\beta).$$

There is also the complex

$$A^\bullet(f) = A^\bullet(X) \oplus A^{\bullet-1}(Z),$$

with

$$d(\alpha,\beta) = (d\alpha, f^*\alpha - d\beta).$$

DEFINITION.

$$H^k(f,G) = H^k(C^\bullet(f,G)).$$

REMARK: If $f\colon Z \to X$ is an embedding, then

$$H^k(f,G) = H^k(X,Z,G).$$

PROPOSITION.
(1) $H^k(A^\bullet(f)) = H^k(f,\mathbf{C})$;
(2) $\mathbf{H}^k(\Omega_f^\bullet) \cong H^k(f,\mathbf{C})$;
(3) The hypercohomology spectral sequence computing $\mathbf{H}^k(\Omega_f^\bullet)$ *has* $E_1^{p,q} = H^q(\Omega^p(f))$ *and degenerates at the* E_2 *term.*

PROOF: (1) follows from the fact that $A^\bullet(f)$ is a fine resolution of $\ker(f^*\colon \mathbf{C}_X \to \mathbf{C}_Z)$. The total complex of the hypercohomology is $A^\bullet(f)$, which proves (2). Finally, we may represent the classes in

$$H^q(\Omega_f^p) = H^q(\Omega_X^p) \oplus H^q(\Omega_Z^{p-1})$$

by harmonic forms on X and Z, which we know are ∂ and $\bar\partial$-closed. If $(\alpha,\beta) \in E_1^{p,q}$, then choosing α,β harmonic, $d_1(\alpha,\beta) = (d\alpha, f^*\alpha - d\beta) = (0, f^*\alpha)$. If $f^*\alpha$ is $\bar\partial$-exact, then by the principle of two types, since it is ∂ and $\bar\partial$-closed, it is $\partial\bar\partial$-exact; hence $f^*\alpha = \bar\partial\partial\gamma$. Then $d_2(\alpha,\beta) = (0,\partial\partial\gamma) = (0,0)$. It now follows easily that all higher d_r are zero. □

We have subcomplexes $\Omega_f^{\geq m}$ and $\Omega_Z^{\bullet-1}$ of $\Omega_f^\bullet$.

DEFINITION. *We define*

$$F^m\mathbf{H}^k(\Omega_f^\bullet) = \operatorname{im}(\mathbf{H}^k(\Omega_f^{\geq m})).$$

We define

$$W_k\mathbf{H}^k(\Omega_f^\bullet) = \mathbf{H}^k(\Omega_f^\bullet),$$
$$W_{k-1}\mathbf{H}^k(\Omega_f^\bullet) = \operatorname{im}(\mathbf{H}^k(\Omega_Z^{\bullet-1}) \to \mathbf{H}^k(\Omega_f^\bullet)),$$
$$W_{k-2}\mathbf{H}^k(\Omega_f^\bullet) = 0.$$

PROPOSITION.

$$\operatorname{Gr}_k^W \mathbf{H}^k(\Omega_f^\bullet) \cong \ker(H^k(X,\mathbf{C}) \to H^k(Z,\mathbf{C})),$$
$$\operatorname{Gr}_{k-1}^W \mathbf{H}^k(\Omega_f^\bullet) \cong \operatorname{coker}(H^{k-1}(X,\mathbf{C}) \to H^{k-1}(Z,\mathbf{C})).$$

Furthermore, the filtration $F^\bullet$ *induces the Hodge filtration on each graded piece.*

PROOF: This is a consequence of what we have done above. □

The point of this proposition is that although $H^k(f,\mathbf{C})$ does not have a Hodge structure (for example, for k odd, the dimension does not need to be even), for each m the m'th graded piece under W each has a Hodge structure of weight m induced by a global filtration of $H^k(f, \mathbf{C})$. As we will see below, this is an example of a mixed Hodge structure.

However, there are other interesting constructions using $H^\bullet(f)$. We will do some of these later in this section. We first need some topological preliminaries:

DEFINITION. *Let M be a smooth oriented manifold of dimension n and Z a smooth oriented manifold of dimension $n-p$. If $f: Z \to M$ is a smooth map, then we have $H_k(M, \mathbf{Z}) \xrightarrow{\cap Z} H_{k-p}(Z, \mathbf{Z})$ defined by intersecting a cycle with $f(Z)$. Under Poincaré duality, this gives a map*

$$\mathrm{Gy}_{Z/M}: H^{n-k}(Z, \mathbf{Z}) \to H^{n-k+p}(M, \mathbf{Z})$$

called the Gysin map.

PROPOSITION. *The composition*

$$H^k(M, \mathbf{Z}) \xrightarrow{f^*} H^k(Z, \mathbf{Z}) \xrightarrow{\mathrm{Gy}_{Z/M}} H^{k+p}(M, \mathbf{Z})$$

is $\cup \psi_Z$.

PROOF: If $\beta \in H^{n-k-p}(M, \mathbf{Z})$, then $\mathrm{Gy}_{Z/M}(f^*\alpha) \cup \beta = f^*\alpha \cup f^*\beta = f^*(\alpha \cup \beta)$. Likewise, $(\psi_Z \cup \alpha \cup \beta)([M]) = f^*(\alpha \cup \beta)([Z])$, and the result follows. □

We recall:

THEOREM (Lefschetz Duality). *Let M be a compact orientable manifold of dimension n and $A \subseteq M$ a closed subset. Then*

$$H_k(M - A, \mathbf{Z}) \cong \bar{H}^{n-k}(X, A, \mathbf{Z}),$$

where

$$\bar{H}^k(X, A, \mathbf{Z}) = \lim\nolimits_{\to} H^k(X, A_\epsilon, \mathbf{Z}),$$

where A_ϵ is a system of open neighborhoods of A converging to A. The map is given by taking a k-cycle α in $X - A$ and choosing ϵ so $\alpha \subseteq X - A_\epsilon$; it then acts on an $(n-k)$-chain β whose boundary is in A_ϵ by intersection.

COROLLARY. *Let X, D be a smooth projective variety and a divisor with normal crossings. Then*

$$H_k(X - D, \mathbf{Z}) \cong H^{2n-k}(X, D, \mathbf{Z}).$$

PROOF: We need to get rid of the bar on the right-hand side of lefschetz duality. It is enough to show that if we put a metric on X, the set D_ϵ of points of distance $< \epsilon$ from D retracts onto D. The only potential problem occurs near singular points. If D locally is given by $z_1 z_2 \cdots z_k = 0$, then the function $|z_1 \cdots z_k|^2$ has non-vanishing gradient on $X - D$, and gives a retraction. □

PROPOSITION/DEFINITION. *If $Z \subseteq M$ is a smooth codimension p oriented submanifold of an n-dimensional oriented smooth manifold M, then there is a natural map*

$$H_k(Z,\mathbf{Z}) \overset{\mathrm{Tube}_{Z/M}}{\longrightarrow} H_{k+p-1}(M-Z,\mathbf{Z})$$

called the **tube map** *that fits into a long exact sequence*

$$H_{k-p+1}(Z,\mathbf{Z}) \to H_k(M-Z,\mathbf{Z}) \to H_k(M,\mathbf{Z}) \overset{\cap Z}{\longrightarrow} H_{k-p}(Z,\mathbf{Z})$$

Furthermore, this sequence is dual to the long exact sequence for the pair (M,Z), where we use Lefschetz

$$H_k(M-Z,\mathbf{Z}) \cong H^{n-k}(M,Z,\mathbf{Z})$$

and Poincaré Duality

$$H_k(M,\mathbf{Z}) \cong H^{n-k}(M,\mathbf{Z})$$
$$H_{k-p+1}(Z,\mathbf{Z}) \cong H^{n-k-1}(Z,\mathbf{Z}).$$

Similarly, there is a tube map

$$H^k(M-Z,\mathbf{Z}) \overset{\mathrm{Tube}_{Z/M}}{\longrightarrow} H^{k-p+1}(Z,\mathbf{Z})$$

that fits into a long exact sequence

$$H^k(M,\mathbf{Z}) \to H^k(M-Z,\mathbf{Z}) \to H^{k-p+1}(Z,\mathbf{Z}) \overset{\mathrm{Gy}_{Z/M}}{\longrightarrow} H^{k+1}(M,\mathbf{Z})$$

that is dual to the long exact sequence for homology of the pair M,Z.

For a divisor D with normal crossings in a smooth n-dimensional projective variety X, one can extend this with some extra work.

We need the following standard description of the cohomology of a blow-up:

PROPOSITION. *Let $Z \subseteq X$ be a smooth codimension p algebraic subvariety of a smooth projective variety X. Let $\epsilon: \hat{X} \to X$ be the blow-up of X at Z, with $\hat{Z}$ the exceptional divisor, and $\epsilon|_Z: \hat{Z} \to Z$ the restriction of ϵ to Z. Since $\hat{Z} = \mathbf{P}(N_{Z/X})$, let $\xi \in H^2(\hat{Z},\mathbf{Z})$ be the tautological divisor, and $\psi_Z \in H^{2p}(X,\mathbf{Z}), \psi_{\hat{Z}} \in H^{2p}(\hat{X},\mathbf{Z})$ the Poincaré dual classes of Z,$\hat{Z}$ respectively. Then*

$$H^*(\hat{X},\mathbf{Z}) \cong (H^{*-2}(\hat{Z},\mathbf{Z}) \oplus H^*(X,\mathbf{Z}))/(\xi^{p-1} = \psi_Z),$$

where the map $H^{k-2}(\hat{Z},\mathbf{Z}) \oplus H^k(X,\mathbf{Z}) \to H^k(\hat{X},\mathbf{Z})$ is $\mathrm{Gy}_{\hat{Z}/\hat{X}} \oplus \epsilon^*$.

COROLLARY. *There is an exact sequence for any k,*

$$0 \to H^k(Z,\mathbf{Z}) \overset{(\cup\xi^{p-1},\mathrm{Gy}_{Z/X})}{\longrightarrow} H^{k+2p-2}(\hat{Z},\mathbf{Z}) \oplus H^{k+2p}(X,\mathbf{Z}) \overset{(\mathrm{Gy}_{\hat{Z}/\hat{X}},\epsilon^*)}{\longrightarrow} H^{k+2p}(\hat{X}) \to 0.$$

COROLLARY. *$\epsilon|_Z$ gives an isomorphism*

$$\ker(H^k(Z,\mathbf{Z}) \xrightarrow{\cup\psi_Z} H^{k+2p}(X,\mathbf{Z})) \cong \ker(H^{k+2p-2}(\hat{Z},\mathbf{Z}) \xrightarrow{\psi_{\hat{Z}}} H^{k+2p}(\hat{X})).$$

Also, ϵ^* gives an isomorphism

$$\mathrm{im}(H^k(Z,\mathbf{Z}) \xrightarrow{\cup\psi_Z} H^{k+2p}(X,\mathbf{Z})) \cong \mathrm{im}(H^{k+2p-2}(\hat{Z},\mathbf{Z}) \xrightarrow{\psi_{\hat{Z}}} H^{k+2p}(\hat{X})).$$

Continuing with the situation of the proposition, we have the Gysin sequence

$$H^{k-2p}(Z,\mathbf{Z}) \to H^k(X,\mathbf{Z}) \to H^k(X-Z,\mathbf{Z}) \to H^{k-2p+1}(Z,\mathbf{Z}) \to H^{k+1}(X,\mathbf{Z})$$

and likewise the Gysin sequence

$$H^{k-2}(\hat{Z},\mathbf{Z}) \to H^k(\hat{X},\mathbf{Z}) \to H^k(\hat{X}-\hat{Z},\mathbf{Z}) \to H^{k-1}(\hat{Z},\mathbf{Z}) \to H^{k+1}(\hat{X},\mathbf{Z}).$$

Now $X - Z = \hat{X} - \hat{Z}$, and we obtain a commutative diagram if we map $H^{k-2p+1}(Z,\mathbf{Z}) \to H^{k-1}(\hat{Z},\mathbf{Z})$ by ξ^{p-1}, and $H^k(X,\mathbf{Z}) \longrightarrow H^k(\hat{X},\mathbf{Z})$ by ϵ^*. We note that since

$$\xi \in F^1H^2(\hat{Z},\mathbf{C}),$$

the map

$$H^{k-2p+1}(Z,\mathbf{C}) \to H^{k-1}(\hat{Z},\mathbf{C})$$

takes

$$F^m H^{k-2p+1}(Z,\mathbf{C}) \to F^{m+p-1} H^{k-1}(\hat{Z},\mathbf{C}).$$

Since $H^k(\hat{X}-\hat{Z},\mathbf{C})$ has a Hodge filtration, we see that an element

$$\phi \in F^m \ker(H^k(Z,\mathbf{Z}) \to H^{k+2p}(X,\mathbf{Z}))$$

lifts to an element

$$\tilde{\phi} \in H^{k+2p-1}(\hat{X}-\hat{Z},\mathbf{Z}).$$

If we mod out by

$$F^{m+p} H^{k+2p-1}(\hat{X}-\hat{Z},\mathbf{C}),$$

we obtain an element

$\nu(\phi) \in$

$$H^{k+2p-1}(\hat{X},\mathbf{C})/(H^{k+2p-3}(\hat{Z},\mathbf{C}) + F^{m+p}H^{k+2p-1}(\hat{X},\mathbf{C}) + H^{k+2p-1}(\hat{X},\mathbf{Z}))$$
$$\cong H^{k+2p-1}(X,\mathbf{C})/(H^{k-1}(Z,\mathbf{C}) + F^{m+p}H^{k+2p-1}(X,\mathbf{C}) + H^{k+2p-1}(X,\mathbf{Z})).$$

This does not depend on the choice of lifting. There are many examples where it gives interesting invariants, some of which we have seen already.

EXAMPLE. *Abel-Jacobi map*

If $Z \in Z_h^p(X)$, then there is a natural

$$\phi \in \ker(H^0(|Z|) \to H^{2p}(X,\mathbf{Z}))$$

which gives rise to a

$$\nu(\phi) \in H^{2p-1}(X, \mathbf{C})/(F^p H^{2p-1}(X, \mathbf{C}) + H^{2p-1}(X, \mathbf{Z})).$$

In fact,

$$\nu(\phi) = AJ_X(Z).$$

It is worth pointing out that if we represent the integral lifting

$$\tilde{\phi} \in H^{2p-1}(\hat{X} - \hat{Z}, \mathbf{Z})$$

as $\tilde{\phi} = \eta_Z + \alpha$, where $\alpha \in H^{2p-1}(\hat{X}, \mathbf{C})$ and

$$\eta_Z \in F^p \mathbf{H}^{2p-1}(\Omega^\bullet_{\hat{X}}(\log(\hat{Z}))),$$

then we may represent η_Z by an element of

$$\oplus_{k \geq p} H^{2p-1-k}(\Omega^k_{\hat{X}}(\log(\hat{Z}))).$$

However,

$$\text{Res}: H^{2p-1-k}(\Omega^k_{\hat{X}}(\log(\hat{Z})) \to H^{k-1,2p-1-k}(\hat{Z}))$$

is zero if $k \neq p$, because

$$H^{2p}(\hat{Z}, \mathbf{C}) = H^{p,p}(Z),$$

and thus we all components of η_Z except the one of type $(p, p-1)$ are in the image of $H^{2p-1}(\hat{X}, \mathbf{C})$, and thus may be incorporated in α. There is thus a unique $\bar{\partial}$-closed $(p, p-1)$ form

$$\eta_Z \in H^{p-1}(\Omega^p_{\hat{X}}(\log(\hat{Z})))$$

satisfying

$$\eta_Z = \tilde{\phi} - \alpha,$$

up to adding elements of $H^{p,p-1}(\hat{X})$. We may regard η_Z as a $(p, p-1)$-form on $X - Z$ having a certain kind of singularity on Z. Now

$$\epsilon^*: H^{2p-1}(X, \mathbf{Z}) \cong H^{2p-1}(\hat{X}, \mathbf{Z}),$$

and if λ is a $(2p-1)$-cycle on X, then

$$\int_\lambda \eta_Z = -\int_\lambda \alpha + \int_\lambda \tilde{\phi},$$

and the last term is integral. Let $Z = \partial U$ on X and let U_ϵ be a neighborhood of U. On $X - U$, we may represent

$$\eta_Z + \alpha - \tilde{\phi} = \partial \tau,$$

where τ is a $(p-1, p-1)$-form which approaches the volume form on normal spheres to U as we approach U. Now to see what element of $F^p H^{2p-1}(X, \mathbf{C})^*$

the class α represents, we note, letting U_ϵ be a neighborhood of U, that for any $\beta \in F^{n-p+1}H^{n-2p+1}(X, \mathbf{C})$,

$$\begin{aligned} <\alpha, \beta> &= \int_X \alpha \wedge \beta \\ &= \lim_{\epsilon \to 0} \int_{X-U_\epsilon} (-\eta_Z + \partial\tau + \tilde{\phi}) \wedge \beta \\ &= \int_X \tilde{\phi} \wedge \beta + \lim_{\epsilon \to 0} \int_{\partial U_\epsilon} \tau \wedge \beta \\ &= \int_X \tilde{\phi} \wedge \beta + \int_U \beta. \end{aligned}$$

We have used that $\eta_Z \wedge \beta = 0$ if $\beta \in F^{n-p+1}H^{n-2p+1}(X, \mathbf{C})$, since it has at least $(n+1)$ differentials of type dz. This means that $\alpha + \tilde{\phi}$ represents $\int_U$ in $H^{2p-1}(X, \mathbf{C})/F^pH^{2p-1}(X, \mathbf{C})$, and then since $\tilde{\phi}$ is integral, α represents $\int_U$ in $J^p(X)$.

DEFINITION. *We will call the form η_Z constructed above the* **residue form** *representing Z.*

EXAMPLE. *Height pairing.*

If Z_1, Z_2 are disjoint algebraic cycles of codimensions p_1, p_2 on an n dimensional smooth projective variety X, such that they are homologically equivalent to zero and $p_1 + p_2 = n + 1$, then if $Z_i = \partial U_i$, the real dimension of U_1 is $2n - 2p_1 + 1 = 2p_2 - 1$, and η_{Z_2} is a $(2p_2 - 1)$-form, so that $\int_{U_1} \eta_{Z_2}$ makes sense. If we change U_1 by a cycle on $X - |Z_2|$, we change $\int_{U_1} \eta_{Z_2}$ by a period of η_{Z_2}, which is information contained in the Abel-Jacobi map of Z_2. If for example $AJ_X(Z_2) = 0$, then the quantity is well-defined modulo $\mathbf{Z}$, hence in $\mathbf{C}/\mathbf{Z}$. This is called the **height pairing**.

EXAMPLE. *Hodge classes on a surface in a threefold.*

Let X be a smooth projective 3-fold containing a smooth surface S. Let γ be a primitive integral (1,1)-class on S. Then

$$\gamma \in F^1(\ker(H^2(S, \mathbf{Z}) \to H^4(X, \mathbf{Z})),$$

and thus gives rise to an element

$$\nu(\gamma) \in H^3(X, \mathbf{C})/(H^1(S, \mathbf{C}) + F^2H^3(X, \mathbf{C}) + H^3(X, \mathbf{Z})) = J^2(X)/\mathrm{im}H^1(S, \mathbf{C}).$$

Now by the Lefschetz (1,1) Theorem we know that γ is represented by a divisor D on S that is homologous to 0 in X.

$$\nu(\gamma) = AJ_X(D);$$

the ambiguity is that we can change D by an element of $\mathrm{Pic}^0(S)$, which accounts for the need to quotient out by the image of $H^1(S, \mathbf{C})$.

EXAMPLE. *1-cycles on curves on a surface.*

Let X be a smooth projective surface and $C \subseteq X$ a smooth curve. Let $\gamma \in H_1(C, \mathbf{Z})$, and $\psi_\gamma \in H^1(C, \mathbf{Z})$ its Poincaré dual. Assume that $\psi_\gamma \mapsto 0$ in $H^1(X, \mathbf{Z})$. Then

$$\nu(\gamma) \in H^2(X, \mathbf{C})/(H^0(C, \mathbf{Z}) + F^1H^2(X, \mathbf{C}) + H^2(X, \mathbf{Z})) = H^{0,2}(X)/H^2(X, \mathbf{Z}).$$

If $\gamma = \partial U$ in X, this is represented by

$$\int_U \in H^{2,0}(X)^*;$$

if we change γ to a homologous cycle on C, the $(2,0)$-forms integrate to zero on the subset of C that is added to U, while if we change U by an integral 2-cycle on X, we don't change the class modulo $H^2(X, \mathbf{Z})$. This kind of construction arises in the work of Blaine Lawson on homotopy groups of spaces of cycles.

We return to the situation of a smooth projective variety X and a normal crossings divisor $D \subseteq X$. Recall that $D_{[i]}$ is the set of points of X having multiplicity $\geq i$ on D. Let $\epsilon_i: \hat{D}_{[i]} \to D_{[i]}$ be the normalization; this is a smooth variety and $\epsilon_i^{-1}(D_{[i+1]})$ is a divisor with normal crossings.

DEFINITION. *On $\hat{D}_{[i]}$, we have a natural local system $\tilde{\mathrm{or}}_i$ which consists of a choice of an ordering of the i irreducible components of $D \cap U$ of a neighborhood U of a point $p \in D_{[i]}$ defining the component of $\hat{D}_{[i]}$ we are on, modulo even permutations. There is a natural isomorphism $\mathrm{or}_i^{\otimes 2} \to 1$. If $j_i: \hat{D}_{[i]} \to X$ is the composition of the normalization and the inclusion, we let $\bar{\mathrm{or}}_i = j_{i*}\mathrm{or}_i$.*

DEFINITION. *We define a filtration $W_\bullet$ on $\Omega_X^\bullet(\log(D))$ by*

$$W_m\Omega_X^p(\log(D)) = \{\omega \in \Omega_X^p(\log(D)) \mid \omega \text{ has a pole of order } \leq m \text{ along } D\}.$$

DEFINITION. *There is a natural map*

$$\mathrm{Res}_i: \Omega_X^\bullet(\log(D))) \to j_{i*}(\Omega_{\hat{D}_{[i]}}^{\bullet - i}(\log(\epsilon_i^* D_{[i+1]})) \otimes \mathrm{or}_i).$$

It is defined, on a component of $\hat{D}_{[i]}$ defined by $z_1 = 0, \ldots, z_i = 0$ in adapted local coordinates (in the proper order) by sending $\alpha \wedge dz_1/z_1 \wedge \cdots \wedge dz_i/z_i \mapsto \epsilon_i^*\alpha$ and everything else to zero. Note that this is well-defined, as we can also obtain it by integrating over a tube, and that it is a map of complexes, since it commutes with d.

PROPOSITION. *Residue gives a natural isomorphism of complexes*

$$\mathrm{Res}_i: \mathrm{Gr}_i^W(\Omega_X^\bullet(\log(D)) \to j_{i*}\Omega_{\hat{D}_{[i]}}^{\bullet - i} \otimes \mathrm{or}_i,$$

where $j_i: \hat{D}_{[i]} \to X$ is the composition of the normalization and the inclusion.

PROOF: There are no poles of order $> i$, so we do not obtain any poles in the image. Poles of lower order are killed. Thus the map is injective. It is surjective, since if p is a point of multiplicity $i + j$ on D, lying over it in $\hat{D}_{[i]}$ we have the $\binom{i+j}{i}$ irreducible local components corresponding to choices of i local components of D. For each subset $I \subseteq \{1, \ldots, i+j\}$, we may extend an α_I to X, and then consider $\sum_I \wedge_{k \in I}(dz_k/z_k) \wedge \alpha_I$, which maps to (α_I) under Res_i. □

COROLLARY. *If D is a smooth codimension 1 subvariety of X, then there is an exact sequence of complexes*

$$0 \to \Omega_X^\bullet \to \Omega_X^\bullet(\log(D)) \to \Omega_D^{\bullet-1} \to 0,$$

where the first map is inclusion and the second map is residue.

Let $j: X - D \to X$ be the inclusion. We would like to compare $j_*\Omega^\bullet_{X-D}$ with $\Omega_X^\bullet(\log(D))$.

DEFINITION. *Let $K^\bullet, d$ be a complex. The τ filtration is defined by*

$$\tau^p K^q = \begin{cases} K^q \text{ if } q < p; \\ \ker(d) \text{ if } q = p; \\ 0 \text{ otherwise.} \end{cases}$$

We note that the identity map $\alpha: (\Omega_X^\bullet(\log(D)), \tau_\bullet) \to (\Omega_X^\bullet(\log(D)), W_\bullet)$ is a map of filtered complexes, since a form of degree $\leq p$ in $\Omega_X^\bullet(\log(D))$ has a pole of order at most p. The natural inclusion $\beta: (\Omega_X^\bullet(\log(D)), \tau_\bullet) \to (j_*\Omega^\bullet_{X-D}, \tau_\bullet)$ is a map of filtered complexes.

THEOREM. *The maps α, β above induce quasi-isomorphisms of filtered complexes, i.e. the maps on the graded pieces are quasi-isomorphisms.*

PROOF: These are purely local statements, so we need only deal with the case of a product of discs and punctured discs. One first shows that there is a "product theorem" to reduce to the case of either the disc or punctured disc, which are easy. The elementary fact one is using is that dz/z^k is exact for $k \neq 1$. □

COROLLARY. $H^k(X - D, \mathbf{C}) \cong \mathbf{H}^k(\Omega_X^\bullet(\log(D))$

DEFINITION. *$F^p H^k(X - D, \mathbf{C})$ will denote the image of*

$$\mathbf{H}^k(\Omega_X^{\geq p}(\log(D))) \to \mathbf{H}^k(\Omega_X^\bullet(\log(D))).$$

$W_m H^k(X - D, \mathbf{C})$ will denote the image of

$$\mathbf{H}^k(W_m\Omega_X^\bullet(\log(D)) \to \mathbf{H}^k(\Omega_X^\bullet(\log(D)).$$

These are called respectively the **Hodge filtration** *and* **weight filtration** *of $H^k(X - D, \mathbf{C})$.*

We will quote the following:

THEOREM (Deligne).
(1)

$$\operatorname{Gr}_m^W H^k(X-D,\mathbf{Q}) \cong \begin{cases} \operatorname{im}(H^{k-m}(\hat{D}_{[m]},\mathbf{Q}\otimes \operatorname{or}_m) \to H^k(X-D,\mathbf{Q})) & \text{for } m\geq 0,\\ 0 & \text{otherwise;}\end{cases}$$

(2) The filtration induced by $F^{\bullet+m}$ on $\operatorname{Gr}_m^W H^k(X-D,\mathbf{C})$ is the Hodge filtration on $\operatorname{im}(H^{k-m}(\hat{D}_{[m]},\mathbf{C}))$;
(3)

$$W_m H^k(X-D,\mathbf{Q}) \cong \begin{cases} \operatorname{im}(H^{k-m}(D_{[m]},\epsilon_{m*}\operatorname{or}_m\otimes q) \to H^k(X-D,\mathbf{Q}) & \text{for } m\geq 0,\\ 0 & \text{otherwise.}\end{cases}$$

(4) If $X-D \cong X'-D'$ are two different ways of writing the same quasi-projective variety, then the weight and Hodge filtrations on $H^\bullet(X-D,\mathbf{C})$ are the same;
(5) The spectral sequence for computing $\mathbf{H}^\bullet(\Omega_X^\bullet(\log(D)))$ with E_1 term

$$E_1^{p,q} = H^q(\Omega_X^p(\log(D))$$

degenerates at the E_1 term;
(6) The spectral sequence for computing $H^q(\Omega_X^p(\log(D)))$ using the filtration $W_\bullet$ degenerates at the E_2 term;
(7) Cup product on $H^\bullet(X-D,\mathbf{C})$ is induced by wedge product on $\Omega_X^\bullet(\log(D))$.

EXAMPLE. *D smooth*

Here,

$$\operatorname{Gr}_0^W H^\bullet(X-D,\mathbf{C}) \cong H^\bullet(X,\mathbf{C})$$

and

$$\operatorname{Gr}_1^W H^\bullet(X-D,\mathbf{C}) \cong H^{\bullet-1}(D,\mathbf{C}).$$

The exact sequence $0 \to W_0 \to W_1 \to W_1/W_0 \to 0$ gives a long exact sequence

$$H^{k-2}(D,\mathbf{C}) \to H^k(X,\mathbf{C}) \to H^k(X-D,\mathbf{C}) \to H^{k-1}(D,\mathbf{C}) \to H^{k+2}(D,\mathbf{C})$$

which is the complexification of the Gysin sequence.

We will denote the integral singular cochains on a space Y by $C^\bullet(Y)$, etc. We note that the composition of the normalization map with ϵ_i gives a map

$$\epsilon_i^* \widehat{D_{[i+1]}} \to \hat{D}_{[i+1]};$$

the variety on the left is just $i+1$ disjoint copies of $\hat{D}_{[i+1]}$, and the map is the identity when restricted to each piece. At a point of multiplicity exactly $i+1$ on D, these correspond to subsets of i elements of the $i+1$ local irreducible components of D going through that point. If we order these components and take an alternating sum of $\pm\mathrm{id}^*$, we obtain a map of complexes

$$\bar{\operatorname{Tr}}_i\colon C^\bullet(\epsilon_i^*\widehat{D_{[i+1]}}) \to C^\bullet(\hat{D}_{[i+1]}).$$

Composing with the composition of the normalization map and the inclusion of $\epsilon_i^* D_{[i+1]} \to \hat{D}_{[i]}$, we obtain a map

$$\mathrm{Tr}_i : C^\bullet(\hat{D}_{[i]}, \mathrm{or}_i) \to C^\bullet(\hat{D}_{[i+1]}, \mathrm{or}_{i+1}),$$

which satisfies the condition

$$\mathrm{Tr}_{i+1} \circ \mathrm{Tr}_i = 0$$

for all $i \geq 0$. Recalling the convention $D_{[0]} = X$, we set

$$\tilde{C}^k(X) = \oplus_{i \geq 0} C^{k-i}(\hat{D}_{[i]}, \mathrm{or}_i),$$

where

$$\delta(\alpha_0, \ldots, \alpha_n) = (\delta\alpha_0, \mathrm{Tr}_0(\alpha_0) - \delta\alpha_1, \ldots, \mathrm{Tr}_{n-1}(\alpha_{n-1}) - \delta\alpha_n).$$

This is a complex. We filter it by

$$W_m \tilde{C}^k(X) = \oplus_{i \geq m} C^{k-i}(\hat{D}_{[i]}, \mathrm{or}_i).$$

A more elegant way to do this is to note that under Tr,

$$C^\bullet(\hat{D}_{[\bullet]}, \mathrm{or})$$

is a bigraded complex and

$$H^k(\tilde{C}^\bullet(X)) \cong \mathbf{H}^k(\mathrm{or}^\bullet),$$

where $\mathrm{or}^\bullet$ is the complex of ± 1-valued sheaves or_i on $\hat{D}_{[i]}$ pushed down to X, and

$$H^k(W_m \tilde{C}^\bullet(X)) \cong \mathbf{H}^k(\mathrm{or}^{\geq m}).$$

PROPOSITION.
(1) $H^k(\tilde{C}^\bullet(X)) \cong H^k(X, D, \mathbf{Z})$;
(2) $H^k(W_m \tilde{C}^\bullet(X))) \cong H^{k-m}(D_{[m]}, \epsilon_{m*}\mathrm{or}_m)$ *for* $m > 0$;
(3) $H^k(\mathrm{Gr}_m^W \tilde{C}^\bullet(X)) \cong H^{k-m}(\hat{D}_{[m]}, \mathrm{or}_i)$.

PROOF: This is basically a massive application of the Mayer-Vietoris sequence. □

We may do the same thing on the level of differential forms, obtaining maps

$$\mathrm{Tr}_i : A^k(\hat{D}_{[i]}) \otimes \mathrm{or}_i \to A^k(\hat{D}_{[i+1]}) \otimes \mathrm{or}_{i+1},$$

and defining a complex

$$\tilde{A}^k(X) = \oplus_{i \geq 0} A^{k-i}(\hat{D}_{[i]}) \otimes \mathrm{or}_i$$

with differential

$$d(\alpha_0, \ldots, \alpha_n) = (d\alpha_0, \mathrm{Tr}_0(\alpha_0) - d\alpha_1, \ldots, \mathrm{Tr}_{n-1}(\alpha_{n-1}) - d\alpha_n).$$

We set

$$W_m \tilde{A}^k(X) = \oplus_{i \geq m} A^{k-i}(\hat{D}_{[i]}) \otimes \mathrm{or}_i.$$

We then have similarly:

PROPOSITION.
(1) $H^k(\tilde{A}^\bullet(X)) \cong H^k(X, D, \mathbf{C})$;
(2) $H^k(W_m\tilde{A}^\bullet(X)) \cong H^{k-m}(D_{[m]}, \mathbf{C} \otimes \epsilon_{m*}\mathrm{or}_m)$ *if* $m > 0$;
(3) $H^k(\mathrm{Gr}_m^W \tilde{A}^\bullet(X)) \cong H^{k-m}(\hat{D}_{[m]}, \mathbf{C} \otimes \mathrm{or}_m)$.

We note that $\tilde{A}^\bullet(X)$ has a Hodge filtration: Let

$$F^p\tilde{A}^\bullet(X) = \oplus_{i\geq 0} F^p A^{\bullet - i}(\hat{D}_{[i]}) \otimes \mathrm{or}_i;$$

this is compatible with the differential. We have further that:

PROPOSITION. *$F^\bullet$ induces the usual Hodge filtration of*

$$H^k(\mathrm{Gr}_m^W \tilde{A}^\bullet(X)) \cong H^{k-m}(\hat{D}_{[m]}, \mathbf{C} \otimes \mathrm{or}_m).$$

Another way to see this is to set

$$W_m\tilde{\Omega}_X^k = \oplus_{i\geq m} \Omega_{\hat{D}_{[i]}}^{k-i} \otimes \mathrm{or}_i$$

with the differential defined above. One then has:

PROPOSITION.
(1) $\mathbf{H}^k(\tilde{\Omega}_X^\bullet) \cong H^k(X, D, \mathbf{C})$;
(2) $\mathbf{H}^k(W_m\tilde{\Omega}_X^\bullet) \cong H^{k-m}(D_{[m]}, \mathbf{C} \otimes \epsilon_{m*}\mathrm{or}_m)$ *if* $m > 0$;
(3) $\mathbf{H}^k(\mathrm{Gr}_m^W \tilde{\Omega}_X^\bullet) \cong H^{k-m}(\hat{D}_{[m]}, \mathbf{C} \otimes \mathrm{or}_m)$;
(4) The hypercohomology spectral sequence for $\mathbf{H}^k(W_m\tilde{\Omega}_X^\bullet)$ *has*

$$E_1^{p,q} = H^q(W_m\tilde{\Omega}_X^p) = \oplus_{i\geq m} H^q(\Omega_{\hat{D}_{[i]}}^{p-i} \otimes \mathrm{or}_i)$$

and degenerates at the E_2 *term;*
(5) The filtration induced by the hypercohomology spectal sequence is $F^\bullet$ *defined above.*

This gives an alternative way to obtain the mixed Hodge structure on $H^\bullet(X, D)$. By Lefschetz duality,

$$H^k(X, D, \mathbf{C}) \cong H_{2n-k}(X - D, \mathbf{C}) \cong H^{2n-k}(X - D, \mathbf{C})^*.$$

Since or_1 is trivial, the weight filtration gives for any D with normal crossings a long exact sequence

$$H^{2n-2}(X, \mathbf{Z}) \to H^{2n-2}(D, \mathbf{Z}) \to H^{2n-1}(X, D, \mathbf{Z}) \to H^{2n-1}(X, \mathbf{Z}) \to 0.$$

Now the dual of $H^{2n-2}(D, \mathbf{Z})$ is $H^0(\hat{D}, \mathbf{Z})$, and the cokernel of the map on the left is dual to the kernel of $H^0(\hat{D}, \mathbf{Z}) \to H^2(X, \mathbf{Z})$ given by taking Poincar'e dual classes of components of $\hat{D}$. We have an exact sequence

$$0 \to H_1(X, \mathbf{Z}) \to H_1(X - D, \mathbf{Z}) \to H_0(\hat{D}) \to H^2(X, \mathbf{Z}).$$

An element ϕ of the kernel of the map on the right, corresponding to a sum of components $\sum_i n_i D_i$ of D that is in $Z_h^1(X)$, lifts to an element $\tilde{\phi} \in H_1(X - D, \mathbf{Z})$.

Now tracing through the Hodge filtration, ϕ belongs to F^1 if we complexify, and hence if we mod out by F^1, we obtain a well-defined element

$$u(\phi) \in H^1(X, \mathbf{C})/(F^1 H^1(X, \mathbf{C}) + H^1(X, \mathbf{Z})) = J^1(X).$$

This is $AJ_X(\sum_i n_i D_i)$.

There is a natural map

$$H_k(D_{[i]}, \epsilon_{i*}\mathrm{or}_i) \to H_{k-1}(D_{[i+1]}, \epsilon_{i+1*}\mathrm{or}_{i+1})$$

constructed as follows. We map

$$H_k(D_{[i]}, \epsilon_{i*}\mathrm{or}_i) \to H_k(D_{[i]}, D_{[i+1]}, \mathrm{or}_i) \cong H_k(\hat{D}_{[i]}, \epsilon_i^* D_{i+1]}, \mathrm{or}_i).$$

Now by boundary we have a map

$$H_k(\hat{D}_{[i]}, \epsilon_i^* D_{i+1]}, \mathrm{or}_i) \to H_{k-1}(\epsilon_i^* D_{[i+1]}, \mathrm{or}_i).$$

Finally, we have

$$\mathrm{Tr}_{i+1}: H_{k-1}(\epsilon_i^* D_{[i+1]}, \mathrm{or}_i) \to H_{k-1}(D_{[i+1]}, \mathrm{or}_{i+1});$$

the composition is the map we want. Dually, this gives a map

$$H^{k-1}(D_{[i+1]}, \mathrm{or}_{i+1} \otimes \mathbf{Q}) \to H^k(D_{[i]}, \mathrm{or}_i \otimes \mathbf{Q}).$$

Finally, there is a map $H^k(D, \mathbf{Z}) \to H^{k+1}(X, D, \mathbf{Z})$ given by coboundary, and hence a map $H^k(\hat{D}, \mathbf{Z}) \to H^{k+1}(X, D, \mathbf{Z})$. The composition of these gives maps

$$H^{k-m}(\hat{D}_{[m]}, \mathbf{Q}) \to H^k(X, D, \mathbf{Q})$$

whose image is $W_m H^k(X, D, \mathbf{Q})$.

DEFINITION. ***A* mixed Hodge structure** *is:*
(1) A real finite-dimensional vector space V *together with a lattice* $\Lambda \subseteq V$*;*
(2) A finite decreasing filtration $F^\bullet V_{\mathbf{C}}$ *on* $V_{\mathbf{C}} = V \otimes \mathbf{C}$*;*
(3) A finite increasing filtration $W_\bullet V$ *defined over* $\mathbf{Q}$ *such that:*
(4) $\mathrm{Gr}_k^W V$*, together with the filtration induced by* $F^\bullet$*, is a Hodge structure of weight* k *for all* k.

REMARK: Here $\mathrm{Gr}_k^W V = W_k V / W_{k+1} V$, and the induced filtration is

$$F^p \mathrm{Gr}_k^W V_{\mathbf{C}} = (F^p V_{\mathbf{C}} \cap W_k V_{\mathbf{C}})/(F^p V_{\mathbf{C}} \cap W_{k+1} V_{\mathbf{C}}).$$

EXAMPLE. *On* $H^k(X - D)$, $F^\bullet$, $W_{\bullet+k}$ *gives a mixed Hodge structure.*

Bibliographic Notes

The Gysin sequence and Lefschetz duality may be found in, for example, Greenberg and Harper, *Algebraic Topology: A First Course*.

The main reference for this section is Deligne, "Théorie de Hodge, II," cited in the last lecture. This is a paper which changed Hodge theory forever. A few weeks after I became Griffiths' PhD student, I flew to Paris. My friend,

Harsh Pittie, met me at the airport and took me directly to the Séminaire Bourbaki, where Deligne was giving a lecture entitled "Travaux de Griffiths" which already incorporated some of these ideas. After the lecture, when Harsh introduced me to one of his French friends as a student of Griffiths, his friend said, "Ah, so then you can explain Deligne's lecture to me." Needless to say, I couldn't. Those who would like help before plunging into Deligne's paper might consult the Cornalba-Griffiths Arcata lectures, or a very short but helpful tour in an appendix on mixed Hodge structures to Dimca, *Singularities and Topology of Hypersurfaces*.

LECTURE 6

In this section, we introduce normal functions.

DEFINITION. *Given a family of projective varieties $\pi: \mathcal{X} \to T$ and π is a submersion, let $X_t = \pi^{-1}(t)$. Let $\mathcal{Z} \subseteq \mathcal{X}$ be a codimension p algebraic cycle such that $Z_t = \mathcal{Z} \cap X_t$ is a codimension p cycle which is homologous to zero for all $t \in T$. Then the* **normal function** *associated to the family of cycles Z_t is*

$$\nu(t) = AJ_{X_t}(Z_t) \in J^p(X_t).$$

DEFINITION. *Given a family of projective varieties $\pi: \mathcal{X} \to T$ and π is a submersion, the* **family of p'th intermediate Jacobians** *associated to $\mathcal{X}$, $\pi_{\mathcal{J}}: \mathcal{J}^p \to T$ is*

$$\mathcal{J}^p = \mathcal{R}^{2p-1}_{\pi*}\mathbf{C}/(\mathcal{R}^{2p-1}_{\pi*}\mathbf{Z} + F^p\mathcal{R}^{2p-1}_{\pi*}\mathbf{C});$$

thus

$$\pi^{p\ -1}_{\mathcal{J}}(t) = J^p(X_t).$$

Since locally we are quotienting a fixed vector space by a fixed lattice and a holomorphically varying linear subspace, $\mathcal{J}^p$ naturally has the structure of a complex manifold, and $\pi_{\mathcal{J}}$ is holomorphic.

DEFINITION. *Let $\nu: T \to \mathcal{J}^p$ be a holomorphic section of $\pi_{\mathcal{J}}: \mathcal{J}^p \to T$. We may choose locally on a small open subset $U \subseteq T$ a lifting $\tilde{\nu}: U \to \mathcal{R}^{2p-1}_{\pi*}\mathbf{C}$ of ν. We will say that ν* **satisfies the infinitesimal condition for normal functions** *if*

$$\nabla\tilde{\nu} \in \Omega^1_T \otimes F^{p-1}\mathcal{R}^{2p-1}_{\pi*}\mathbf{C}.$$

This condition is independent of the local lifting, since if $\hat{\nu}: U \to \mathcal{J}^p$ is another lifting of ν, then

$$\hat{\nu} - \tilde{\nu} = f + \lambda,$$

where $f \in F^p\mathcal{R}^{2p-1}_{\pi*}\mathbf{C}$ and $\lambda \in \mathcal{R}^{2p-1}_{\pi*}\mathbf{Z}$. Now

$$\nabla f \in \Omega^1_T \otimes F^{p-1}\mathcal{R}^{2p-1}_{\pi*}\mathbf{C}$$

by the infinitesimal period relations, and $\nabla\lambda = 0$, since integral classes are locally constant. Thus

$$\nabla\hat{\nu} - \nabla\tilde{\nu} \in F^{p-1}\mathcal{R}^{2p-1}_{\pi*}\mathbf{C},$$

and the concept is well-defined.

DEFINITION. *A* **normal function** *for the family $\pi: \mathcal{X} \to T$ is a holomorphic section $\nu: T \to \mathcal{J}^p$ which satisfies the infinitesimal condition for normal functions.*

REMARK: For many purposes, if $T = \bar{T} - D$, it is necessary to also add a condition on the behavior of normal functions near D. We can get away without this condition, which is rather delicate.

THEOREM. *The normal function associated to an analytic family of algebraic cycles homologically equivalent to zero is a normal function.*

PROOF: Let

$$\nu(t) = AJ_{X_t}(Z_t).$$

The first thing we must do is to show that ν is holomorphic. Since we are working locally, we may assume T is the unit disc, and that we have written $Z_t = \partial U_t$ with $U_t \subseteq X_t$ depending continuously on t. Since $\mathcal{X}$ is diffeomorphic to $X_0 \times T$, we use De Rham's Theorem on $\mathcal{X}$ to find a closed form

$$\phi \in A^{2n-2p+1}(\mathcal{X})$$

representing any given class in $H^{2n-2p+1}(X_0, \mathbf{C})$. Let t be joined to 0 by a path γ, and let

$$\mathcal{U} = \cup_{s\in\gamma} U_s.$$

Let

$$\mathcal{Z}_\gamma = \cup_{s\in\gamma} Z_s.$$

Then

$$\partial\mathcal{U} = U_t - U_0 + \mathcal{Z}_\gamma.$$

By Stokes' Theorem,

$$\int_{U_t} \phi = \int_{U_0} \phi - \int_{\mathcal{Z}_\gamma} \phi + \int_{\mathcal{U}} d\phi.$$

Since ϕ is closed, the last term is zero. Taking limits and recalling the notation Y from Lecture 3, we see that

$$\frac{\partial}{\partial t}\left(\int_{U_t} \phi\right)(0) = -\int_{Z_0} < \phi, Y >;$$

$$\frac{\partial}{\partial \bar{t}}\left(\int_{U_t} \phi\right)(0) = -\int_{Z_0} < \phi, \bar{Y} > .$$

We note that

$$\frac{\partial}{\partial t}\left(\int_{U_t} \phi\right)(0) = 0$$

if

$$\phi \in F^{n-p+2} H^{2n-2p+1}(X_0, \mathbf{C}),$$

while

$$\frac{\partial}{\partial \bar{t}}\left(\int_{U_t} \phi\right)(0) = 0$$

if

$$\phi \in F^{n-p+1} H^{2n-2p+1}(X_0, \mathbf{C}).$$

Now let $\tilde{\phi}$ be a holomorphic local section of

$$F^{n-p+1}\mathcal{R}^{2n-2p+1}_{\pi *}\mathbf{C}$$

agreeing with ϕ at 0. We may write

$$\tilde{\phi} = \phi + t\psi + \cdots.$$

Now

$$\frac{\partial}{\partial t}(\int_{U_t} \tilde{\phi})(0) = \frac{\partial}{\partial t}(\int_{U_t} \phi)(0) + \int_{U_0} \psi$$

and

$$\frac{\partial}{\partial \bar{t}}(\int_{U_t} \tilde{\phi})(0) = \frac{\partial}{\partial \bar{t}}(\int_{U_t} (\phi))(0).$$

We thus conclude that

$$\frac{\partial}{\partial \bar{t}}(\int_{U_t} \tilde{\phi})(0) = 0$$

if

$$\phi \in F^{n-p+1}H^{2n-2p+1}(X_0, \mathbf{C}),$$

and this is the statement that ν is a holomorphic section of $\mathcal{J}^p$. Since

$$\int_{U_t} \in H^{2n-2p+1}(X_t, \mathbf{C})^*$$

represents a lifting $\tilde{\nu}$ of ν to $\mathcal{R}^{2p-1}_{\pi *}\mathbf{C}$, our computation for $\frac{\partial}{\partial t}(\int_{U_t} \phi)(0)$ gives that

$$\nabla \tilde{\nu} \in \Omega^1_T \otimes F^{p-1}\mathcal{R}^{2p-1}_{\pi *}\mathbf{C},$$

which is the infinitesimal condition for normal functions. □

REMARK: What we have actually shown is that we may choose a lifting $\tilde{\nu}$ of ν to $\mathcal{R}^{2p-1}_{\pi *}\mathbf{C}$ such that

$$\nabla_{\frac{\partial}{\partial t}} \tilde{\nu} = (\phi \mapsto \int_{Z_t} < \phi, Y >);$$

$$\nabla_{\frac{\partial}{\partial \bar{t}}} \tilde{\nu} = (\phi \mapsto \int_{Z_t} < \phi, \bar{Y} >).$$

DEFINITION. *Let $\pi: \mathcal{X} \to T$ be a family of smooth projective varieties with π a submersion. The Gauss-Manin connection ∇ gives maps*

$$\Omega^k_T \otimes F^p\mathcal{R}^{2p-1}_{\pi *}\mathbf{C} \to \Omega^{k+1} \otimes F^{p-1}\mathcal{R}^{2p-1}_{\pi *}\mathbf{C}.$$

If

$$K^k_p = \Omega^k_T \otimes F^{p-k}\mathcal{R}^{2p-1}_{\pi *}\mathbf{C},$$

then $K^\bullet_p, \nabla$ is a complex, which we will call the **p'th Koszul complex** *associated to the family $\mathcal{X} \to T$.*

DEFINITION. *Let $\nu: T \to \mathcal{J}^p$ be a normal function for the family $\pi: \mathcal{X} \to T$. Let $\tilde{\nu}: U \to \mathcal{R}^{2p-1}_{\pi *}\mathbf{C}$ be a local lifting of ν on $U \subseteq T$. Then*

$$\nabla \tilde{\nu} \in \Omega^1_T \otimes F^{p-1}\mathcal{R}^{2p-1}_{\pi *}\mathbf{C} = K^1_p.$$

Since $\nabla^2 = 0$, $\nabla\tilde{\nu}$ maps to zero in K_p^2. If $\hat{\nu}$ is another lifting of ν, then

$$\hat{\nu} - \tilde{\nu} = f + \lambda,$$

where

$$f \in F^p\mathcal{R}_{\pi*}^{2p-1}\mathbf{C} = K_p^0$$

and

$$\lambda \in \mathcal{R}_{\pi*}^{2p-1}\mathbf{Z}.$$

Thus

$$\nabla\hat{\nu} - \nabla\tilde{\nu} = \nabla f \in (\nabla K_p^0).$$

We thus conclude that $\nabla\tilde{\nu}$ determines a well-defined element

$$\delta(\nu) \in H^1(K_p^\bullet),$$

called **Griffiths' infinitesimal invariant** *of the normal function ν.*

DEFINITION. *A normal function $\nu: T \to \mathcal{J}^p$ is* **locally constant** *if at every point it has a local lifting $\tilde{\nu}: U \to \mathcal{R}_{\pi*}^{2p-1}\mathbf{C}$ such that $\nabla\tilde{\nu} = 0$, i.e. $\tilde{\nu}$ is constant on U.*

PROPOSITION. *A normal function ν is locally constant if and only if $\delta(\nu) = 0$.*

PROOF: If ν is locally constant, then there is a lifting $\tilde{\nu}$ of ν such that $\nabla\tilde{\nu} = 0$, which of course implies that $\delta(\nu) = 0$. If instead we assume that $\delta(\nu) = 0$, then there is a local lifting $\tilde{\nu}$ such that $\nabla\tilde{\nu} = \nabla f$ for some $f \in F^p\mathcal{R}_{\pi*}^{2p-1}\mathbf{C}$; however, $\hat{\nu} = \tilde{\nu} - f$ is another lifting of ν satisfying $\nabla\hat{\nu} = 0$, so ν is locally constant. □

DEFINITION. *The complex $K_p^\bullet$ is filtered by the subcomplexes $K_{p,m}^\bullet$ with*

$$K_{p,m}^k = \Omega_T^k \otimes F^{m-k}\mathcal{R}_{\pi*}^{2p-1}\mathbf{C}$$

for all $m \geq p$. Let $\bar{K}_{p,m}^\bullet = K_{p,m}^\bullet / K_{p,m+1}^\bullet$.

REMARK: We note that

$$\bar{K}_{p,m}^k = \Omega_T^k \otimes H^{m-k,2p-1+m-k}(X_t),$$

and the differentials are just the derivative of the period map. We note that

$$H^1(K_p^\bullet) = 0 \quad \text{if } H^1(\bar{K}_{p,m}^\bullet) = 0 \text{ for all } m \geq p.$$

There are several ways to rephrase the above concepts that are more elegantly and invariantly expressed. We know

$$H^k(\mathcal{X}, \mathbf{C}) = \mathbf{H}^k(\Omega_{\mathcal{X}}^\bullet).$$

Now $\Omega_{\mathcal{X}}^\bullet$ is filtered by

$$\operatorname{im}(\pi^*\Omega_T^p \otimes \Omega_{\mathcal{X}}^{\bullet-p});$$

the graded pieces are

$$\pi^*\Omega_T^p \otimes \Omega_{\mathcal{X}/T}^{\bullet-p}.$$

There is thus a spectral sequence which abuts to

$$\mathbf{H}^\bullet(\Omega^\bullet_{\mathcal{X}})$$

whose E_1 term is

$$E_1^{p,q} = \mathbf{H}^{p+q}(\pi^*\Omega^p_T \otimes \Omega^{\bullet-p}_{\mathcal{X}/T}).$$

By the Leray spectral sequence, if T is a ball, then

$$E_1^{p,q} = H^0(\Omega^p_T \otimes \mathbf{R}^{p+q}_{\pi *}(\Omega^{\bullet-p}_{\mathcal{X}/T})) \cong H^0(\Omega^p_T \otimes \mathcal{R}^q_{\pi *}\mathbf{C}).$$

The d_1 maps for this spectral sequence are just ∇, and the filtration on $E_1^{p,q}$ induced by the filtration of $\mathbf{R}^{p+q}_{\pi *}(\Omega^{\bullet-p}_{\mathcal{X}/T})$ is just the Hodge filtration, shifted, so that the graded pieces of $E_1^{\bullet,2p-1}$, d_1 are just the Koszul complexes $\bar{K}^\bullet_{p,m}$ defined above.

Another way to look at normal functions is using Deligne cohomology. If we have a Deligne class

$$\Phi \in H^{2p}_{\mathcal{D}}(\mathcal{X})$$

whose cohomological part

$$\Phi_c \in H^{p,p}_{\mathbf{Z}}(\mathcal{X})$$

belongs to the kernel of $H^{2p}(\mathcal{X}, \mathbf{C}) \to H^{2p}(X_t, \mathbf{C})$ (this condition is independent of t, of course), then the restriction map $\nu(t) = \Phi|_{X_t} \in H^{2p}_{\mathcal{D}}(X_t)$ lies in $J^p(X_t)$. We will call this map $\nu: T \to \mathcal{J}^p$ the **normal function associated to the Deligne class Φ**.

PROPOSITION. *The normal function associated to a Deligne class Φ is a normal function.*

PROOF: We may shrink T to a ball without loss of generality, since the restriction map factors through the Deligne cohomology of the smaller family. Now

$$H^{2p}(\mathcal{X}, \mathbf{Z}) \cong H^{2p}(X_t, \mathbf{Z}),$$

so we may assume that

$$\Phi_c = 0.$$

Now the complex $\mathbf{Z} \to \Omega^{<p}_{\mathcal{X}}$ has sub-complex $\Omega^{<p}_{\mathcal{X}}[-1]$ with quotient $\mathbf{Z}$, and we have an exact sequence

$$H^{2p-1}(\mathcal{X}, \mathbf{Z}) \to \mathbf{H}^{2p-1}(\Omega^{<p}_{\mathcal{X}}) \to H^{2p}_{\mathcal{D}}(\mathcal{X}) \to H^{2p}(\mathcal{X}, \mathbf{Z}).$$

Thus Φ comes from a class in

$$\mathbf{H}^{2p-1}(\Omega^{<p}_{\mathcal{X}})/H^{2p-1}(\mathcal{X}, \mathbf{Z}).$$

Now $\Omega^{<p}_{\mathcal{X}/T}$ is a quotient complex of $\Omega^{<p}_{\mathcal{X}}$, so we map to a class in

$$\mathbf{H}^{2p-1}(\Omega^{<p}_{\mathcal{X}/T})/H^{2p-1}(\mathcal{X}, \mathbf{Z}),$$

and the map to $J^p(X_t)$ factors through this. More specifically,

$$\mathbf{H}^{2p-1}(\Omega^{<p}_{\mathcal{X}/T}) \cong H^0(T, \mathbf{R}^{2p-1}_{\pi*}(\Omega^{<p}_{\mathcal{X}/T})) \cong H^0(T, \mathcal{R}^{2p-1}_{\pi*}\mathbf{C}/F^P\mathcal{R}^{2p-1}_{\pi*}\mathbf{C}),$$

and this latter object is just holomorphic sections of $\mathcal{J}^p$ over T.

We need to get the infinitesimal condition for normal functions as well. The point is that $\Omega^{<p}_{\mathcal{X}}$ is filtered by

$$\operatorname{im}(\pi^*\Omega^i_T \otimes \Omega^{<p-i}_{\mathcal{X}}),$$

and the graded quotients are

$$\pi^*\Omega^i_T \otimes \Omega^{<p-i}_{\mathcal{X}/T}[-i].$$

Taking the associated spectral sequence for this filtration, we compute the group $\mathbf{H}^{2p-1}(\Omega^{<p}_{\mathcal{X}})$ as the abutment of a spectral sequence where the d_1 differential is

$$H^0(\mathcal{R}^{2p-1}_{\pi*}\mathbf{C}/F^p\mathcal{R}^{2p-1}_{\pi*}\mathbf{C}) \to H^0(\Omega^1_T \otimes \mathcal{R}^{2p-1}_{\pi*}\mathbf{C}/F^{p-1}\mathcal{R}^{2p-1}_{\pi*}\mathbf{C})$$

given by the map induced by ∇. Those holomorphic sections of $\mathcal{J}^p$ which survive to the E_2 term are those which satisfy the infinitesimal condition for normal functions. □

REMARK: Of course, if $\mathcal{Z} \subseteq \mathcal{X}$ is a codimension p algebraic cycle, the normal function for $\mathcal{Z}$ is just the normal function associated to $\psi_{\mathcal{Z}} \in H^{2p}_{\mathcal{D}}(\mathcal{X})$. This in turn provides an alternate proof of the earlier proposition about normal functions of cycles.

Note that given an integral (p,p)-class, one may lift it (non-canonically) to a Deligne class, and then get a normal function. In the literature preceding the introduction of Deligne cohomology, this subtlety in how a Hodge class gives rise to a normal function looks awkward.

EXAMPLE. *The difference of two g^1_3's on a curve of genus 4.*

The first example of $\delta\nu$ was computed by Griffiths even before the correct formulation of the invariant was known. On the moduli of curves of genus 4, we have on a general curve two g^1_3's corresponding to the rulings of the smooth quadric containing the canonical curve. Locally, we may take the difference of the g^1_3's (or look at the subset of $\mathbf{P}(H^0(Q, \mathcal{O}_Q(3)))$ corresponding to smooth curves) and obtain a normal function ν, whose infinitesimal invariant we may think of as

$$\delta\nu \in H^1(H^0(K_X) \to H^0(K^2_X) \otimes H^0(K_X)^*).$$

Now

$$Q\colon H^0(K_X)^* \to H^0(K_X),$$

and if we take

$$H^0(K^2_X) \otimes H^0(K_X)^* \xrightarrow{1\otimes Q} H^0(K_X)^2 \otimes H^0(K_X)$$

and then compose with the map

$$H^0(K^2_X) \otimes H^0(K_X) \to H^0(Q, \mathcal{O}_Q(3)),$$

then the image of $H^0(K_X)$ maps to zero, so that we obtain a well-defined element of $H^0(\mathcal{O}_Q(3))$ from $\delta\nu$. This is the cubic V such that

$$Q \cap V = \phi_K(X).$$

One notices from this that the infinitesimal invariant can be non-zero, and that it contains sufficient information to allow one to reconstruct the curve.

Bibliographic Notes

Normal functions go back to Poincaré. The papers that I have found useful are Griffiths, "A theorem concerning the differential equations satisfied by normal functions associated to algebraic cycles," Am. J. Math. 101 (1979), 94-131 and the papers of Zucker, El-Zein and Zucker, and Griffiths on normal fucntions in Griffiths, *Topics in Transcendental Algebraic Geometry*.

Griffiths infinitesimal invariant comes out of an idea of Griffiths in "Infinitesimal variation of Hodge structures (III): determinantal varieties and the infinitesimal invariant of normal functions," Comp. Math. 50 (1983) 267-324, where the case of the difference of two g_4^1's on curves of genus 4 also may be found. In that paper, Griffiths did not have quite the right form of the invariant, so that he had a necessary but not sufficient condition for a normal function to be locally constant. The full infinitesimal invariant was found by myself and Claire Voisin independently for our work on the Abel-Jacobi map of a general 3-fold of degree ≥ 6 presented in Lecture 7.

LECTURE 7

For varieties whose Hodge groups have a simple explicit algebraic description, e.g. complete intersections, cyclic branched coverings of $\mathbf{P}^n$, etc., one can hope to apply algebraic tools and obtain Hodge-theoretic results. In this lecture, we give some illustrations for hypersurfaces in $\mathbf{P}^n$.

We first state the algebraic results that we will need:

PROPOSITION. *Let $d > 0$ be an integer. For any integer $c \geq 0$, there is a unique expression*

$$c = \binom{k_d}{d} + \binom{k_{d-1}}{d-1} + \cdots + \binom{k_\delta}{\delta},$$

such that $k_d > k_{d-1} > \cdots > k_\delta \geq \delta > 0$.

PROOF: We proceed by induction on d. The case $d = 1$ is automatic, since $k_1 = c, \delta = 1$ is the only choice.

Now, for any d, choose k_d such that

$$\binom{k_d}{d} \leq c < \binom{k_d+1}{d}.$$

If we now write $\hat{c} = c - \binom{k_d}{d}$ as

$$\hat{c} = \binom{k_{d-1}}{d-1} + \cdots + \binom{k_\delta}{\delta}$$

applying the theorem for $d-1$, we see that $k_{d-1} < k_d$, since otherwise $c \geq \binom{k_d}{d} + \binom{k_d}{d-1} = \binom{k_d+1}{d}$. Thus there exists at least one expression for c of the form claimed in the theorem. If there is a second expression for c as

$$c = \binom{l_d}{d} + \cdots + \binom{l_\epsilon}{\epsilon}$$

with $l_d > l_{d-1} > \cdots > l_\epsilon \geq \epsilon > 0$, then $k_d \geq l_d$, and if equality holds, then by induction, we may apply uniqueness of representation to $\hat{c}$ to get that the two expressions are identical. If $k_d > l_d$, then we note that

$$c \leq \binom{l_d}{d} + \binom{l_d-1}{d-1} + \cdots + \binom{l_d-(d-1)}{1} < \binom{l_d+1}{d} \leq \binom{k_d}{d} \leq c,$$

which is a contradiction. So the expression claimed by the theorem is unique.□

DEFINITION. *Let $d > 0$ be an integer. For any integer $c \geq 0$, the unique expression*

$$c = \binom{k_d}{d} + \binom{k_{d-1}}{d-1} + \cdots + \binom{k_\delta}{\delta},$$

such that $k_d > k_{d-1} > \cdots > k_\delta \geq \delta > 0$ is called the **d'th Macaulay representation** *for c. We will call $k_d, k_{d-1}, \ldots, k_\delta$ the* **d'th Macaulay coefficients** *of c.*

LEMMA. *The usual order on integers coincides with the lexicographic order on their d'th Macaulay representations, i.e. if $c, \hat{c}$ have d'th Macaulay coefficients $k_d, k_{d-1}, \ldots$ and $l_d, l_{d-1}, \ldots$ respectively, then $c \leq \hat{c}$ if and only if $k_d \leq l_d$ or $k_d = l_d$ and $k_{d-1} \leq l_{d-1}$, or $k_d = l_d$, $k_{d-1} = ld - 1$, and $k_{d-2} \leq l_{d-2}$, etc.*

PROOF: This is easy. □

THEOREM. *Let I be a homogeneous polynomial ideal in $S = \mathbf{C}[x_1, \ldots, x_n]$. Let $h_{S/I}(d) = \dim(S_d/I_d)$ be the Hilbert function of S/I. Let*

$$h_{S/I}(d) = \binom{k_d}{d} + \cdots + \binom{k_\delta}{\delta}$$

be the d'th Macaulay representation of $h_{S/I}(d)$.
(1) (Macaulay's bound)

$$h_{S/I}(d+1) \leq \binom{k_d+1}{d+1} + \binom{k_{d-1}+1}{d} + \cdots + \binom{k_\delta+1}{\delta+1};$$

(2) For a general hyperplane H, if $I_H \subset S_H$ is the restriction of I to the hyperplane H, then

$$h_{S_H/I_H}(d) \leq \binom{k_d-1}{d} + \cdots + \binom{k_\delta-1}{\delta},$$

where $\binom{a}{b} = 0$ is $a < b$;
(3) (Gotzmann's Persistence Theorem) If equality holds in Macaulay's bound, then

$$h_{S/I}(d+m) = \binom{k_d+m}{d+m} + \cdots + \binom{k_\delta+m}{\delta+m}$$

for all $m \geq 0$.

A reference for the proof of this is given at the end of the lecture.

Another useful result is:

THEOREM. *Let $S = \mathbf{C}[x_1, \ldots, x_n]$. Let $W \subseteq S_d$ be a base-point free linear subspace with $\dim(S_d/W) = c$. The Koszul complex*

$$\wedge^{p+1} W \otimes S_{k-d} \to \wedge^p W \otimes S_k \to \wedge^{p-1} W \otimes S_{k+d}$$

is exact at the $\wedge^p W$ term if $k \geq p + c + d$.

A simple proof of this result appears in [].

We now give three applications of these results to Hodge theory.

DEFINITION. *The* **Noether-Lefschetz locus** *in degree d is*

$\mathcal{NL}_d = \{F \in \mathbf{P}(H^0(\mathbf{P}^3, \mathcal{O}_{\mathbf{P}^3}(d))) \mid$

F represents a nonsingular surface S with $\mathrm{Pic}(\mathbf{P}^3) \to \mathrm{Pic}(S)$ not surjective$\}$.

THEOREM. *Let Σ be an irreducible component of $\mathcal{NL}_d$. Then*

$$\mathrm{codim}(\Sigma) \geq d - 3.$$

PROOF: Let $S \in \mathcal{NL}_d$ and $\lambda \in H^{1,1}_{\mathbf{Z}}(S)$ denote a primitive integral class. The condition that λ remains infinitesimally of type $(1,1)$ to first order in tangent direction η is that $\nabla_\eta \lambda$ is of type $(1,1)$, which since we are deforming an integral (and therefore real) class, is equivalent to saying that $(\nabla_\eta \lambda)^{0,2} = 0$.

Let F be the homogeneous polynomial of degree d defining S. We know that

$$H^{2-q,q}_{\mathrm{pr}}(S) \cong R_{d-4+qd},$$

where $R_k = S_k/J_k$, where J is the Jacobi ideal of F. The map

$$H^1(\Theta_S) \otimes H^{1,1}_{\mathrm{pr}}(S) \to H^{0,2}(S)$$

given by $(\eta, \lambda) \mapsto (\nabla_\eta \lambda)^{0,2}$ is just the multiplication map

$$R_d \otimes R_{2d-4} \to R_{3d-4}.$$

Let $T \subseteq R_d$ be the Zariski tangent space to the component of $\mathcal{NL}_d$ keeping λ of type $(1,1)$. Then $T \otimes \lambda \mapsto 0$. If we dualize and look at

$$H^1(\Theta_S) \otimes H^{2,0}(S) \to H^{1,1}_{\mathrm{pr}}(S)^* \cong H^{1,1}_{\mathrm{pr}}(S),$$

then

$$T \otimes R_{d-4} \mapsto (\lambda)^\perp.$$

In particular, the map

$$T \otimes R_{d-4} \to R_{2d-4}$$

is not surjective. If $\tilde{T} \subseteq S_d$ is the preimage of T under the projection $S_d \to R_d$, then *a fortiori* the multiplication map

$$\tilde{T} \otimes S_{d-4} \to S_{2d-4}$$

is not surjective. However,

$$\tilde{T} \supseteq J_d,$$

and hence $\tilde{T}$ is base-point free since S is smooth.

We may now use either Gotzmann's Persistence Theorem or case $p = 0, k = 2d - 4$ of the vanishing theorem above for Koszul cohomology. The latter result says that if $c \leq d - 4$, then the multiplication map is surjective, and since we know it is not, we conclude $c \geq d - 3$. Since c is the codimension of $\tilde{T}$ and hence of T, we conclude that the codimension of Σ is at least $d-3$. To use Gotzmann's result, we note that if $c \leq d-4$, then we may write it as

$$\binom{d}{d}+\cdots+\binom{5}{5}.$$

Now in the next degree, the codimension of the linear space it generates is at most

$$\binom{d+1}{d+1}+\cdots+\binom{6}{6}=d-4.$$

If equality holds, then it holds for all higher degrees, and this would force $\tilde{T}$ to have a basse-point. If equality fails, then we may continue the argument to see that the codimension drops in each degree, and thus the ideal generated by $\tilde{T}$ has codimension at most $d-4-k$ in degree $d+k$, and hence codimension ≤ 0 in degree $2d-4$, contradicting the surjectivity of multiplication. □

THEOREM (Donagi Symmetrizer Lemma). *Let $X \subseteq \mathbf{P}^{n+1}$ be a smooth hypersurface defined by a homogeneous polynomial F of degree d. Let J be the Jacobi ideal of F and $R_k = S_k/J_k$. Then the Koszul complex*

$$R_{k-a} \to S_a^* \otimes R_k \to \wedge^2 S_a^* \otimes R_{k+a}$$

is exact at the middle term provided $k \leq (n+2)(d-2)-d+1$ and $k \leq (n+2)(d-2)-a-1$.

PROOF: Let $e=(n+2)(d-2)$. By Macaulay's Theorem, $R_k^* \cong R_{e-k}$, with the duality induced by multiplication. If we dualize the sequence of the Theorem, we have

$$\wedge^2 S_d \otimes R_{e-k-a} \to S_d \otimes R_{e-k} \to R_{e-k+a}.$$

If we consider the complex

$$C_p = \wedge^p S_d \otimes R_{e-k+a-pa},$$

we are trying to show

$$H_1(C_\bullet)=0.$$

If we let

$$A_p = \wedge^p S_d \otimes J_{e-k+a-pa}$$

and

$$B_p = \wedge^p S_d \otimes S_{e-k+a-pa},$$

then we have an exact sequence of complexes

$$0 \to A_\bullet \to B_\bullet \to C_\bullet \to 0.$$

It thus suffices to show

$$H_1(B_\bullet)=0$$

and

$$H_0(A_\bullet)=0.$$

The vanishing theorem for Koszul cohomology states that $H_1(B_\bullet) = 0$ provided

$$e - k + a - a \geq a + 1,$$

i.e.

$$k \leq e - a - 1.$$

Since J is generated in degree $d - 1$,

$$H_0(A_\bullet) = 0$$

if

$$e - k \geq d - 1,$$

i.e.

$$k \leq e - d + 1.$$

This holds under the hypothesis on k. □

COROLLARY. *If k satisfies the inequalities of the theorem, then*

$$\{u \in \mathrm{Hom}(S_d, R_k) \mid u(P)Q = u(Q)P \text{ for all } P, Q \in S_d\} = R_{k-d}.$$

PROOF: In terms of the proof above, the only issue is that $H_0(C_\bullet) = 0$, and this follows because it $H_0(B_\bullet) = 0$. □

THEOREM (Generic Torelli for Projective Hypersurfaces). *For smooth projective hypersurfaces of degree d in $\mathbf{P}^{n+1}$, the period map has degree one on hypersurfaces/projective equivalence except possibly when (d, n) is either $(3, 2)$, $(4, 4m)$, $(6, 6m + 1)$, or $(m, mk - 2)$.*

PROOF: The heart of the proof is that if we know the multiplication map

$$R_a \otimes R_b \rightarrow R_{a+b},$$

,for some $a < b$, then we may construct both R_{b-a} and the natural map

$$R_{b-a} \rightarrow \mathrm{Hom}(S_a, R_b)$$

using the symmetrizer construction, provided that $b \leq e - d + 1$ and $b \leq e - a - 1$.

Now, choose q so that $0 \leq d - n - 2 + qd < d$. Outside of the exceptional ranges, the symmetrizer lemma is true, if we take

$$a = d - n - 2 + qd, \quad b = d.$$

We may thus replace a by $b - a$ and b by a. Since we are lowering b and $a + b$, once the inequalities are true, they stay true. We thus eventually get down to having the map

$$R_f \otimes R_f \rightarrow R_{2f},$$

where $f = (d, n+2)$. If $2f < d - 1$, then we have the map $S_f \times S_f \rightarrow S_{2f}$. We may identify the $2f$'th powers of linear forms, because they have $\binom{2f}{f}$ preimages, and from this we may identify the standard action of $GL(n + 2, \mathbf{C})$ on these spaces. Reversing the symmetrizer construction, we obtain the maps $S_k \rightarrow R_k$ for all multiples k of f, and hence from just the Hodge groups alone and the derivative of the period map we may reconstruct the Jacobi ideal in all degrees divisible by f. If $2f \geq d-1$, then we must contend with identifying quadrics of rank 4 in J_{2f}, and seeing that there are few enough that we can still recognize powers of linear forms, and use this to once again recover the Jacobi ideal. Finally, we need to know that we can recognize a hypersurface up to projective equivalence from its Jacobi ideal. This follows from:

THEOREM (Mather-Yau). *If $f, g \in S_d$ have the same Jacobi ideal, then they are related by a projective transformation.*

Once we have this, we are done. □

Here is an application of these ideas by Claire Voisin and I to the Abel-Jacobi map.

THEOREM. *Let $X \subseteq \mathbf{P}^{2n+2}$ be a general $(2n+1)$-fold of degree d. For $d > 2 + 3/m$, the image of the Abel-Jacobi map*

$$AJ_X: Z_h^{n+1}(X) \to J^{n+1}(X)$$

has image contained in the torsion subgroup of $J^{n+1}(X)$.

PROOF: Let B parametrize the space of smooth $(2n+1)$-folds in $\mathbf{P}^{2n+2}$. Given a general X and a codimension n algebraic cycle Z on X, on a branched cover $\pi: \tilde{B} \to B$, we may find an analytic family of cycles $Z_{\tilde{t}}$ so that $Z_{\tilde{t}} = Z$ for some $\tilde{t} \in \pi^{-1}(0)$. If ν is the normal function assoicated to this family of cycles, then the graded pieces of the infinitesimal invariant of ν lie in the cohomology at the middle term of the complexes

$$H^{n+1+k,n-k}(X_t) \to T_B^* \otimes H^{n+k,n+1-k}(X_t) \to \wedge^2 T_B^* \otimes H^{n-1-k,n+2+k}(X_t)$$

for $k \leq 0$. Dualizing these complexes, we obtain

$$\wedge^2 T_B \otimes H^{n+2+k,n-1-k}(X_t) \to T_B \otimes H^{n+1+k,n-k}(X_t) \to H^{n+k,n-k+1}(X_t).$$

Taking $T_B = S_d$, we may rewrite these as

$$\wedge^2 S_d \otimes R_{d-(2n+3)+(n-1-k)d} \to S_d \otimes R_{d-(2n+3)+(n-k)d} \to R_{d-(2n+3)+(n+1-k)d}.$$

By the vanishing theorem for Koszul cohomology, combined with the same argument we used for the Donagi symmetrizer lemma, this vanished provided $(n-k)d > 2n+3$, which holds for all $k \leq 0$ provided $d > 2 + 3/n$. Thus the infinitesimal invariant of ν vanishes, and ν is locally constant.

We may thus locally write ν on each open subset $U_\alpha \subseteq \tilde{T}$ as $\nu = c_\alpha + f_\alpha$, where c_α is a constant vector in $H^{2n+1}(X, \mathbf{C})$ and $f_\alpha \in F^{n+1}H^{2n+1}(X_t, \mathbf{C})$ for all t. However, on $U_\alpha \cap U_\beta$,

$$\gamma_{\alpha\beta} = c_\alpha - c_\beta \in F^{n+1}H^{2n+1}(X_t, \mathbf{C}).$$

However, this means that

$$\nabla_\eta \gamma_{\alpha\beta} \in F^{n+1}$$

for all directions η, and this translates into having an element of $R_{2n+3+nd}$ that goes to zero in $\mathrm{Hom}(R_d, R_{2n+3+(n+1)d})$, and by Macaulay's Theorem this forces

$$\gamma_{\alpha\beta} = 0.$$

Thus the locally constant lifting of ν is unique, modulo integral classes. Thus

$$\nu(0) \in H^{2n+1}(X_0, \mathbf{C})/H^{2n+1}(X_0, \mathbf{Z})$$

is invariant under the monodromy group.

If $\phi \in H^{2n+1}(X, \mathbf{Z})$ is a vanishing cycle, then for the element of the monodromy group giving rise to this cycle, we have that

$$\nu(0) \mapsto \nu(0) + N < \phi, \nu(0) > \phi,$$

where the N comes from the fact that we might have had to take a branched cover. By the argument above, $N < \phi, \nu(0) > \phi \in H^{2n+1}(X, \mathbf{Z})$. The vanishing cycles have finite index in $H^{2n+1}(X, \mathbf{Z})$, so it follows that for some integer M, $< \nu(0), \lambda > \in \mathbf{Z}$ for all $\lambda \in H^{2n+1}(X, \mathbf{Z})$. This forces $\nu(0) \in H^{2n+1}(X, \mathbf{Q})$, and hence $\nu(0) \in J^{n+1}(X_0)$ is a torsion point. □

The kind of vanishing theorem we have used goes through for "sufficiently ample" line bundles or vector bundles. An extremely useful embodiment of this idea is the following result of Nori, which we will use heavily in the next lecture in proving the Nori Connectedness Theorem: Let

$$E = \oplus_{i=1}^{h} \mathcal{O}_X(a_i),$$

and assuming E is generated by global sections, let M_E be defined by the exact sequence

$$0 \to M_E \to H^0(X, E) \otimes \mathcal{O}_X \to E \to 0.$$

THEOREM. *For any fixed coherent sheaf $\mathcal{F}$ on X, there is a constant $N(\mathcal{F}, c)$ such that if all the $a_i \geq N(\mathcal{F}, c)$, then $H^a(X, \mathcal{F} \otimes E^{\otimes b} \otimes M_E^{\otimes c}) = 0$ if $a > 0$ and $b > 0$ and $c \geq 0$.*

PROOF: Let $X^m = \times_{i=1}^{m} X$, and $p_0, \ldots, p_{m-1}$ the projections on the factors. For any multi-index $I \subseteq \{1, \ldots, c\}$, define a multi-diagonal

$$\Delta_I = \{(x_0, \ldots, x_c) \in X^{c+1} \mid x_i = x_0 \text{ if } i \in I\}.$$

Let $Z[I]$ denote the free group generated by I. Set

$$C^q = \sum_{\#(I)=q} \wedge^q Z[I] \otimes \mathcal{O}_{\Delta_I}.$$

This may be made into a complex by noting that if $J = I \cup \{j\}$ with $j \notin I$, then $\wedge j$ gives an isomorphism

$$\lambda_{I,j} : \wedge^q Z[I] \to \wedge^{q+1} Z[J].$$

Let

$$r_{I,j} : \mathcal{O}_{\Delta_I} \to \mathcal{O}_{\Delta_J}$$

denote the restriction map. Then

$$\delta^q : C^q \to C^{q+1}$$

is defined by

$$\delta(\sum_{\#(I)=q} \phi_I \otimes f_I) = \sum_{\#(I)=q} \sum_{j \notin I} \lambda_{I,j}(\phi_I) \otimes r_{I,j}(f_I).$$

One may check that $\mathcal{H}^k(C^\bullet) = 0$ for $k > 0$. Let

$$Z^q = \ker(\delta^q).$$

Let

$$\mathcal{E} = p_1^* E \otimes \cdots \otimes p_c^* E.$$

Then

$$p_{0*}(C^0 \otimes \mathcal{E}) = \otimes_{i=1}^c H^0(X,E) \otimes \mathcal{O}_X;$$

while

$$p_{0*}(C^1 \otimes \mathcal{E}) \cong \sum_{i=1}^{c} \otimes^{c-1} H^0(X,E) \otimes E;$$

etc. This may be put more elegantly by saying that if $K^\bullet$ is the complex

$$H^0(X,E) \otimes \mathcal{O}_X \to E,$$

then

$$p_{0*}(C^\bullet \otimes \mathcal{E}) = (K^\bullet)^{\otimes c}.$$

Thus

$$\mathcal{H}^q(C^\bullet \otimes \mathcal{E}) = \begin{cases} 0 & \text{if } q > 0; \\ H^0(K^\bullet)^{\otimes c} \cong M_E^{\otimes c} & \text{if } q = 0. \end{cases}$$

One has that

$$p_{0*}(Z^q \otimes \mathcal{E}) = \ker(p_{0*}(C^q \otimes \mathcal{E}) \to p_{0*}(C^{q+1} \otimes \mathcal{E}))$$

for all $q \geq 0$. Thus

$$p_{0*}(Z^0 \otimes \mathcal{E}) = M_E^{\otimes c}.$$

Now if we compute

$$H^a(X^{c+1}, p_0^*(\mathcal{F} \otimes E^{\otimes b}) \otimes Z^0 \otimes \mathcal{E}),$$

we have that on each fiber of the projection p_0, we have a fixed sheaf tensored with something containing arbitrarily high twists on each factor of the fiber X^c, thus all the higher direct images vanish and

$$H^a(X^{c+1}, p_0^*(\mathcal{F} \otimes E^{\otimes b}) \otimes Z^0 \otimes \mathcal{E}) \cong H^a(X, \mathcal{F} \otimes E^{\otimes b} \otimes M_E^{\otimes c}).$$

However, the group on the left involves a tensor with an arbitrarily high power of the ample bundle

$$p_0^* \mathcal{O}_X(1) \otimes p_1^* \mathcal{O}_X(1) \otimes \cdots \otimes p_c^* \mathcal{O}_X(1),$$

and therefore vanishes.

REMARK: Ein and Lazarsfeld have vanishing theorems which give explicit bounds for a very ample line bundle L on what powers of K_X and L are needed to get this kind of result.

Bibliographic Notes

A fairly simple proof of Macaulay's bound for the growth of an ideal and Gotzmann's Persistence Theorem may be found in my paper "Restrictions of linear series to hyperplanes, and some results of Macaulay and Gotzmann," 76-86 in Ballico and Ciliberto, *Algebraic Curves and Projective Geometry, Proceedings, Trento 1988.* A different proof will appear in my "Notes on Generic Initial Ideals."

The idea of using Koszul cohomology in for Hodge-theoretic computations goes back to Lieberman, Peters, and Wilsker, "A theorem of local Torelli type," Math. Ann 231 (1977), 39-45.

The vanishing theorem for Koszul cohomology is proved in my lecture notes, "Koszul cohomology and geometry," in Cornalba, Gomez-Mont, and Verjovsky, *Lectures on Riemann Surfaces, ICTP, Trieste, Italy*, which also treats some of the Hodge-theoretic applications.

The idea of using infinitesimal computations to prove (an eventually to improve) the Noether-Lefschetz Theorem arose in Carlson, Green, Griffiths, and Harris, "Infinitesimal variation of Hodge structure (I)," Comp. Math. 50 (1983), 109-205. The estimate given here is from Green, "A new proof of the explicit Noether-Lefschetz Theorem," J. Diff. Geom. 27 (1988), 155-159. There have been many subsequent discoveries about the Noether-Lefschetz locus, by Voisin, Ellingsrud-Peskine, Ciliberto-Harris-Miranda, and Kim, among others.

The Donagi Symmetrizer Lemma has a history similar to Griffiths' proof of infinitesimal Torelli for hypersurfaces—Donagi proved it first for the Fermat hypersurfaces, and then he and I together proved it for all smooth hypersurfaces in the appropriate range of degrees, once again using Macaulay's Theorem. The brilliant idea of using the symetrizer to reconstruct the Jacobi ideal is due to Donagi, and appears in his paper "Generic Torelli for projective hypersurfaces," Comp. Math. 50 (1983), 325-353. Our slight improvement is Donagi and Green, "A new proof of the symmetrizer lemma and a stronger weak Torelli theorem for projective hypersurfaces," J. Diff. Geom. 20 (1984), 459-461.

The results on the Abel-Jacobi map follow the presentation of my paper, "Griffiths' infinitesimal invariant and the Abel-Jacobi map," J. Diff. Geom. 29 (1989), 545-555; as mentioned in the text, this result is due independently to Voisin, who has carried it much further in work described in her lectures in this volume.

LECTURE 8

A beautiful generalization of the ideas in the foregoing section is due to Nori; this is the goal toward which we have been working in these lectures . Let X be a smooth projective variety of dimension $n+h$, and $E = \oplus_{i=1}^h \mathcal{O}_X(a_i)$. Let $S = \prod_{i=1}^h \mathbf{P}(H^0(X, \mathcal{O}_X(a_i)))$, T a variety and $\pi: T \to S$ a smooth morphism, i.e. π_* is surjective at all points of T. We let $A = X \times S$ and

$$B \subseteq A = \{(x, f_1, \dots, f_h) \mid f_i(x) = 0 \text{ for all } i\}.$$

Let $A_T = A \times_S T$, $B_T = B \times_S T$.

THEOREM (Nori Connectedness Theorem). *For any number c, there is a number $N(c)$ such that if $a_i \geq N(c)$ for all i, then*
(1) $F^k H^{n+k}(A_T, B_T, \mathbf{C}) = 0$ for all $k \leq c$;
(2) $H^k(A_T, B_T, \mathbf{Q}) = 0$ for all $k \leq 2n$.

We first remark that (2), which is the part of Nori's theorem most useful for applications, is a fairly easy consequence of (1). We have not constructed the mixed Hodge structure on relative cohomology groups of quasi-projective varieties, but such a mixed Hodge structure exists and fits into a long exact sequence of mixed Hodge structures. The point is that if we write $A_T = \bar{A}_T - D$ where D is a divisor with normal crossings, we can arrange that $B_T = \bar{B}_T - (D \cap \bar{B}_T)$ with $D \cap \bar{B}_T$ a divisor with normal crossings in $\bar{B}_T$. Now take the mapping cone of

$$\Omega^\bullet_{\bar{A}_T}(\log(D)) \to \Omega^\bullet_{\bar{B}_T}(\log(D \cap \bar{B}_T))$$

as in Lecture 5 and define the height and weight filtrations as in that chapter. In particular,

$$\mathrm{Gr}^W_m H^{n+k}(A_T, \mathbf{C}) = 0 \qquad \text{for } m < n+k$$

and likewise

$$\mathrm{Gr}^W_m H^{n+k-1}(B_T, \mathbf{C}) = 0 \qquad \text{for } m < n+k-1.$$

. It follows that $\mathrm{Gr}^W_m H^{n+k}(A_T, B_T, \mathbf{C}) = 0$ for $m < n+k-1$. However, $\mathrm{Gr}^W_m H^{n+k}(A_T, B_T, \mathbf{C})$ has a Hodge structure of weight m, and thus if this group is non-zero, we must have $F^p \mathrm{Gr}^W_m H^{n+k}(A_T, B_T, \mathbf{C}) \neq 0$ for some $p \geq m/2$. It follows that $F^p H^{n+k}(A_T, B_T, \mathbf{C}) \neq 0$ for some $p \geq (n+k-1)/2$. By part (1), if $H^{n+k}(A_T, B_T, \mathbf{Q}) \neq 0$, we must have $k-1 \geq (n+k-1)/2$, which is equivalent to $n+k \geq 2n+1$. □

We next remark that the choice of S is somewhat arbitrary; for some purposes, it might be more natural to take $S = H^0(X, E)$ or $S = \mathbf{P}(H^0(X, E))$ or S the moduli space of complete intersections of type $(a_1, \dots, a_h)$, and the following two lemmas allow us to do this.

LEMMA. *Let $i: X \to Y$ be an embedding, and $f: Y \to T$ a morphism. Let $g: U \to T$ be a smooth morphism. Then there is a natural filtration of the bundle $\Omega^p_{Y\times_T U, X\times_T U}$ by $p_X^*\Omega^i_{Y,X} \otimes p_U^*\Omega^{p-i}_{U/T}$ for $0 \le i \le p$, where $p_X: X \times_T U \to X$ and $p_U: X \times_T U \to U$ are the natural projections.*

PROOF: This is completely local, so we may take $U = T \times F$ with $g: T \times F \to T$ projection on the first factor. Now $X \times_T U = X \times F$ and $Y \times_T U = Y \times F$. Now the lemma is an easy local computation. □

The next step is the following Lemma of Nori:

LEMMA. *Let $g: U \to T$ be a smooth morphism, and q_0, b_0 integers. Consider the following statements:*
(1) $\mathcal{R}^q_{p_U *}(\Omega^b_{A_U,B_U}) = 0$ for $q \le q_0$, $q + b \le q_0 + b_0$;
(2) $\mathcal{R}^q_{p_T *}(\Omega^b_{A_T,B_T}) = 0$ for $q \le q_0$, $q + b \le q_0 + b_0$.
Then (2) implies (1). If g is also surjective, then (1) and (2) are equivalent.

PROOF: The preceding Lemma implies that $\Omega^b_{A_U,B_U}$ has a natural filtration with graded pieces $p_{A_T}^*\Omega^{b-i}_{A_T,B_T} \otimes p_U^*\Omega^i_{U/T}$ for $0 \le i \le b$, where $p_{A_T}: A_U \to A_T$ and $p_U: A_U \to U$ are the projections. A local computation shows that

$$\mathcal{R}^q_{p_U *}(p_{A_T}^*\Omega^b_{A_T,B_T}) \cong g^*\mathcal{R}^q_{p_T *}(\Omega^b_{A_T,B_T}).$$

If (2) holds, then using the filtration above we see that it is enough to show

$$\mathcal{R}^q_{p_U *}(p_U^*\Omega^i_{U/T} \otimes p_{A_T}^*\Omega^{b-i}_{A_T,B_T}) = 0$$

for all $q \le q_0$, $q + b \le q_0 + b_0$, $0 \le i \le b$, and this follows from (2).

If g is surjective, to prove that (1) implies (2), we note that (2) is local in U, and thus we may shrink U and T so that the filtration becomes a direct sum decomposition

$$\Omega^b_{A_U,B_U} \cong \oplus_{i=0}^b (p_U^*\Omega^i_{U/T} \otimes p_{A_T}^*\Omega^{b-i}_{A_T,B_T}).$$

Now

$$\mathcal{R}^q_{p_U *}(\Omega^b_{A_U,B_U}) \cong \oplus_{i=0}^b (\Omega^i_{U/T} \otimes g^*\mathcal{R}^q_{p_T *}(\Omega^{b-i}_{A_T,B_T})).$$

If (1) holds, then we get that $g^*\mathcal{R}^q_{p_T *}(\Omega^b_{A_T,B_T}) = 0$ for $q \le q_0$, $q + b \le q_0 + b_0$, and since g is surjective, this implies (2). □

The proof of the main theorem is a lengthy sequence of vanishings. For any embedding $i: X \to Y$, it is true that

$$H^k(Y, X, \mathbf{C}) \cong \mathbf{H}^k(\Omega^\bullet_{Y,X}).$$

The image of

$$\mathbf{H}^k(\Omega^{\ge p}_{Y,X}) \to \mathbf{H}^k(\Omega^\bullet_{Y,X})$$

defines a filtration $G^\bullet$ on $H^k(Y, X, \mathbf{C})$. Now there is a map of complexes from

$$\Omega^\bullet_{\bar{A}_T}(\log(D)) \to \Omega^\bullet_{\bar{B}_T}(\log(D \cap \bar{B}_T))$$

to

$$\Omega^\bullet_{A_T} \to \Omega^\bullet_{B_T},$$

and therefore a map of the mapping cones. It follows that the Hodge filtration $F^\bullet \subseteq G^\bullet$, and therefore it is enough to show that

$$\mathbf{H}^{n+k}(\Omega^{\geq k}_{A_T,B_T}) = 0$$

for $k \leq c$. This of course would be implied by showing that

$$H^a(\Omega^b_{A_T,B_T}) = 0$$

if $b \geq k$, $a + b = n + k$, $k \leq c$. This in turn is implied by knowing that

$$H^{a-q}(\mathcal{R}^q_{p_T*}\Omega^b_{A_T,B_T}) = 0$$

for $b \geq k$, $a + b = n + k$, $0 \leq q \leq a$, $k \leq c$, where $p_T\colon A_T \to T$ is the projection. This would follow from knowing

$$\mathcal{R}^q_{p_T*}\Omega^b_{A_T,B_T} = 0$$

if $q \leq n$ and $q + b \leq n + c$.

Now if $p_A\colon A_T \to A$ and $p_B\colon B_T \to B$ are the canonical projections, then $\Omega^1_{A_T/A} \cong p_T^*\Omega^1_{T/S}$ and similarly for B. Now $\Omega^\bullet_{A_T}$ is filtered with graded pieces

$$\Omega^i_{A_T/A} \otimes p_A^*\Omega^{\bullet-i}_A \cong p_T^*\Omega^i_{T/S} \otimes p_A^*\Omega^{\bullet-i}_A,$$

and there is a compatible filtration on $\Omega^\bullet_{B_T}$, with the result that $\Omega^\bullet_{A_T,B_T}$ is filtered with graded pieces $p_T^*\Omega^i_{T/S} \otimes p_A^*\Omega^{\bullet-i}_{A,B}$. To show the vanishing we want, it is enough to show that

$$\mathcal{R}^q_{p_T*}(p_T^*\Omega^i_{T/S} \otimes p_A^*\Omega^{b-i}_{A,B}) = 0$$

for $q \leq n$, $q+b \leq n+c$, $0 \leq i \leq b$. Using the computation made in the preceding Lemma, we have that

$$\mathcal{R}^q_{p_T*}(p_T^*\Omega^i_{T/S} \otimes p_A^*\Omega^{b-i}_{A,B}) \cong \Omega^i_{T/S} \otimes g^*\mathcal{R}^q_{p_S*}(\Omega^{b-i}_{A,B}),$$

and thus it suffices to prove

$$\mathcal{R}^q_{p_S*}\Omega^b_{A,B} = 0$$

for $q \leq n$ and $q + b \leq n + c$.

At this point, we now invoke the Lemma. If

$$U = \prod_{i=1}^{h} H^0(X, \mathcal{O}_X(a_i)) = H^0(X, E)$$

and

$$U' = \prod_{i=1}^{h}(H^0(X, \mathcal{O}_X(a_i)) - 0),$$

then the natural projection $U' \to S$ is a smooth surjective morphism, so we may replace S by U'. Again by the Lemma, since the inclusion $U' \to U$ is a smooth morphism, we may replace U' by U. We are thus reduced to proving

$$\mathcal{R}^q_{p_U*}(\Omega^b_{A_U,B_U}) = 0$$

for $q \leq n$, $b \leq n+c$.

Let $p_X: A_U \to X$ and $\pi_X: B_U \to X$ be the projections. $\Omega^b_{A_U}$ is filtered by $p_X^*\Omega^i_X \otimes \Omega^{b-i}_{A_U/X}$ and similarly for $\Omega^b_{B_U}$, with the result that $\Omega^b_{A_U,B_U}$ is filtered by $p_X^*\Omega^i_X \otimes \Omega^{b-i}_{A_U,B_U/X}$. We are thus reduced to showing

$$\mathcal{R}^q_{p_U*}(p_X^*\Omega^i_X \otimes \Omega^{b-i}_{A_U,B_U/X}) = 0$$

for $q \leq n$, $q+b \leq n+c$, $0 \leq i \leq q$, $s \in U$.

If $E = \oplus_{i=1}^h \mathcal{O}_X(a_i)$ and M_E is defined by the exact sequence

$$0 \to M_E \to H^0(X,E) \otimes \mathcal{O}_X \to E \to 0,$$

then we may take $\Omega^1_{A_U/X} \cong H^0(X,E)^* \otimes \mathcal{O}_A$ and $\Omega^1_{B_U/X} \cong \pi_X^* M_E^*$. It follows that there is an exact sequence

$$0 \to \ker(\wedge^p H^0(X,E)^* \otimes \mathcal{O}_A \to p_X^* \wedge^p M_E^*) \to \Omega^p_{A_U,B_U/X} \to p_X^* \wedge^p M_E^* \otimes \mathcal{I}_{B_U/A_U} \to 0.$$

We thus need to show that

$$\mathcal{R}^q_{p_U*}(p_X^*\Omega^i_X \otimes \ker(\wedge^{b-i} H^0(X,E)^* \otimes \mathcal{O}_A \to p_X^* \wedge^{b-i} M_E^*)) = 0$$

for $q \leq n$, $q+b \leq n+c$, $0 \leq i \leq b$, and

$$\mathcal{R}^q_{p_U*}(p_X^*\Omega^i_X \otimes p_X^* \wedge^{b-i} M_E^* \otimes \mathcal{I}_{B_U/A_U}) = 0$$

for all $q \leq n$, $q+b \leq n+c$, $0 \leq i \leq b$.

We attack the second vanishing first. By the Koszul resolution for $\mathcal{I}_{A_U/B_U}$,

$$0 \to p_X^* \wedge^h E^* \to \cdots \to p_X^* E^* \to \mathcal{I}_{B_U/A_U} \to 0,$$

it is enough for this part to show that

$$\mathcal{R}^{q+r-1}_{p_U*}(p_X^*\Omega^i_X \otimes p_X^* \wedge^{b-i} M_E^* \otimes p_X^* \wedge^r E^*)$$

for all $q \leq n$, $q+b \leq n+c$, $0 \leq i \leq b$, $0 < r \leq h$. Since the bundles in question are $\mathcal{O}_U$-flat, it is enough to know the pointwise vanishing

$$H^q(X \times \{s\}, \Omega^i_X \otimes \wedge^{b-i} M_E^* \otimes \wedge^r E^*) = 0$$

for all $q \leq n$, $q+b \leq n+c$, $0 \leq i \leq b$, $0 < r \leq h$ and all $s \in U$. By Serre Duality, this is equivalent to

$$H^{n+h+1-q-r}(X, \Omega^{n+h-i}_X \otimes \wedge^{b-i} M_E \otimes \wedge^r E) = 0$$

with the same inequalities. However, $r \leq h$, $q \leq n$, implies that $n+h+1-q-r > 0$. We recall from the previous section the result:

THEOREM. *For any fixed coherent sheaf $\mathcal{F}$ on X, there is a constant $N(\mathcal{F}, c)$ such that if all the $a_i \geq N(\mathcal{F}, c)$, then $H^a(X, \mathcal{F} \otimes E^{\otimes b} \otimes M_E^{\otimes c}) = 0$ if $a > 0$ and $b > 0$ and $c \geq 0$.*

We may invoke this result (noting that $r > 0$) to conclude the vanishing we need. The fact that we have an upper bound for the tensor power of M_E which occurs allows us to get a single N that works for all of the cases we need.

The other vanishing we need goes as follows: there is a filtration on the bundle $\ker(\wedge^{b-i} H^0(E)^* \otimes \mathcal{O}_A \to p_X^* \wedge^{b-i} M_E^*)$ with graded pieces $p_X^* \wedge^r E^* \otimes p_X^* \wedge^{b-i-r} M_E^*$ with $0 < r \leq b - i$. The vanishing we need (using again the trick of pointwise reduction) would be implied by

$$H^q(\Omega_X^i \otimes \wedge^r E^* \otimes \wedge^{b-i-r} M_E^*) = 0$$

for $q \leq n$, $q + b \leq n + c$, $0 \leq i \leq b$, $0 < r \leq b - i$. By Serre Duality, this is equivalent to

$$H^{n+h-q}(X, \Omega_X^{n+h-i} \otimes \wedge^r E \otimes \wedge^{b-i-r} M_E) = 0$$

with the same inequalities, and this follows from the Theorem just quoted. This completes the proof of the Nori Connectedness Theorem. □

Nori's first application is:

THEOREM. *Let X be a smooth projective variety of dimension $n + h$ and*

$$E = \oplus_{i=1}^h \mathcal{O}_X(a_i).$$

Let $s \in H^0(X, E)$ be a general section and X_s the smooth dimension n variety it defines. Let $i_s \colon X_s \to X$ be the inclusion. If the a_i are all sufficiently large, then if $Z \in Z^d(X)$ whose cycle class $\psi_X(Z)$ is non-zero in $H^{2d}(X, \mathbf{Q})$, then $i_s^* Z \in Z^d(X_s)$ is not algebraically equivalent to 0.

This is a special case of the following more general theorem he proved, which we state after some preliminary results.

DEFINITION. *For any smooth variety X, let $I_r^p(X) \subseteq J^p(X)$ denote the image in $J^p(X)$ of the maximal integral sub Hodge-structure of weight $2r + 1$ in $H^{2p-1}(X, \mathbf{C})$. We may extend this by taking a filtration $I_r H_{\mathcal{D}}^{2p}(X)$ to be $I_r^p(X)$ if $r \geq 0$ and $H_{\mathcal{D}}^{2p}(X)$ if $r < 0$. We extend this to cycles by defining*

$$I_r CH^p(X) = \{Z \in CH^p(X) \mid \psi_Z \in I_r H_{\mathcal{D}}^{2p}(X)\},$$

and set

$$I_r(CH^p(X) \otimes \mathbf{Q}) = (I_r CH^p(X)) \otimes \mathbf{Q} \subseteq CH^p(X) \otimes \mathbf{Q}.$$

DEFINITION. *For any smooth variety X, let $A_r CH^d(X)$ be the group generated by all cycles of the form*

$$(p_X)_*(\alpha \cap p_W^* \beta)$$

where W is a smooth projective variety and

$$\alpha \in CH^{d+r}(X \times W)$$

and

$$\beta \in (Z_h)_r(W).$$

Note that $A_0CH^d(X)$ is the cycles algebraically equivalent to 0 on X, that it is an increasing filtration, that it pulls back under morphisms and pushes forward under proper maps. $\oplus_d A_r CH^d(X)$ is an ideal in $CH^\bullet(X)$.

LEMMA. *In the definition above, if we work rationally, it is enough to take W to have dimension $2r+1$.*

PROOF: The class $\beta \in CH_r(W) \otimes \mathbf{Q}$ can be represented by a linear combination of smooth disjoint cycles $\sum_i q_i Z_i$. We may now find, for $k >> 0$, a section in $H^0(W, \mathcal{O}_W(k))$ that defines a smooth hypersurface W' containing the Z_i provided that $\dim(W) > 2r+1$ by a general position argument. By the Lefschetz Theorem, $H_{2r}(W', \mathbf{Z}) \to H_{2r}(W, \mathbf{Z})$ is injective, so the cycle class of $\sum_i q_i Z_i$ in W' is also zero in cohomology. This allows us to reduce dimension until we are $\leq 2r+1$. □

In general, if $\alpha \in Z^k(X \times W)$, then we obtain a natural map

$$a_\alpha \colon Z_r(W) \to Z^{k-r}(X)$$

by

$$\beta \mapsto p_{X*}(p_W^* \beta \cap \alpha).$$

There is a similar map on the level of Deligne cohomology, where if $w = \dim(W)$,

$$b_\alpha \colon H_{\mathcal{D}}^{2w-2r}(W) \to H_{\mathcal{D}}^{2k-2r}(X)$$

is defined by

$$\lambda \mapsto p_{X*}(p_W^* \lambda \cup \psi_\alpha).$$

These maps are related by

$$b_\alpha(\psi_\beta) = \psi_{a_\alpha(\beta)}.$$

In particular,

$$\psi_{a_\alpha(\beta)} \in \operatorname{im}(H_{\mathcal{D}}^{2w-2r}(W)).$$

In particular, if $\beta \in (Z_h)_r(W)$, then

$$\psi_{a_\alpha(\beta)} \in \operatorname{im}(J^{w-r}(W) \to J^{k-r}(X)).$$

Since we may take $\dim(W) = 2r+1$, we see that:

PROPOSITION. *For all $r \geq 0$,*

$$A_r CH^p(X) \subseteq I_r CH^p(X).$$

REMARK: We actually know a bit more, since the sub-Hodge structure actually is the Hodge structure of a variety.

DEFINITION. *We define the* **rational Deligne cohomology** *groups by*

$$H^{2p}_{\mathcal{D}}(X,\mathbf{Q}) = \mathbf{H}^{2p}(\mathbf{Q} \to \Omega^{<p}_X).$$

REMARK: There is a natural exact sequence

$$0 \to J^p(X)/\text{torsion} \to H^{2p}_{\mathcal{D}}(X,\mathbf{Q}) \to H^{p,p}_{\mathbf{Q}}(X) \to 0.$$

The term on the right is what comes up in the Hodge conjecture, and one presumes that the rational Deligne cohomology groups are the correct thing to consider in similar situations. As far as I know, there is no conjecture as to what the image of the rational cycle class map

$$Z^p(X) \otimes \mathbf{Q} \to H^{2p}_{\mathcal{D}}(X,\mathbf{Q})$$

is; it would be interesting to have such a conjecture.

We now state Nori's main corollary:

THEOREM. *Let X, X_s, i_s be as in the previous theorem. Then for all $0 \leq r < n-p$,*
(1) The map

$$i_s^*: CH^p(X) \otimes \mathbf{Q} \to CH^p(X_s) \otimes \mathbf{Q}$$

has

$$(i_s^*)^{-1}(A_r CH^p(X_s) \otimes \mathbf{Q}) \subseteq I_r(CH^p(X) \otimes \mathbf{Q}).$$

(2)

$$\ker(i_s^*) \subseteq \ker(\psi: CH^p(X) \otimes \mathbf{Q} \to H^{2p}_{\mathcal{D}}(X,\mathbf{Q})).$$

REMARK: This captures information about AJ_X in $J^p(X)/\text{torsion}$.
PROOF: (1) If $Z \in CH^p(X)$, we may piece together our W's and cycles after taking a branched cover and deleting the branch locus; thus we get a smooth morphism $g: T \to S$ so that there is a family $\mathcal{W} \to T$ of smooth varieties of dimension $\leq 2r+1$ with fibers W_t so that there are cycles $\alpha^\nu_t \in CH^{p+r}(X_{g(t)} \times W_t)$ and $\beta^\nu_t \in (Z_h)_r(W_t)$ satisfying

$$i^*_{g(t)} Z = \sum_\nu p_{X_{g(t)}*}(\alpha^\nu_t \cap p^*_{W_t} \beta^\nu_t)$$

which patch together to give cycles

$$\alpha^\nu \in CH^{p+r}(B_T \times_T \mathcal{W})$$

and

$$\beta^\nu \in CH_{r+\dim(T)}(\mathcal{W}).$$

Now

$$\psi_{\alpha^\nu} \in H^{2p+2r}_{\mathcal{D}}(B_T \times_T \mathcal{W}, \mathbf{Q}).$$

It follows from the Nori connectedness theorem (since $A_T \times_T \mathcal{W} = A_{\mathcal{W}}$) that

$$H^k_{\mathcal{D}}(A_T \times_T \mathcal{W}, B_T \times_T \mathcal{W}, \mathbf{Q}) = 0$$

if $k \leq 2p+2r-1$ and $p+r<n$, which implies $k<2n$. So

$$H_{\mathcal{D}}^{2p+2r}(A_T \times_T \mathcal{W}, \mathbf{Q}) \to H_{\mathcal{D}}^{2p+2r}(B_T \times_T \mathcal{W}, \mathbf{Q})$$

is an isomorphism, so ψ_{α^ν} comes from a class $\mu^\nu \in H_{\mathcal{D}}^{2p+2r}(A_T \times_T \mathcal{W}, \mathbf{Q})$. Now

$$\pi_X^*(Z) = \sum_\nu p_{B_T *}(\alpha^\nu \cap p_{\mathcal{W}}^* \beta^\nu).$$

Taking cycle classes and pulling up to $A_T \times_T \mathcal{W}$,

$$p_X^* \psi_Z = \sum_\nu p_{A_T *}(\mu^\nu \cap p_{\mathcal{W}}^* \psi_{\beta^\nu}).$$

If $i_t \colon X \to A_T$ is $x \mapsto (x,t)$, then pulling back by i_t gives

$$\psi_Z = \sum_\nu i_t^* p_{A_T *}((j_t^* \mu^\nu) \cup p_{W_t}^* \psi_{\beta_t^\nu}),$$

where $j_t \colon X \times W_t \to A_T \times_T \mathcal{W}$ is $(x,w) \mapsto (x,t) \times w$. From this we conclude that

$$\psi_Z \in \operatorname{im}(J^{w-r}(W_t)).$$

Once again, we may take $\dim W_t = 2r+1$, from which we conclude that

$$\psi_Z \in I_r H_{\mathcal{D}}^{2p}(X),$$

and hence

$$Z \in I_r(CH^p(X) \otimes \mathbf{Q}).$$

(2) The same argument proceeds, except that now instead of knowing $\dim(W_t) \leq 2r+1$, we know that W_t is a projective space. Since

$$\psi_Z \in \operatorname{im}(J^{w-r}(W_t)) = 0,$$

we are done. □

Another application of Nori's connectedness theorem is the following result of Stefan Müller-Stach and myself:

THEOREM. *Let X be a smooth projective variety of dimension $n+h$ and X_s the zero locus of a general section $s \in H^0(X,E)$), where*

$$E = \oplus_{i=1}^h \mathcal{O}_X(a_i)$$

with the a_i chosen sufficiently large. Let

$$\psi_s \colon CH^p(X_s) \otimes \mathbf{Q} \to H_{\mathcal{D}}^{2p}(X_s, \mathbf{Q})/\operatorname{im}(I_0 H_{\mathcal{D}}^{2p}(X, \mathbf{Q}))$$

be the cycle map for X_s and

$$\phi_s \colon CH^p(X) \otimes \mathbf{Q} \to H_{\mathcal{D}}^{2p}(X_s, \mathbf{Q})/\operatorname{im}(I_0 H_{\mathcal{D}}^{2p}(X, \mathbf{Q}))$$

the composition of ψ_s and i_s^*; both computed modulo $\operatorname{im}(I_0 H_{\mathcal{D}}^{2p}(X,\mathbf{Q}))$. Then

$$\operatorname{im}(\phi_s) = \operatorname{im}(\psi_s)$$

provided $p<n$.

PROOF: Since s is general, for any $Z_s \in Z^p(X_s)$, there is a cycle $\mathcal{Z} \in Z^p(B_T)$ for some smooth morphism $g: T \to S$ such that for some $t \in p^{-1}(s)$, $Z_s = \mathcal{Z} \cap p_T^{-1}(t)$. Now $\psi_{\mathcal{Z}} \in H^{2p}_{\mathcal{D}}(B_T, \mathbf{Q})$ must come from a class $\alpha \in H^{2p}_{\mathcal{D}}(A_T, \mathbf{Q})$ provided that $2p + 1 \leq 2n$, i.e. $p < n$. We know that for any t,

$$\psi_{p_T^{-1}(t) \cap \mathcal{Z}} = i_t^* \alpha,$$

where $i_t: X \to X \times T = A_T$ is the inclusion taking $x \mapsto (x, t)$.

Now we have a map

$$T \to H^{2p}_{\mathcal{D}}(X)$$

given by $t \mapsto i_t^* \alpha$. Since T is connected, this is constant when projected to the integral part. We therefore get a map $Z_0(T) \to J^p(X)$ given by

$$\sum_j n_j t_j \mapsto \sum_j n_j i_{t_j}^* \alpha.$$

This factors $J^{\dim(T)}(T)$, which is an abelian variety. If we mod out by $I_0 J^p(X)$, then this map is zero. It therefore follows that $i_t^* \alpha$ is independent of t, modulo $I_0 H^{2p}_{\mathcal{D}}(X, \mathbf{Q})$; call it $\bar{\alpha}$.

If $j_s: X_s \to X$ is the inclusion and $k_t = i_t \circ j_{g(t)}$, and if $m_t: X_{g(t)} \to B_T$ is the inclusion as $\pi_T^{-1}(t)$, and if $q: B_T \to A_T$ is the inclusion, then $q \circ m_t = k_t$. Now

$$\begin{aligned} \psi_{Z_t} &= m_t^* \psi_{\mathcal{Z}} \\ &= k_t^* \alpha \\ &= j_{g(t)}^* i_t^* \alpha \\ &= j_{g(t)}^* \bar{\alpha} \end{aligned}$$

if we work modulo the image of $I_0 H^{2p}_{\mathcal{D}}(X, \mathbf{Q})$. It therefore follows that

$$\psi_{Z_{t_1}} = \psi_{Z_{t_2}} \quad \text{mod im}(I_0 H^{2p}_{\mathcal{D}}(X, \mathbf{Q}))$$

if $g(t_1) = g(t_2)$. We thus conclude that the cycle class of $\sum_{t \in g^{-1}(s)} Z_t$ is $\deg(g)$ times the cycle class of Z_t for any $t \in g^{-1}(s)$, once again modulo $\text{im}(I_0 H^{2p}_{\mathcal{D}}(X, \mathbf{Q}))$. If we push $\mathcal{Z}$ down from B_T to B, then after dividing out by $\deg(g)$ (this is one of the places where it is essential that we work rationally), we have a cycle $\bar{\mathcal{Z}}$ on B such that (using the same notation for the maps)

$$\psi_{X_s \cap \bar{\mathcal{Z}}} = \psi_{Z_s}.$$

We now want to patch together our cycles on X_s to give a cycle on X.

LEMMA. *Let Y be a smooth projective variety and and $V \cdot \subseteq H^0(Y, L)$. a pencil. Let $W \subset Y \times \mathbf{P}^1$ be defined by*

$$W = \{(y, s) \in Y \times \mathbf{P}(V) \mid s(y) = 0\}.$$

Assume W is smooth. Let $Y_s = \operatorname{div}(s)$ and Λ the base locus of V. Let $\pi_1: W \to Y$ and $\pi_2: W \to \mathbf{P}^1$ be the projections, which we assume are dominant. Let $Z \in Z^p(W)$, and $Z_s = \pi_{1*}(Z \cap \pi_2^*\{s\})$. Then

$$\pi_{1*}(Z) \cap Y_s = Z_s + \pi_{1*}((Z \cap (\Lambda \times \mathbf{P}^1))$$

in $CH^p(Y)$. Note that the last term is a codimension $(p-1)$ cycle in Λ.

PROOF: We may replace Z by anything rationally equivalent to it. We first note that

$$\begin{aligned}\pi_1^* Y_s &= \{(y,t) \mid s(y) = 0, t(y) = 0\}\\ &= (Y_s \times \{s\}) + (\Lambda \times \mathbf{P}^1).\end{aligned}$$

Now

$$\begin{aligned}\pi_{1*}(Z) \cap Y_s &= \pi_{1*}(Z \cap \pi_1^* Y_s)\\ &= \pi_{1*}(Z \cap (Y_s \times \{s\}) + Z \cap (\Lambda \times \mathbf{P}^1))\\ &= \pi_{1*}(Z \cap \pi_2^*\{s\}) + \pi_{1*}(Z \cap (\Lambda \times \mathbf{P}^1))\\ &= Z_s + \pi_{1*}(Z \cap (\Lambda \times \mathbf{P}^1)).\end{aligned}$$

□

We may generalize this lemma to:

LEMMA. *Let Y be a smooth projective variety and $V_1, \ldots V_h$ pencils, with*

$$V_i \subseteq H^0(Y, L_i).$$

Let

$$W \subset Y \times (\mathbf{P}^1)^h$$

be defined by

$$W = \{(y, s_1, \ldots, s_h) \mid s_1(y) = 0, \ldots, s_h(y) = 0\}.$$

Assume that W is smooth and that the projection

$$\pi_2: W \to (\mathbf{P}^1)^h$$

is dominant. Let $\Lambda_i \subset Y$ be the base locus of V_i. Let

$$Y_s = \pi_{1*}(\pi_2^*\{s\})$$

for $s = (s_1, \ldots, s_h)$. Let $Z \in Z^p(W)$ and

$$Z_s = \pi_{1*}(Z \cap \pi_2^*\{s\}).$$

Then

$$\pi_{1*} Z \cap Y_s = Z_s + \sum_{i=1}^{h} G_s^i,$$

where

$$G_s^i \in Z^{p-1}(X_s \cap \Lambda_i).$$

Returning to the proof of the theorem, we may choose general pencils

$$V_i \subseteq H^0(X, \mathcal{O}_X(a_i))$$

so that if we restrict B to $W \subset X \times (\times_i \mathbf{P}(V_i))$, then W satisfies the conditions of the lemma. We may intersect $\bar{\mathcal{Z}}$ with W to obtain $Z \in Z^p(W)$. Then up to rational equivalence in X_s, we have that

$$\pi_{1*} Z \cap X_s = Z_s + \sum_{i=1}^{h} G_s^i,$$

where $G_s^i \in Z^{p-1}(X_s \cap \Lambda_i)$. Since we chose a general pencil, we may use the theorem inductively on p (since we are decreasing both p and n by 1 and hence not disturbing that hypothesis, and since although h is allowed to go up, it does so by at most n) to say that the rational cycle class of G_s^i in $X_s \cap \Lambda_i$ is the same as that of $Z^i \cap (X_s \cap \Lambda_i)$, where $Z^i \in Z^{p-1}(X)$. Now $X_s \cap \Lambda_i = X_s \cap D_i$, where $D_i \in |V_i|$, and $Z^i \cap (X_s \cap \Lambda_i) = (Z^i \cap D_i) \cap X_s$. The Gysin map

$$H_{\mathcal{D}}^{2p-2}(X_s \cap D_i, \mathbf{Q}) \to H_{\mathcal{D}}^{2p}(X_s, \mathbf{Q})$$

takes the cycle class of G_s^i as a codimension $p-1$ cycle of $X_s \cap \Lambda_i$ to the cycle class of G_s^i as a codimension p cycle on X_s, and hence this is the image of the cycle class of $Z^i \cap D_i$ under

$$i_s^* \colon H_{\mathcal{D}}^{2p}(X, \mathbf{Q}) \to H_{\mathcal{D}}^{2p}(X_s, \mathbf{Q}).$$

The upshot is that ψ_{Z_s} is the image of

$$\psi_{\pi_{1*}Z - \sum_{i=1}^{h} Z^i}$$

under i_s^*, always working modulo $\mathrm{im}(I_0 H_{\mathcal{D}}^{2p}(X, \mathbf{Q}))$. □

Bibliographic Notes

The Nori Connectedness Theorem appears in M. Nori, "Algebraic cycles and Hodge theoretic connectivity," Inv. Math. 111 (1993), 349-373. My result with Müller-Stach will appear in M. Green and S. Müller-Stach, "Algebraic cycles on a general complete intersection of high multi-degree of a smooth projective variety," preprint.

I have cheated a bit in this section by using Deligne cohomology for quasi-projective varieties (known as Deligne-Beilinson cohomology), which I have not shown how to construct; this may be found in Esnault and Viehweg, "Deligne-Beilinson cohomology," in Rapoport, Schappacher and Schneider, *Beilinson's Conjectures on Special Values of L-Functions*.

Algebraic cycles and algebraic aspects of cohomology and K–theory

J.P. Murre
Rijksuniversiteit Leiden, The Netherlands

These are the notes of the lectures delivered at the C.I.M.E. meeting in Torino, June 93. I have tried to keep the written version as much as possible in the informal style of the lectures. The content of the eight lectures is grouped into seven chapters. In my lectures, with the exception of chapters 4 and 5, the emphasis was on the *algebraic* methods in studying algebraic cycles; as such the lectures complement those of C. Voisin and M. Green. The chapters are as follows:

1. Algebraic cycles. Basic notions.
2. The Chow ring and the Grothendieck group of coherent sheaves.
3. The Chow ring and higher algebraic K–theory.
4. Introduction to the Deligne–Beilinson cohomology.
5. The Hodge–Conjecture.
6. Applications of the theorem of Merkurjev and Suslin to the theory of algebraic cycles of codimension two.
7. Grothendieck's theory of motives.

I thank the C.I.M.E. foundation and especially Fabio Bardelli for inviting me to lecture at this conference and for his careful and skilful planning of the scientific program. It has been a pleasure and a privilege to collaborate with Claire Voisin and Mark Green. Furthermore I want to thank Alberto Albano for his help during the conference and the participants for their interest and valuable comments. Also I like to thank Mrs. Ooms from our Department in Leiden for the careful typing of the manuscript and Chris Peters for his help to type the more complicated diagrams.

Chapter I. Algebraic cycles. Basic Notions

1.1. Notations and conventions.

Let k be a field, algebraically closed, of arbitrary characteristic. Let $\mathcal{V}'$ be the category of algebraic varieties defined over k and let $\mathcal{V}$ be the subcategory of *smooth*, *quasi-projective* varieties defined over k. In these lectures we always assume, unless stated explicitly otherwise, that our varieties are smooth and quasi-projective, i.e., that we are working in $\mathcal{V}$ (however, for technical reasons, we sometimes have to work in $\mathcal{V}'$). If $X \in \mathcal{V}'$ is irreducible and of dimension d we write X_d if we want to indicate the dimension.

For this chapter the general references are [F] and [B], for most of the proofs we refer to [F] Chap. 1.

M. Green et al.: LNM 1594, A. Albano and F. Bardelli (Eds.), pp. 93–152, 1994.

Ia. Algebraic Cycles and Equivalence Relations.

1.2. Algebraic cycles.
Let $X \in \mathcal{V}'$ be of dimension d and $0 \le i \le d$. Let $\mathcal{Z}^i(X)$ be the group of algebraic cycles on X of codimension i; i.e., $Z \in \mathcal{Z}^i(X)$ is a finite formal sum $Z = \sum_\alpha n_\alpha Z_\alpha$ with $Z_\alpha \subset X$ an irreducible subvariety of codimension i and $n_\alpha \in \mathbb{Z}$. For $X_d \in \mathcal{V}'$ we also sometimes write $\mathcal{Z}_{d-i}(X) = \mathcal{Z}^i(X)$. Furthermore put $\mathcal{Z}(X) = \oplus_i \mathcal{Z}^i(X)$.

Examples.

a. $i = 1, \mathcal{Z}^1(X)$ is the group of Weil divisors $\mathrm{Div}(X)$ on X.
b. $i = d, \mathcal{Z}_o(X) = \mathcal{Z}^d(X)$ is the group of zero cycles on X.
c. for $f \in k(X)$ a rational function on $X, \mathrm{div}(f) \in \mathcal{Z}^1(X)$ denotes the divisor of zeros minus the divisor of poles of f (see [F] 1.2 and 1.3)

1.3. Adequate (= "good") equivalence relations. Let $X \in \mathcal{V}$ and $Y \in \mathcal{V}$. Let $\sim$ be an equivalence relation between algebraic cycles on smooth, quasi-projective varieties.

Definition. The equivalence relation $\sim$ is *adequate* if it satisfies the following conditions:

RI. $\mathcal{Z}^i_\sim(X) = \{Z \in \mathcal{Z}^i(X); Z \sim 0\}$ is a subgroup of $\mathcal{Z}^i(X)$ and $\sim$ is defined via the cosets of this subgroup.
RII. $Z \sim 0$ or $X \Rightarrow Z \times Y \sim 0$ on $X \times Y$.
RIII. If Z_1 and Z_2 are cycles such that $Z_1 \cdot Z_2$ is defined then $Z_1 \sim 0$ implies $Z_1 \cdot Z_2 \sim 0$.
RIV. If $Z \sim 0$ on $X \times Y$ then $pr_X(Z) \sim 0$ on X.
RV. ("moving lemma") Let $Z \in \mathcal{Z}^i(X)$ and let $W_j (j \in J)$ be a finite set of subvarieties on X then there exists a $Z' \in \mathcal{Z}^i(X)$ such that $Z' \sim Z$ and such that $Z' \cdot W_j$ is defined for all $j \in J$.

Theorem. *Let $\sim$ be an adequate equivalence relation. Put $C^i_\sim(X) = \mathcal{Z}^i(X)/\mathcal{Z}^i_\sim(X)$ and $C_\sim(X) = \oplus_i C^i_\sim(X)$. Then:*

1. $C_\sim(X)$ is a ring with respect to addition and intersection of cycles (so-called "intersection ring").

2. Let $W \in \mathcal{Z}^a(X_d \times Y_e)$. Then this defines a map

$$W : C^i_\sim(X) \to C^j_\sim(X) \quad (j = i + a - d)$$

defined by $Z \mapsto W(Z) := pr_Y\{W \cdot (Z \times Y)\}$. Moreover W is an additive homomorphism.

3. Let $\phi : X \to Y$ be a morphism of varieties. Let Γ_ϕ be its graph and ${}^t\Gamma_\phi$ its transpose. Then ϕ defines a map $\phi^* : C^i_\sim(Y) \to C^i_\sim(X)$ given by $Z \mapsto \phi^*(Z) := pr_X\{{}^t\Gamma_\phi \cdot (X \times Z)\}$ and ϕ^* is a homomorphism of rings.

We leave the proof (which is not difficult) as an exercise to the reader (see [Sa 2] prop. 6 and 7).

1.4. Examples of good equivalence relations.

1.4.1. Rational equivalence. ([F], Chap. 1) Let $X_d \in \mathcal{V}'$ (i.e. not necessary smooth) and $Z \in \mathcal{Z}^i(X)$.

Definition 1. Z is called rationally equivalent to zero if there exists a finite collection $\{Y_\lambda, f_\lambda\}$, with Y_λ irreducible subvarieties of X of codimension $(i-1)$ and f_λ rational functions on Y_λ, such that $Z = \sum_\lambda \operatorname{div}(f_\lambda)$.

Example: $i = 1$. A divisor D is rationally equivalent to zero if $D = \operatorname{div}(f)$, with f a rational function on X. Hence for divisors rational equivalence is the same as linear equivalence.

Put $\mathcal{Z}^i_{\text{rat}}(X) = \{Z \in \mathcal{Z}^i(X); Z \text{ rationally equivalent to zero}\}$.

Definition 2. Put $CH^i(X) = \mathcal{Z}^i(X)/\mathcal{Z}^i_{\text{rat}}(X)$ and $CH(X) = \oplus_i CH^i(X)$.

$CH^i(X)$ is called the i–th *Chow group* of X and $CH(X)$ is called the *Chow ring* of X because if X is smooth, quasi–projective then rational equivalence is an adequate equivalence relation and hence $CH(X)$ is a ring with respect to intersection of cycles.

We also write sometimes $CH_{d-i}(X) = CH^i(X)$. Moreover put $CH^i_{\mathbb{Q}}(X) = CH^i(X) \otimes \mathbb{Q}$.

We have the following equivalent formulation for rational equivalence:

Lemma 1. *Let $X \in \mathcal{V}$ and $Z \in \mathcal{Z}^i(X)$. Then the following are equivalent:*

1. $Z \sim 0$ rational equivalence,

2. $\exists Y \in \mathcal{Z}^i(\mathbb{P}_1 \times X)$ and two points a and $b \in \mathbb{P}_1$ such that $Y(a) = Z$ and $Y(b) = 0$, where for $t \in \mathbb{P}_1$ we put $Y(t) = pr_X\{Y \cdot (t \times X)\}$.

The following property is important:

Lemma 2. *Let $X_d \in \mathcal{V}'$ and Y a subvariety of X and put $U = X - Y$. Then for $0 \le q \le d$ the following sequence is exact*

$$CH_q(Y) \xrightarrow{i_*} CH_q(X) \xrightarrow{j^*} CH_q(U) \to 0$$

where $i : Y \hookrightarrow X$ and $j : U \hookrightarrow X$ are the inclusions (note that neither X nor Y need to be smooth; i_* and j^* are the obvious maps).

Remark. For the proofs of these properties we refer to [F], Chap. 1. Rational equivalence was introduced (in a rigorous way) in 1956 by Samuel [Sa 1] and by Chow [Cho] and treated in detail in the Chevalley seminar ([Che]).

1.4.2. Algebraic equivalence. Let $X \in \mathcal{V}$ and $Z \in \mathcal{Z}^i(X)$.

Definition. Z is called algebraically equivalent to zero if there exists a couple (T, Y), with $T \in \mathcal{V}$ and connected, $Y \in \mathcal{Z}^i(T \times X)$ and two points $t_1, t_0 \in T$ such that $Y(t_1) = Z$ and $Y(t_0) = 0$.

Remarks.

Remarks.

1. In the above definition we may take for T a smooth irreducible curve or an abelian variety.
2. The notion of algebraic equivalence was, over arbitrary algebraically closed fields, introduced by Weil in [We] for which we refer to a proof of remark 1.

1.4.2 bis. τ–equivalence.

Definition. Z is called τ–equivalent to zero if nZ is algebraically equivalent to zero for some $n \neq 0$.

1.4.3. Numerical equivalence. Let always $X_d \in \mathcal{V}$ and $Z \in \mathcal{Z}^i(X)$.

Definition. Z is numerically equivalent to zero if for all $Y \in \mathcal{Z}^{d-i}(X)$ such that $Z \cdot Y$ is defined we have that the intersection number $\#(Z \cdot Y) = 0$.

1.4.4. Homological equivalence. Fix a Weil (="good") cohomology theory $H(X)$ ("good" means that the cohomology groups are vectorspaces over a field of characteristic zero, of finite dimension, satisfying Poincaré duality, the Künneth formula, behave functorially, that there exists a cycle map, and moreover that the Lefschetz hyperplane theorem and the hard Lefschetz theorem hold).

Examples:

1. If $k = \mathbb{C}$ one can take $H(X_{an}, \mathbb{Q})$.
2. For k arbitrary one can take $H_{et}(X_{\overline{k}}, \mathbb{Q}_\ell)$ with ℓ a prime number, $\ell \neq p =$ char(k) (of course in our case we did already assume that $k = \overline{k}$, i.e. that k is algebraically closed, however we want to exphasize that for a "good" cohomology theory one must always go to the (separable) algebraic closure).

Let $\gamma : \mathcal{Z}^i(X) \to H^{2i}(X)$ be the cycle map.

Definition. Z is homologically equivalent to zero if $\gamma(Z) = 0$.

Remark. A–priori this depends on the choice of the cohomology theory, however at least in characteristic zero it does not depend on this choice thanks to the comparison theorems between the different cohomology theories.

1.4.5. For a good equivalence relation on cycles put $\mathcal{Z}^i_\sim(X) = \{Z \in \mathcal{Z}^i(X); Z \sim 0\}$. Then one has the following relations:

$$(1) \qquad \mathcal{Z}^i_{\text{rat}}(X) \subseteq \mathcal{Z}^i_{\text{alg}}(X) \subseteq \mathcal{Z}^i_\tau(X) \subseteq \mathcal{Z}^i_{\text{hom}}(X) \subseteq \mathcal{Z}^i_{\text{num}}(X) \subseteq \mathcal{Z}^i(X)$$

1.4.6. Conjecture. It is an old and crucial conjecture that $\mathcal{Z}^i_{\text{hom}}(X) = \mathcal{Z}^i_{\text{num}}(X)$.

Remark. For the above see [F], Chap. 19, part 19.3. Note that in our case we did take a field as coefficients in the cohomology theory.

Ib. Some principal results.

1.5. Survey of some results for divisors (see lecture 1 of Claire Voisin [Vo]). Always $X \in \mathcal{V}$ and $k = \overline{k}$.

1.5.1. Picard variety.

Theorem. $\mathcal{Z}^1_{\text{alg}}(X)/Z^1_{\text{rat}}(X) \xrightarrow{\sim} \text{Pic}^0(X)(k)$, *where the* $\text{Pic}^0(X)$ *is an abelian variety, the so-called Picard variety of* X.

Remark. For arbitrary closed fields this was proved by Matsusaka (51,52), Weil (54), Chow (54), Chevalley (58), Seshadri (60) and in fact one can obtain much more precise results as was shown by Grothendieck in his theory of the Picard functor (61) [Gr2]. To be precise: in the above one should write $\text{Pic}^0_{\text{red}}(X)(k)$, where $\text{Pic}^0_{\text{red}}$ is the reduced part of the Picard scheme.

1.5.2. Homologous versus numerical equivalence.

Theorem. (Matsusaka, 57) $\mathcal{Z}^1_r(X) = \mathcal{Z}^1_{\text{hom}}(X) = \mathcal{Z}^1_{\text{num}}(X)$.

1.5.3. Néron–Severi group.

Theorem. (Severi 40, Néron 52). *The group* $\mathcal{Z}^1(X)/\mathcal{Z}^1_{\text{alg}}(X)$ *is a finitely generated group. (This group is denoted by* $NS(X)$ *and is called the Néron–Severi group).*

1.6. Some of the main "facts" for cycles of codimension larger than one.

It has been discovered during the last 25 years that the situation in codimension larger than one is entirely different from the situation for divisors. This becomes quite clear from the results below (see also the report [H 1]).

1.6.1. Mumford's result for 0–dimensional cycles.

Let $X_d \in \mathcal{V}$, recall that there exists an abelian variety, the so–called *Albanese variety* $\text{Alb}(X)$, and the albanese map $\alpha(X) : X \to \text{Alb}(X)$ which has the universal property for morphisms into abelian varieties:

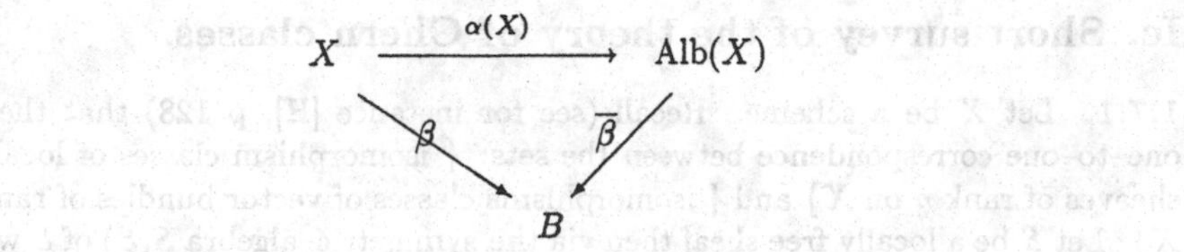

Now take a 0–dimensional cycle $\mathfrak{a} = \sum n_i p_i$ (where $p_i \in X$) and of degree zero (i.e. $\sum n_i = 0$); in other words $\mathfrak{a} \in \mathcal{Z}^d_{\text{num}}(X) = \mathcal{Z}^d_{\text{alg}}(X)$. Put $CH^d_{(0)}(X) = \mathcal{Z}^d_{\text{alg}}(X)/\mathcal{Z}^d_{\text{rat}}(X)$, then we clearly get a homomorphism

$$\alpha : CH^d_{(0)}(X) \longrightarrow \text{Alb}(X)$$

Definition. $\mathfrak{a} \in \mathcal{Z}^d_{\text{alg}}(X)$ is called *albanese equivalent* to zero if $\alpha(\mathfrak{a}) = 0$.

Theorem. ([Mum], Mumford 69). *Let $k = \mathbb{C}$ and let $X = S$ be a smooth, projective surface with $p_g(S) \neq 0$ (recall $p_g(S) = \dim H^0(S, \Omega_S^2)$). Then albanese equivalence $\neq$ rational equivalence. In fact the group $\mathcal{Z}^d_{\text{alb}}(S)/\mathcal{Z}^d_{\text{rat}}(S)$ is "very large"; more precisely, this group can not be parametrized by an algebraic variety.*

Remarks.

1. The result of Mumford has been extended by Roitman and by Bloch. For all of this see the lectures 5 and 6 by Claire Voisin [Vo].
2. **Conjecture** ([B], Bloch). If $p_g(S) = 0$ then on S albanese equivalence and rational equivalence coincide .

   Bloch was lead to this conjecture by considerations from algebraic K–theory (see Chap. III, 3.12).

1.6.2. Griffiths theorem ([Gri], 69): *There exists a three–dimensional smooth projective variety X for which $\mathcal{Z}^2_r(X) \neq \mathcal{Z}^2_{\text{hom}}(X)$.*

Remarks.

1. Somewhat less precise this means: in codimension > 1 algebraic equivalence $\neq$ homological equivalence.
2. For $i > 1$ one introduces the *Griffiths group*

$$Gr^i(X) = \mathcal{Z}^i_{\text{hom}}(X)/\mathcal{Z}^i_{\text{alg}}(X)$$

3. For all of this see lecture 7 of C. Voisin [Vo].

1.6.3. The theorem of Clemens ([Cl], 82): *There exists a smooth, projective three-dimensional variety X such that* $\dim_{\mathbb{Q}}(Gr^2(X) \otimes \mathbb{Q}) = \infty$ *(In fact one can take for X a generic quintic 3–fold).*

See lecture 8 of Claire Voisin [Vo].

Ic. Short survey of the theory of Chern classes.

1.7.1. Let X be a scheme. Recall (see for instance [H], p 128) that there is a one–to–one correspondence between the sets: { isomorphism classes of locally free sheaves of rank n on X} and { isomorphisms classes of vector bundles of rank n on X}. Let $\mathcal{E}$ be a locally free sheaf then via the symmetric–algebra $S(\mathcal{E})$ of $\mathcal{E}$ we get a vector bundle $\mathbb{V}(\mathcal{E}) \to X$ over X; conversely from a vector bundle $\pi : V \to X$ over X we get via the local sections a locally free sheaf $\mathcal{S}(V/X)$ on X; moreover starting from $\mathcal{E}$ we get

$$\mathcal{E} \to \{\mathbb{V}(\mathcal{E}) \to X\} \to \mathcal{S}(\mathbb{V}(\mathcal{E})/X) = \mathcal{E}^*$$

where $\mathcal{E}^*$ is the dual of $\mathcal{E}$.

In view of the above it is sufficient to work with locally free sheaves.

1.7.2. Let $X \in \mathcal{V}$. Let $\mathcal{L}$ be an invertible sheaf (= loc.free sheaf of rank 1) on X.

Lemma. *Let* Pic(X) *be defined as the group of isomorphism classes of invertible sheaves on* X. *Then*

$$CH^1(X) \cong \text{Pic}\,(X) \cong H^1(X, \mathcal{O}_X^*).$$

Proof: see [H], Chap. II-6.

If $D \in \text{Div}(X)$ and $[D] \in CH^1(X)$ then $[D]$ determines up to isomorphism an invertible sheaf $\mathcal{O}_X(D)$ (denoted by $\mathcal{L}(D)$ in Hartshorne, see p 144) and every invertible sheaf of $\mathcal{L}$ is of this type.

Definition. Given $\mathcal{L} \in \text{Pic}(X)$, put $c_1(\mathcal{L}) = [D]$. $c_1(\mathcal{L})$ is called the first Chern class of $\mathcal{L}$.

1.7.3. Let $X \in \mathcal{V}$ and let $\mathcal{E}$ be a locally free sheaf on X of rank r on X. Let $\mathbf{V}(\mathcal{E})$ be the corresponding vector bundle on X and $\pi : \mathbb{P}(\mathcal{E}) = \mathbb{P}(\mathbf{V}(\mathcal{E})) \to X$ the associated projective bundle ([H], Chap. II, 7, page 162) and $\mathcal{O}(1) = \mathcal{O}_{\mathbb{P}(\mathcal{E})}(1)$ the canonical invertible sheaf on $\mathbb{P}(\mathcal{E})$; put $\xi = c_1(\mathcal{O}(1)) \in CH^1(\mathbb{P}(\mathcal{E}))$. Consider the homomorphism

$$\pi^* : CH(X) \to CH(\mathbb{P}(\mathcal{E})).$$

The basis for the theory of Chern classes is the following ([Gr 1]):

Theorem. *The* $CH(\mathbb{P}(\mathcal{E}))$ *is, by means of* π^*, *a free* $CH(X)$*-module with basis* $1, \xi, \xi^2, \ldots, \xi^{r-1}$.

For the proof one reduces, by using lemma 2 of 1.4.2, to the case where $\mathcal{E}$ is free (see for instance [El], p 49).

Corollary. We have in $CH(\mathbb{P}(\mathcal{E}))$ a relation of the form

$$\xi^r - \pi^*(a_1)\xi^{r-1} + \pi^*(a_2)\xi^{r-2} + \ldots + (-1)^r \pi^*(a_r) = 0$$

with $a_i \in CH^i(X)$ and the a_i are uniquely determined.

Definition of the Chern classes:

For $\mathcal{E}$ locally free of rank r on X put

$$\begin{aligned} c_i(\mathcal{E}) &= a_i \quad 1 \le i \le r \\ c_0(\mathcal{E}) &= 1 \\ c_i(\mathcal{E}) &= 0 \qquad i > r \end{aligned}$$

This defines the Chern classes $c_i(\mathcal{E}) \in CH^i(X)$ for every locally free sheaf $\mathcal{E}$ on X. Note that it agrees with the definition in 17.2 if $\mathcal{E} = \mathcal{L}$ is an invertible sheaf, for then $\mathbb{P}(\mathcal{L}) = X$ and the relation (in the corollary) becomes $\xi - c_1(\mathcal{L}) = 0$ in $CH(X)$, i.e. $c_1(\mathcal{L}) = \xi = [D]$ and this agrees with the definition in 17.2.

1.7.4. Properties. (see [H], app. A and [F], Chap.3).

The Chern classes satisfy the following properties:

C1. If $\mathcal{E} = \mathcal{O}_X(D)$ then $c_1(\mathcal{E}) = [D]$.

C2. Functoriality: if $f : X' \to X$ is a morphism and $\mathcal{E}$ locally free on X then $c_i(f^*(\mathcal{E}) = f^* c_i(\mathcal{E})$.

C3. Put $c_t(\mathcal{E}) = 1 + c_1(\mathcal{E})t + c_2(\mathcal{E})t^2 + \dots$, with $c_t(\mathcal{E}) \in CH(X)[[t]]$ (the formal power series ring). If $0 \to \mathcal{E}' \to \mathcal{E} \to \mathcal{E}'' \to 0$ is an exact sequence then we have $c_t(\mathcal{E}) = c_t(\mathcal{E}') \cdot c_t(\mathcal{E}'')$.

1.7.5. Chern character.

Always $\mathcal{E}$ is a locally free sheaf of rank r on X and $X \in \mathcal{V}$. Write formally $c_t(\mathcal{E}) = \prod_{i=1}^{r}(1 + \alpha_i t)$ and define the Chern character of $\mathcal{E}$ as follows:

$$ch(\mathcal{E}) := \sum_{i=1}^{r} e^{\alpha_i},$$

then $ch(\mathcal{E}) \in CH_{\mathbb{Q}}(X)$ because, via the elementary symmetric functions, $ch(\mathcal{E})$ can be expressed as polynomials in the $c_i(\mathcal{E})$ with coefficients in $\mathbb{Q}$. In fact (see [H], p. 432) $ch(\mathcal{E}) = r + c_1 + \frac{1}{2}(c_1^2 - 2c_2) + \frac{1}{3!}(c_1^3 - 3c_1c_2 + 3c_3) + \dots$

Properties. (see [F], p.56)

ch 1. : $0 \to \mathcal{E}' \to \mathcal{E} \to \mathcal{E}'' \to 0$ exact $\Longrightarrow ch(\mathcal{E}) = ch(\mathcal{E}') + ch(\mathcal{E}'')$
ch 2. : $ch(\mathcal{E}_1 \otimes \mathcal{E}_2) = ch(\mathcal{E}_1) \cdot ch(\mathcal{E}_2)$

Remark. (The "splitting-principle").

The above "formal" calculations can effectively be realized provided we go to a suitable associated variety. Namely there exists a morphism $f : X' \to X$ with $X' \in \mathcal{V}$ such that

1. $f^* : CH(X) \to CH(X')$ is an injection,
2. $f^*(\mathcal{E}) = \mathcal{E}'$ has a filtration $\mathcal{E}' = \mathcal{E}'_0 \supset \mathcal{E}'_1 \supset \dots \supset \mathcal{E}'_r = (0)$ such that $\mathcal{E}'_{i-1}/\mathcal{E}'_i \simeq \mathcal{L}'_i$ with $\mathcal{L}'_i$ an invertible sheaf on X'.

Then on X' we have $f^*(c_t(\mathcal{E})) = c_t(\mathcal{E}') = \prod_i c_t(\mathcal{L}'_i) = \prod_i(1 + c_1(\mathcal{L}'_i)t)$.

1.7.6. Todd class. ([H], App.A and [F], Chap. 3).

Again let $\mathcal{E}$ be a locally free $\mathcal{O}_X$-module of rank r. As before write formally $c_t(\mathcal{E}) = \prod_{i=1}^{r}(1 + \alpha_i t)$.

Definition.

$$td(\mathcal{E}) = \prod_{i=1}^{r} \frac{\alpha_i}{1 - e^{-\alpha_i}}.$$

Again $td(\mathcal{E}) \in CH_{\mathbb{Q}}(X)$ can be expressed as a polynomial in the Chern classes $c_i(\mathcal{E})$, in fact it starts as follows:

$$td(\mathcal{E}) = 1 + \frac{1}{2}c_i + \frac{1}{12}(c_1^2 + c_2) + \frac{1}{24}c_1c_2 + \dots$$

Property: If $0 \to \mathcal{E}' \to \mathcal{E} \to \mathcal{E}''$ is exact then $td(\mathcal{E}) = td(\mathcal{E}') \cdot td(\mathcal{E}'')$

Also note for the following that $(td(\mathcal{E}))^{-1} = 1 + \dots$ makes sense and is an element of $CH_{\mathbb{Q}}(X)$.

Chapter II. The Chow ring and the Grothendieck group of coherent sheaves.

In this chapter we discuss the relation between the *Chow ring* and the *Grothendieck group* (or in fact: ring) of coherent sheaves. We keep the assumptions and conventions of chapter I, in particular we assume tacitly (unless explicitly stated otherwise) that $X \in \mathcal{V}$, i.e., that X is smooth and quasi-projective and defined over an algebraically closed field k of arbitrary characteristic.

The main references for this chapter are [BS] and SGA 6, Chap.0, appendix RRR (i.e., the original manuscripts of Grothendieck), see further also [Ma 2], [F] Chap. 15 and [FL].

IIa. Definition and properties of the Grothendieck group of coherent sheaves.

2.1. Definitions.

Let $X \in \mathcal{V}'$ (i.e., arbitrary variety, or even arbitrary scheme). Let $\mathcal{S}(X)$ be the category of coherent $\mathcal{O}_X$-modules on X and $\mathcal{E}(X)$ the subcategory of locally free $\mathcal{O}_X$-modules of finite rank. For $\mathcal{F} \in \mathcal{S}(X)$ let $[\mathcal{F}]$ denote its isomorphism class, similarly for $\mathcal{E} \in \mathcal{E}(X)$ we have $[\mathcal{E}]$. Put $\mathbb{Z}[\mathcal{S}(X)] :=$ free group generated by the isomorphism classes $[\mathcal{F}]$ with $\mathcal{F} \in \mathcal{S}(X)$ and $\mathbb{Z}[\mathcal{E}(X)]$: free group generated by the isomorphism classes $[\mathcal{E}]$ with $\mathcal{E} \in \mathcal{E}(X)$. Now define

(1) $$K_0(X) := \mathbb{Z}[\mathcal{S}(X)]/\text{subgroup of relations}$$

where the *relations* are defined as follows: for every exact sequence $\mathcal{O} \to \mathcal{F}' \to \mathcal{F} \to \mathcal{F}'' \to 0$ consider the element $[\mathcal{F}] - [\mathcal{F}'] - [\mathcal{F}'']$ in $\mathbb{Z}[\mathcal{S}(X)]$, the relations consist of the subgroup generated by the elements of this type.

Entirely similar one defines

(2) $$K^0(X) := \mathbb{Z}[\mathcal{E}(X)]/\text{subgroup of relations}$$

with the relations coming from exact sequences $0 \to \mathcal{E}' \to \mathcal{E} \to \mathcal{E}'' \to 0$. $K_0(X)$ is called the *Grothendieck group of coherent sheaves* and $K^0(X)$ the *Grothendieck group of locally free sheaves* or of vector bundles.

2.2. Properties and remarks.

Denote by $cl(\mathcal{F})$ the image of $[\mathcal{F}]$ in $K_0(X)$ and $cl(\mathcal{E})$ the image of $[\mathcal{E}]$ in $K^0(X)$.

a. $K^0(X)$ is a ring by defining $cl(\mathcal{E}_1) \cdot cl(\mathcal{E}_2) := cl(\mathcal{E}_1 \otimes \mathcal{E}_2)$ (and note that this is well-defined in $K^0(X)$).
b. $K_0(X)$ is a $K^0(X)$-module: $cl(\mathcal{E}) \cdot cl(\mathcal{F}) := cl(\mathcal{E} \otimes \mathcal{F})$.
c. $\exists$ a homomorphism $\alpha : K^0(X) \to K_0(X)$, namely $\alpha(cl(\mathcal{E})) = cl(\mathcal{E})$.

Theorem. *If X is smooth and quasi-projective then α is an isomorphism (and then we often write $K_0(X) = K^0(X) = K(X)$).*

For the proof we refer to [BS], the essential points being that due to quasi projectivity one can obtain a resolution of a coherent sheaf by locally free sheaves and next if X is of dimension d and if we have a resolution

$$0 \to \mathcal{G} \to \mathcal{E}_{d-1} \to \mathcal{E}_{d-2} \to \ldots \to \mathcal{E}_1 \to \mathcal{E}_0 \to \mathcal{F} \to 0$$

then by the syzygy theorem of Hilbert if X is smooth we have $\mathcal{G} \in \mathcal{E}(X)$; writing $\mathcal{G} = \mathcal{E}_d$ one can then define an "inverse" map $\beta : K_0(X) \to K^0(X)$ by $\beta(cl(\mathcal{F})) = \sum_i (-1)^i cl(\mathcal{E}_i)$.

d.Functorial properties. Let $f : X \to Y$ be a morphism. Define $f^* : K^0(Y) \to K^0(X)$ via $\mathcal{E} \mapsto f^*(\mathcal{E})$ and if f is *proper* define $f_* : K_0(X) \to K_0(Y)$ via $\mathcal{F} \mapsto \sum_i (-1)^i R^i f_*(\mathcal{F})$. Note that both homomorphims are well-defined, the f_* because of the long exact sequence of the higher direct images in case $0 \to \mathcal{F}' \to \mathcal{F} \to \mathcal{F}'' \to 0$ exact (and f proper is used to ensure that the $R^i f_*(\mathcal{F})$ are coherent).

e. Let $i : Y \hookrightarrow X$ be a subvariety of X and put $j : U = X \backslash Y \to X$. Then the following sequence is exact

$$K_0(Y) \xrightarrow{i_*} K_0(X) \xrightarrow{j^*} K_0(U) \to 0$$

where j^* is the restriction (see [BS] or [H] p. 148).

f. There exists an *isomorphism* $p^* : K_0(X) \to K_0(X \times \mathbf{A}_1)$ where $\mathbf{A}_1$ is the affine line and $p : X \times \mathbf{A}_1 \to X$ the projection ("homotopy"-property). In this case p^* is defined for any projection $p : X \times Y \to X$ because p is flat. For the proof of the property see [BS], prop. 8.

g. The λ-structure on $K^0(X)$. Let $\mathcal{E} \in \mathcal{E}(X)$ be of rank r then the exterior product $\Lambda^i(\mathcal{E}) \in \mathcal{E}(X)$ and is of rank $\binom{r}{i}$.

Lemma. *If* $0 \to \mathcal{E}' \to \mathcal{E} \to \mathcal{E}'' \to 0$ *is an exact sequence of locally free sheaves then we have in* $K_0(X)$

$$cl(\Lambda^i \mathcal{E}) = \sum_{j=0}^{i} (-1)^j cl(\Lambda^j(\mathcal{E}')) \cdot cl(\Lambda^{i-j}(\mathcal{E}'')).$$

Proof. (cf [H], p 127 or [Ma 2], prop. 3.9).

This comes from the fact that we have on $\Lambda^i(\mathcal{E})$ a filtration

$$\Lambda^i(\mathcal{E}) = \Lambda^i_0(\mathcal{E}) \supset \Lambda^i_1(\mathcal{E}) \supset \ldots \supset \Lambda^i_j(\mathcal{E}) \supset \ldots \supset \Lambda^i_i(\mathcal{E}) = \Lambda^i(\mathcal{E}')$$

where $\Lambda^i_j(\mathcal{E})$ is locally generated by elements of type $a_1 \wedge \ldots \wedge a_j \wedge b_{j+1} \wedge \cdot \wedge b_i$ with $a_q \in \mathcal{E}'$.

Corollary = Theorem. $\exists$ *uniquely maps* $\lambda^i : K^0(X) \to K^0(X)$ *determined by the condition* $\lambda^i(cl(\mathcal{E})) = cl(\Lambda^i(\mathcal{E}))$ $(i = 0, 1, \ldots)$.

Proof: From the lemma it follows that there are well-defined maps

$$\lambda_t : K^0(X) \to K^0(X)[[t]]$$

into the multiplicative group of units in the formal power series defined by $\lambda_t(cl(\mathcal{E})) = \sum_i cl(\Lambda^i(\mathcal{E}))\cdot t^i$ with $cl(\Lambda^0(\mathcal{E})) = \mathcal{O}_X (= 1$ in the ring $K^0(X))$. The map is indeed well-defined because by the lemma we have $\lambda_t(\mathcal{E}) = \lambda_t(\mathcal{E}')\cdot\lambda_t(\mathcal{E}'')$. Next for an element $\xi \in K^0(X)$ write $\lambda_t(\xi) = \sum_{i=0}^{\infty} \lambda^i(\xi)t^i$ and this defines $\lambda^i(\xi)$.

h. Chern classes. (cf. [BS], §6). Let now $X \in \mathcal{V}$. We have seen (Chap.I, 17.4) that for $0 \to \mathcal{E}' \to \mathcal{E} \to \mathcal{E}'' \to 0$ exact we have $c_t(\mathcal{E}) = c_t(\mathcal{E}')\cdot c_t(\mathcal{E}'')$ in $CH(X)[[t]]$.

Hence for $\xi \in K_0(X)$ the element $c_t(\xi)$ is well-defined and writing $c_t(\xi) = 1 + c_1(\xi)t + \ldots + c_i(\xi)t^i + \ldots$ it follows that $c_i(\xi)$ is well-defined as an element in $CH^i(X)$. Moreover proceding as in Chap.I, 17.5 we can define the *Chern character* on $K^0(X) = K_0(X) = K(X)$

$$ch : K(X) \to CH_{\mathbb{Q}}(X)$$

and we have the properties (see Chap.I, 17.5):

$$ch(\xi_1 + \xi_2) = ch(\xi_1) + ch(\xi_2)$$
$$ch(\xi_1 \cdot \xi_2) = ch(\xi_1)\cdot ch(\xi_2)$$

for $\xi_1, \xi_2 \in K(X)$, i.e. *ch is a ring homomorphism.*

i. Todd classes. Similarly, due to the property $td(\mathcal{E}) = td(\mathcal{E}')\cdot td(\mathcal{E}'')$ the Todd class $td(\xi) \in CH_{\mathbb{Q}}(X)$ makes sense for $\xi \in K(X)$ and $td(\xi_1 + \xi_2) = td(\xi_1)\cdot td(\xi_2)$. (see Chap. I, 17.6).

IIb. $K_0(X)$ and algebraic cycles.

Let X_d be smooth and quasi-projective and defined over k.

2.3. Topological filtration.
Let $\mathcal{F} \in \mathcal{S}(X)$; consider Supp$(\mathcal{F})$, i.e., the set of points $x \in X$ for which the stalk $\mathcal{F}_x \neq 0$. Supp$(\mathcal{F})$ is a closed algebraic subset of X. Introduce on $K_0(X)$ the *topological filtration*

$$K_0(X) = F^0 \supset F^1 \supset \ldots \supset F^d \supset F^{d+1} = 0$$

where $F^i_{\text{top}}K_0(X)$ is the subgroup generated by $cl(\mathcal{F})$ with $\dim \text{Supp}(\mathcal{F}) \leq d - i$. Let $\mathcal{Z}(X)$ be the group of algebraic cycles on X (see Chap.I).

2.4. Proposition. *$\exists$ surjective homomorphism $\gamma : \mathcal{Z}(X) \to K_0(X)$ defined as follows: if $Z \hookrightarrow X$ is a closed irreducible subvariety then $\gamma(Z) = cl(\mathcal{O}_Z)$ and further γ is defined by linearity.*

Proof. We have only to see the surjectivity. For this it suffices to see $\gamma : \mathcal{Z}^i(X) \to F^i \pmod{F^{i+1}}$ is surjective. Now let $\mathcal{F} \in \mathcal{S}(X)$ with $\dim \text{Supp}(\mathcal{F}) = d - i = q$.

Then consider the cycle

$$Z(\mathcal{F}) = \sum_{Y \subset X} \ell_{\mathcal{O}_{X,y}}(\mathcal{F}_y) \cdot Y$$

where Y runs through the irreducible subvarieties of dimension q in X, where $y \in Y$ is the generic point of Y, $\mathcal{F}_y$ is the stalk of $\mathcal{F}$ in y and $\ell(-)$ denotes the length as $\mathcal{O}_{X,y}$-module. Then one sees easily that $\gamma(Z(\mathcal{F})) \equiv cl(\mathcal{F})$ modulo F^{i+1}.

2.5. Consider now the homomorphism φ^i induced by γ as follows

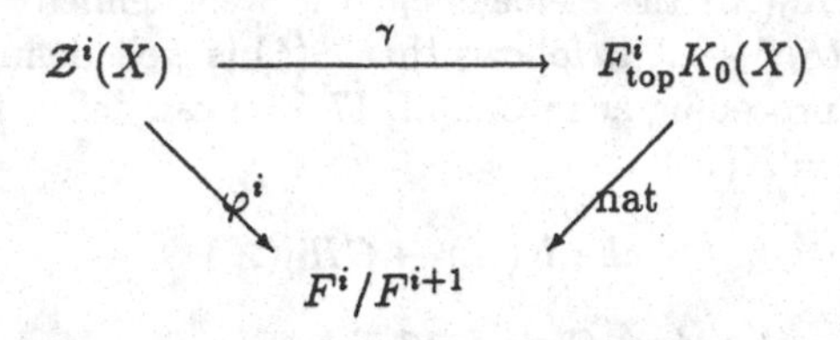

2.6. Properties of γ.
Let $X \in \mathcal{V}$.

a. Let $Z_1 \in \mathcal{Z}^i(X), Z_2 \in \mathcal{Z}^j(X)$ and let $Z_1 \cdot Z_2$ be defined. Then $\gamma(Z_1) \cdot \gamma(Z_2) \equiv \gamma(Z_1 \cdot Z_2) \bmod F^{i+j+1}$.
Indication of proof: There are three main ingredients:

α. $cl(\mathcal{F}) \cdot cl(G) = \sum_{q \geq 0} (-1)^q cl(Tor_q^{\mathcal{O}_X}(\mathcal{F}, G))$.

β. $Tor_0(\mathcal{O}_{Z_1}, \mathcal{O}_{Z_2}) = \mathcal{O}_{Z_1} \otimes_{\mathcal{O}_X} \mathcal{O}_{Z_2} \equiv \mathcal{O}_{Z_1 \cap Z_2} \pmod{F^{i+j+1}}$ (and in fact $=$ if Z_1 and Z_2 intersect transversally).

γ. $\dim \mathrm{Supp} Tor_q(\mathcal{O}_{Z_1}, \mathcal{O}_{Z_2}) < d - (i+j)$ if $q > 0$.
For details see SGA 6, p. 49 and following.

b. Let $f : X' \to X$, let $Z \in \mathcal{Z}^i(X)$ be such that $f^*(Z)$ is defined. Then

$$\gamma'_X(f^*(Z)) \equiv f^*(\gamma_X(Z)) \bmod F^{i+1}K(X').$$

Indication of proof:

α. Special case: $f : X' \hookrightarrow X$ closed immersion. Then $f^*(Z) = Z \cdot X' = Z'$ (say). Then $\gamma'_X(f^*(Z)) = cl(\mathcal{O}_{Z'}) = cl(\mathcal{O}_Z \otimes \mathcal{O}_{X'}) \equiv f^*(cl(\mathcal{O}_Z)) \pmod{F^{i+1}} = f^*(\gamma_X(Z)) \pmod{F^{i+1}}$.

β. In the general case let $X'' = \Gamma_f$ be the graph of f, put $j : X'' \hookrightarrow X \times X'$, let $p_1 : X \times X' \to X$ and $p'_1 = p_1 \bullet j : X'' \to X$. Since p_1 is flat we have in fact $\gamma_{X \times X'}(p_1^*(Z)) = p_1^*(\gamma_X(Z))$. Next using the special case we have $\gamma''_X(j^* \cdot p_1^*(Z)) \equiv j^* \gamma_{X \times X'}(p_1^*(Z)) \pmod{F^{i+1}K(X'')} = j^* \cdot p_1^*(\gamma_X(Z)) = p_1'^*(\gamma_X(Z))$. Since $X'' \xrightarrow{\sim} X'$ is an isomorphism, this proves the formula. For details see SGA 6, p.53.

2.7. Theorem. (see SGA6, p.58 and [F], p 285). *Let X be smooth, quasi-projective. Let $Z_1, Z_2 \in \mathcal{Z}^i(X)$ and let Z_1 be rationally equivalent to Z_2. Then*

$$\gamma(Z_1) \equiv \gamma(Z_2) \mod F^{i+1}_{\text{top}}K(X).$$

Proof.

We have $Y \in \mathcal{Z}^i(X \times \mathbb{A}_1)$ and two points $a, b \in \mathbb{A}_1$ such that $Y(a) = Z_1$ and $Y(b) = Z_2$ (Chap I. 1.4.1 lemma 1). Now consider $f_a : X \to X \times \mathbb{A}_1$ defined by $f_a(x) = (x, a)$ and similarly $f_b : X \to X \times \mathbb{A}_1$. Then $Z_1 = f_a^*(Y)$ and $Z_2 = f_b^*(Y)$. By 2.6 property b we have $\gamma_X(Z_1) = \gamma_X(f_a^*(Y)) \equiv f_a^*\gamma_{X\times\mathbb{A}_1}(Y) \pmod{F^{i+1}K(X)}$ and similarly $\gamma_X(Z_2) \equiv f_b^*\gamma_{X\times\mathbb{A}_1}(Y) \pmod{F^{i+1}K(X)}$. So it suffices to see

$$f_a^* = f_b^* : K(X \times \mathbb{A}_1) \to K(X).$$

Consider $p : X \times \mathbb{A}_1 \to X$, then we have $id_X = p \bullet f_a$, and hence

$$id : K(X) \xrightarrow{p^*} K(X \times \mathbb{A}_1) \xrightarrow{f_a^*} K(X).$$

However by 2.2f we know that p^* is an isomorphism hence we have $f_a^* = (p^*)^{-1} = f_b^*$.

2.8. Corollary *(see SGA6, p 59). For X smooth, quasi-projective we have $F^i_{\text{top}}K(X) \cdot F^j_{\text{top}}K(X) \subset F^{i+j}_{\text{top}}K(X)$. In particular it follows that $Gr^{\cdot}_{\text{top}}K(X)$ has a ring structure.*

Proof.

We proceed by decreasing induction on i and j. $F^iK(X)$ is generated by the $\gamma(Z)$ with $Z \hookrightarrow X$ of codimension i. By 2.6c we have $\gamma(Z_1) \cdot \gamma(Z_2) \equiv \gamma(Z_1 \cdot Z_2)$ mod F^{i+j+1} for $Z_1 \in \mathcal{Z}^i(X)$ and $Z_2 \in \mathcal{Z}^j(X)$ provided $Z_1 \cdot Z_2$ is defined, however by theorem 2.7 we may always assume that $Z_1 \cdot Z_2$ is defined, hence $F^i \cdot F^j \subseteq F^{i+j}$.

2.9. Corollary *(see SGA6, p 59 cor 2). Let X be smooth, quasi-projective. Then*

1. $\exists$ surjective ring homomorphism

$$\varphi : \oplus_i CH^i(X) \longrightarrow \oplus F^i_{\text{top}}/F^{i+1}_{\text{top}} = Gr^{\cdot}_{\text{top}}K(X)$$

where $\varphi = \oplus\varphi^i$ with φ^i from 2.5, i.e., determined by $\varphi(Z) = cl(\mathcal{O}_Z)$ mod F^{i+1} for $Z \hookrightarrow X$ closed, irreducible of codimension i.

2. φ is functorial, i.e., if $f : X \to Y$ is a morphism of smooth, quasi-projective *varieties then the following diagram is commutative*

$$\begin{array}{ccc} CH(Y) & \xrightarrow{f^*} & CH(X) \\ \downarrow{\scriptstyle \varphi_Y} & & \downarrow{\scriptstyle \varphi_X} \\ Gr^{\bullet}_{\text{top}}K(Y) & \xrightarrow[Gr^{\bullet}(f^*)]{} & Gr^{\bullet}_{\text{top}}K(X) \end{array}$$

Proof.

φ is well-defined by 2.5 and 2.7, φ is a ring homomorphism as follows from the definition of the ring structure in $Gr^{\cdot}_{\text{top}}K(X)$ as is seen in the proof of Corollary 2.8. Finally $Gr^{\cdot}(f^*)$ is defined as follows from 2.6b and the commutativity follows also from 2.6b.

IIc. The Chow ring and $K(X)$.

2.10. Let X be smooth and quasi-projective, defined over k. We have seen in 2.2h that the Chern character defines a ring homomorphism $ch : K(X) \to CH_{\mathbb{Q}}(X)$. Consider now the commutative diagram

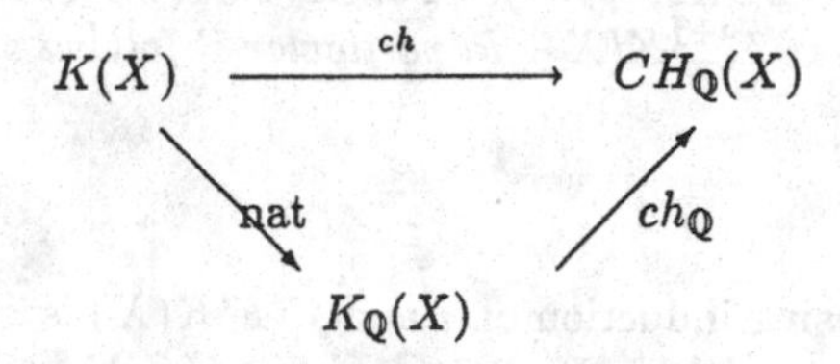

where $K_{\mathbb{Q}}(X) := K(X) \otimes \mathbb{Q}$.

Main theorem *(Grothendieck; see SGA6, p 673-674 and [F], p 294-295).*
Let X be smooth, quasi-projective defined over k. Then

1. $ch_{\mathbb{Q}} : K_{\mathbb{Q}}(X) \to CH_{\mathbb{Q}}(X)$ *is a ring isomorphism.*
2. $Gr^{\cdot}(ch) : Gr^{\cdot}_{\text{top}}(K(X)) \to \oplus CH^i_{\mathbb{Q}}(X)$ *is defined and is a ring homomorphism of graded rings which becomes an isomorphism after tensoring with* $\mathbb{Q}$.
3. $CH^i(X) \xrightarrow{\varphi^i} Gr^i_{\text{top}}(K(X)) \xrightarrow{Gr^i(ch)} CH^i_{\mathbb{Q}}(X)$ *is the natural inclusion.*

Outline of the proof:
The proof uses a special case of the Riemann–Roch–Grothendieck theorem. We proceed along several lemma's.

2.11. Lemma. *Let X_d be smooth, quasi-projective and let $f : Y \hookrightarrow X$ be a closed immersion of codimension i. Then*

$$ch(cl(\mathcal{O}_Y)) = f_*([Y] + \alpha)$$

where $\alpha \in \bigoplus_{j=0}^{q-1} CH_j(Y) \otimes \mathbb{Q}$ *where* $\underline{q} = d - i$.

2.12. Lemma $\Longrightarrow$ theorem

Firstly it shows $ch(F^i_{\text{top}}K(X)) \subset \oplus_{j\geq i}CH^j_{\mathbb{Q}}(X)$, hence $Gr^{\cdot}(ch)$ is defined. Moreover we see that

$$CH^i(X) \xrightarrow{\varphi^i} F^i/F^{i+1} \xrightarrow{Gr^i(ch)} CH^i_{\mathbb{Q}}(X)$$

is the natural inclusion $[Z] \to cl(\mathcal{O}_Z) \to [Z]$. Hence tensoring with $\mathbb{Q}$ both $\varphi^i_{\mathbb{Q}}$ and $Gr^i(ch_{\mathbb{Q}})$ become additive isomorphisms because φ^i is surjective. But then also $ch_{\mathbb{Q}}$ itself must be injective (we knew already that it is surjective), hence, since it is a ring homomorphism it is a ring isomorphism and this shows that the lemma implies the theorem.

2.13. Step 1 of the proof of the lemma: Reduction to the case $f : Y \hookrightarrow X$ with Y also smooth and (hence) f a regular embedding (i.e, given by a regular sequence of parameters). Namely, it suffices to replace X by an open set $U \subset X$ for which this is the case such that $\dim S < d - i$, where $S = X - U$. We have an exact sequence (Chap I. 1.4.1 lemma 2)

$$CH_j(S) \to CH_j(X) \to CH_j(U) \to 0$$

and hence we have $CH_q(X) \xrightarrow{\sim} CH_q(U)$ for $q = d - i$; moreover we have a similar sequence for the K-groups (2.2e)

2.14. Step 2: Lemma 1 follows now from following key-lemma:

Lemma 2 *$f : Y \to X$ closed regular imbedding, with X and Y smooth, quasi-projective. Then*

$$ch(cl(\mathcal{O}_Y)) = f_*(Y \cdot (tdN_{Y/X})^{-1})$$

where $N_{Y/X}$ is the normal bundle of Y in X and td is the Todd class.

Proof.

First note that lemma 2 $\Longrightarrow$ lemma 1 because the RHS in the above equation is of the type required in lemma 1 (see Chap.I, 1.7.6). Lemma 2 itself is a special case of the Riemann-Roch-Grothendieck theorem ([F], 15.2). Namely apply the RRG-theorem to an element $\xi \in K(Y)$, then we get

$$ch(f_*\xi) \cdot tdT_X = f_*\{ch(\xi) \cdot td(T_Y)\}$$

where T_X, respectively T_Y, is the tangent bundle of X, respectively Y. Take $\xi = cl(\mathcal{O}_Y)$ and use the exact sequence $0 \to T_Y \to f^*T_X \to N_{Y/X} \to 0$ and the well-known multiplicative behaviour of the Todd class by exact sequences (see Chap. I, 1.7.6). We get

$$ch(cl(\mathcal{O}_Y)) \cdot (tdT_X) = f_*\{Y \cdot (tdN_{Y/X})^{-1} \cdot f^*(tdT_X)\}.$$

Applying the well-known projection formula on the RHS we get that the RHS equals $f_*\{Y \cdot (tdN_{Y/X})^{-1}\} \cdot tdT_X$ and "dividing out" tdT_X (as is allowed, see Chap. I 1.7.6) this completes the proof.

2.15. Remark.

By the main theorem (see 2.10) the ring homomorphism

$$\varphi : CH(X) \to Gr^{\cdot}_{\text{top}}(K(X))$$

becomes an isomorphism after tensoring with $\mathbb{Q}$. However φ itself does not need to be an isomorphism as is shown by an example in SGA6, XIV 4.7, page 679-680.

IId. A formula for a subvariety.

2.16. Theorem (see [F], p 298 and SGA6, p 674). *Let X_d be smooth and quasi-projective, defined over k. Let $\alpha \in F^q_{\text{top}}K(X)$. Then we have for the Chern classes:*

$$c_0(\alpha) = 1, \;\; c_i(\alpha) = 0 \text{ for } 0 < i < q$$

and c_q induces an additive homomorphism (still denoted by c_q)

$$c_q : F^q/F^{q+1} \to CH^q(X)$$

such that the composition with the homomorphism φ (from 2.5)

$$CH^q(X) \xrightarrow{\varphi^q} F^q/F^{q+1} \xrightarrow{c_q} CH^q(X)$$

gives

$$c_q \bullet \varphi^q = (-1)^{q-1}(q-1)!id.$$

$$\varphi^q \bullet c_q = (-1)^{q-1}(q-1)!id.$$

2.17. Outline of proof of 2.16.

Concerning the last two formula's first note that, since φ is surjective (see 2.4), it suffices to prove the first one. The proof of 2.16 proceeds along several steps.

Step 1. Since F^q_{top} is generated by elements of type $\gamma(Z)$ with $Z \to X$ a closed immersion of an irreducible subvariety of codimension q and since $c_t(\alpha + \beta) = c_t(\alpha) \cdot c_t(\beta)$ it suffices to prove the assertions for elements $\alpha = \gamma(Z)$ with Z as above. In that case one has the following more precise fact:

2.18. Corollary *(cf [Gr 1], p.151). Let $i : Z \to X$ be a closed immersion of an irreducible subvariety of codimension q.*
Then:

$$c_0(\gamma(Z)) = 1, c_j(\gamma(Z)) = 0 \text{ for } 0 < j < q$$

$$c_q(\gamma(Z)) = (-1)^{q-1}(q-1)![Z]$$

where $[Z]$ is the class of Z in $CH^q(X)$.

2.19. Outline of proof of 2.18 (cf. [Hi], p. 53).

Step 2. Reduction to the case that $Z = H_1 \cdot H_2 \cdot \ldots H_q \cdot X$, where $H_j (1 \leq j \leq q)$ are hypersurfaces in the ambient projective space of X and intersecting transversally on X.

This reduction goes itself in several steps. First we choose hypersurfaces such that $H_1 \cdot H_2 \ldots H_q \cdot X = Z + Z_1$ for some Z_1 of codimension q. Next by omitting a closed set S of codimension larger than q, and using Lemma 2 of Chap. I, 1.4.1, we can assume that $Z \cap Z_1 = \emptyset$ and that the intersection is transversal on X, hence both Z and Z_1 smooth. Finally by blowing up X along the smooth Z_1 and using $CH(X) \subset CH(B_{Z_1}(X))$ we can reduce to the case $Z_1 = \emptyset$.

Step 3. Let now $Z = H_1 \cdot H_2 \ldots H_q \cdot X$ with the H_i intersecting transversally on X. Let J_r be the ideal sheaf of H_r on X and put $J = \oplus_r J_r$. Now we have a global Koszul resolution for $\mathcal{O}_Z$ (cf [BS], §10 c):

$$0 \to \Lambda^q J \to \ldots \to \Lambda^2 J \to J \to \mathcal{O}_X \to \mathcal{O}_Z \to 0.$$

Moreover $\Lambda^s J = \Lambda^s(\oplus_r J_r) \cong \oplus_\alpha (J_{\alpha_1} \otimes J_{\alpha_2} \otimes \ldots \otimes J_{\alpha_s})$. From this we get in $K(X)$ that

$$\gamma(Z) = \sum_{s=0}^{q} (-1)^s cl(\Lambda^s J) = \sum_{s=0}^{q} (-1)^s \{\sum_\alpha cl(J_{\alpha_1} \otimes \ldots \otimes J_{\alpha_s})\}$$

In $CH(X)[[t]]$ we have $c_t(J_r) = 1 - h_r t$, where $h_r = H_r \cdot X$ is the divisor on X cut out by H_r and next $c_t(J_{\alpha_1} \otimes \ldots \otimes J_{\alpha_s}) = 1 - h_{\alpha_1} t - h_{\alpha_2} t - \quad - h_{\alpha_s} t$. Therefore we get in $CH(X))[[t]]$ that

$$c_t(\gamma(Z)) = \prod_{s=0}^{q} \{\prod_\alpha c_t(J_{\alpha_1} \otimes \ldots \otimes J_{\alpha_s})\}^{(-1)^s} = \prod_{s=0}^{q} \{\prod_\alpha (1 - h_{\alpha_1} t - \ldots - h_{\alpha_s} t)\}^{(-1)^s}.$$

That means that the proof of 2.18 (and hence of 2.16) reduces to the following completely formal:

Step 4:

Lemma. *In the power series ring* $\mathbb{Z}[[x_1, \ldots, x_q]]$ *we have* $\prod_{s=0}^q \{\prod_\alpha (1 - x_{\alpha_1} - \ldots - x_{\alpha_s})\}^{(-1)^s} = 1 + (-1)^{q-1}(q-1)! x_1 x_2 \ldots x_q +$ *higher order terms.*

We omit the proof of this (Hint: take of both sides the logarithm).

Chapter III. The Chow ring and higher algebraic K–theory.

In this chapter we shall say something about the relation between the Chow ring and higher algebraic K–theory; it is only intended as a first introduction to this subject. A good reference for K_0, K_1 and K_2 is the book of Milnor [Mi]. The basic reference for higher algebraic K–theory is the fundamental paper [Q] of Quillen, see also the book [Sr] and the lectures [LP, et al]. For the relation between the theory of algebraic cycles and algebraic K–theory see Chap.4 of the book of Bloch [B] and again [Q] and [Sr].

IIIa. Some remarks on algebraic K–theory.

3.1. K–groups of rings. For every $n \in \mathbb{Z}, n \geq 0$ there is a covariant functor

$$K_n : \{ \text{ category of commutative rings } \} \to \{ \text{ category of abelian groups } \}.$$

For the general construction see [Q] and [Sr]. Here we say only something about the cases $n = 0$ and $n = 1$. Let R be a commutative ring (with unit element).

3.1.1. Case $n = 0$. The $K_0(R)$ is the Grothendieck group of R constructed similarly as in Chap.2. Namely the generators of $K_0(R)$ are the isomorphism classes [P] of the finitely generated projective R-modules P; the relations are generated by the elements $[P_1 \oplus P_2] - [P_1] - [P_2]$.

3.1.1. Case $n = 1$. The $K_1(R)$ is constructed (see [Mi]) via the general linear group $GL(R)$ as follows. Firstly $GL(R) := \varinjlim GL(n, R)$ where the transition morphisms $GL(n, R) \to GL(n + 1, R)$ are given by

$$A \to \begin{pmatrix} A & 0 \\ 0 & 1 \end{pmatrix}$$

and then $K_1(R) := GL(R)/[GL(R), GL(R)]$.

3.2. K–groups of local rings. For local rings one can be more explicit for the cases $n = 0, 1$ and 2. Let R be a commutative *local ring.*

3.2.1. Case $n = 0$. A projective module over a local ring is free, so only the rank matters. Therefore it follows immediately from the definition of K_0 that $K_0(R) \simeq \mathbb{Z}$ if R is a local ring.

3.2.2. Case $n = 1$. From the description of $K_1(R)$ we have for any commutative ring a commutative diagram (where R^* is the group of units of R)

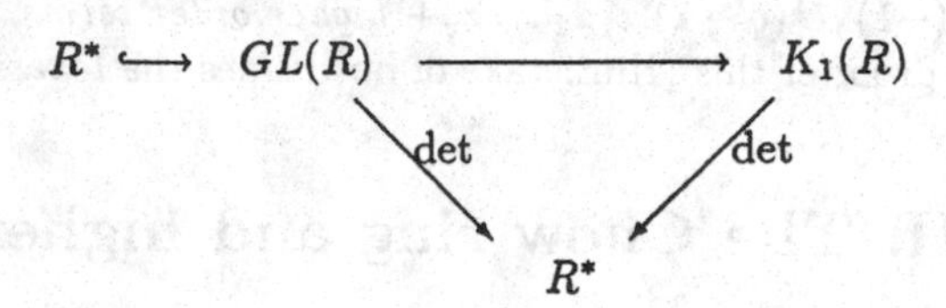

Theorem 1.(see [Mi], p.28)*For a local ring* $K_1(R) = R^*$.

3.2.3. Case $n = 2$. In this case there is a bilinear pairing (see for instance [Mi], §8 and [B], 4.4)

$$R^* \otimes_{\mathbb{Z}} R^* \longrightarrow K_2(R)$$
$$a, b \longmapsto \{a, b\}$$

and this map is *surjective* for a *local ring.* The elements $\{a, b\}$ are called *Steinberg symbols*; hence $K_2(R)$ is generated by Steinberg symbols. Moreover these symbols satisfy the following relations

i.$\{a_1 a_2, b\} = \{a_1, b\}\{a_2, b\}$

$\{a, b_1 b_2\} = \{a, b_1\}\{a, b_2\}$ bilinearity,
ii. $\{1, 1-a\} = 1$ Steinberg relation,
iii. $\{a, -a\} = 1$
iv. $\{a, b\} = \{b, a\}^{-1}$
v. $\{a, 1\} = 1$

Theorem 2. (Matsumoto, [Mi], p.93). *If $R = F$ is a field then all the relations in $K_2(F)$ are generated by the above relations i. and ii.*

Theorem 3. (Van der Kallen, [vdK 1], p. 309). *The same is true for a commutative local ring provided the residue field has at least 6 elements.*

3.2.4. Milnor K-theory. For higher K_n things become much more complicated but one can define the so-called *Milnor K-groups* by means of generators and relations similar as in the case $n = 2$. For instance if $R = F$ is a field, define

$$K_n^M(F) := F^* \otimes \ldots \otimes F^* / \{(a_1 \otimes \ldots \otimes a_n); a_i + a_j = 1 \text{ for some } i \neq j\},$$

and there are well-known relations between $K_n^M(F)$ and $K_n(F)$ (Suslin; see for instance the paper of Loday in [LP et al], p.52).

3.3. The Zariski sheaves $\mathcal{K}_{n,X}$.

Now let X be a variety defined over a field k say, or even more general let X be a scheme. Since the $K_n(R)$ is a covariant functor for the ring R, we get, if we take for $R = \Gamma(U, \mathcal{O}_X)$ with U a *Zariski open set* of X, a presheaf on X:

$$U \mapsto K_n(\Gamma(U, \mathcal{O}_X))$$

and hence by sheafification a *Zariski sheaf* $\mathcal{K}_{n,X}$ which we sometimes also denote shortly by $\mathcal{K}_n$ if X is fixed. Since K_n commutes with inductive limits we have for the stalks at the points $x \in X$ that $\mathcal{K}_{n,X,x} = K_n(\mathcal{O}_{X,x})$. Therefore we have $\mathcal{K}_{0,X} = \mathbb{Z}, \mathcal{K}_{1,X} = \mathcal{O}_X^*$ and the stalks of $\mathcal{K}_{2,X}$ are generated by Steinberg symbols $\{a, b\}$ with $a, b \in \mathcal{O}_X^*$ and for which we have well-known relations by the theorem of Van der Kallen.

3.4. K-groups of schemes. Quillen has defined, in a very general way, K-groups for so-called "exact categories" (see [Q] and [Sr], and for an introduction also [B], Chap.4). If X is a Noetherian scheme then this applies in particular to the abelian categories $\mathcal{E}(X)$ of locally free sheaves of finite rank and $\mathcal{S}(X)$ of coherent sheaves on X. Then one defines $K^n(X) := K_n(\mathcal{E}(X))$ and $K_n(X) := K_n(\mathcal{S}(X))$. From the inclusion $\mathcal{E}(X) \subset \mathcal{S}(X)$ one gets a homomorphism $K^n(X) \to K_n(X)$. In case X is *regular* (and quasi-compact) then this is in fact an *isomorphism* $K^n(X) \xrightarrow{\sim} K_n(X)$; this follows, like in Chapter 2, from the fact that for every $\mathcal{F} \in \mathcal{S}(X)$ there is a finite resolution by means of locally free sheaves ([Q], p.116). Also $K^0(X)$ coincides with the Grothendieck group $K^0(X)$ from Chapter 2 and similarly for $K_0(X)$. Furthermore in case $X = \operatorname{Spec} R$ with R a commutative ring (with unit) then $K_n(X) = K_n(R)$ (see [Sr], p. 58).

3.5. Filtration by means of support. Consider in $\mathcal{S}(X)$ the topological filtration, i.e., the filtration by means of support:

$$\mathcal{S}(X) = \mathcal{S}^0 \supset \mathcal{S}^1 \supset \ldots \supset \mathcal{S}^p \supset \ldots \supset \mathcal{S}^d \supset \mathcal{S}^{d+1} = 0 \tag{1}$$

where $\mathcal{S}^p = \mathcal{S}^p(X)$ consists of the subcategory of $\mathcal{S}(X)$ consisting of sheaves $\mathcal{F}$ with codimension $(\mathrm{Supp}(\mathcal{F})) \geq p$ and where we have assumed that X has dimension d. Also let X^p be the set of points on X of codimension p, i.e.

$$X^p = \{x \in X; \text{codimension } \{\overline{x}\} = p\}$$

where $\{\overline{x}\}$ is the Zariski closure of x in X (i.e. x is the "generic" point of the irreducible subscheme $\{\overline{x}\}$ of X and $\{\overline{x}\}$ has codimension p in X). Consider the inclusion $\mathcal{S}^{p+1}(X) \subset \mathcal{S}^p(X)$, then it follows from the general theorems of Quillen (see [Q], §5, thm.5 and thm 4, cor.1) that there is an exact sequence of abelian groups

$$\ldots \to K_n(\mathcal{S}^{p+1}) \to K_n(\mathcal{S}^p) \to \bigoplus_{x \in X^p} K_n(k(x)) \to K_{n-1}(\mathcal{S}^{p+1}) \to \ldots \tag{2}$$

where $k(x) = \mathcal{O}_{X,x}/M_{X,x}$ is the residue field of x. Writing down the same sequence for the inclusion $\mathcal{S}^{p+2} \subset \mathcal{S}^{p+1}$ and comparing terms one gets homomorphisms

$$\bigoplus_{x \in X^p} K_n(k(x)) \to \bigoplus_{y \in X^{p+1}} K_{n-1}(k(y))$$

where the homomorphisms go from x to those y for which $y \in \{\overline{x}\}$. Combining all these homomorphisms one gets finally a complex of abelian groups:

$$\begin{aligned} 0 \to K_n(\mathcal{S}(X)) \to \bigoplus_{x \in X^0} K_n(k(x)) \to \bigoplus_{x \in X^1} K_{n-1}(k(x)) \to \ldots \\ \ldots \to \bigoplus_{x \in X^{n-1}} K_1(k(x)) \to \bigoplus_{x \in X^n} K_0(k(x)) \to 0. \end{aligned} \tag{3}$$

3.6. Gersten–Quillen resolution. Let, as above, X be a noetherian scheme. We sheafify the complex (3) as follows. For $x \in X^p$ consider on the closed subscheme $\{\overline{x}\}$ of X the *constant* sheaf $K_{n-p}(k(x))$; let $j_x : \{\overline{x}\} \hookrightarrow X$ be the natural inclusion, then $j_{x*}K_{n-p}(k(x))$ is the sheaf on X obtained by extending the above constant sheaf on $\{\overline{x}\}$ by zero outside of $\{\overline{x}\}$. In this way we get from the complex (3) of abelian groups the following *complex of Zariski sheaves* on X:

$$\begin{aligned} 0 \to \mathcal{K}_{n,X} \to \bigoplus_{x \in X^0} j_{x*}K_n(k(x)) \to \bigoplus_{x \in X^1} j_{x*}K_{n-1}(k(x)) \to \ldots \\ \ldots \to \bigoplus_{x \in X^{n-1}} j_{x*}K_1(k(x)) \to \bigoplus_{x \in X^n} j_{x*}K_0(k(x)) \to 0. \end{aligned} \tag{4}$$

The following result was conjectured by Gersten and proved by Quillen:

Theorem 4. (Quillen, [Q], §7, Thm. 5.11) *Let X be a smooth variety defined over a field k. Then the above sequence (4) of Zariski sheaves is exact.*

For the proof see [Q]. Note that this theorem is giving, on a smooth variety, a "concrete" flasque (= "flabby") resolution for the sheaf $\mathcal{K}_{n,X}$.

IIIb. Application to algebraic cycles. Bloch's formula.

3.7. Bloch's formula.

Theorem 5. ([Q], §7 Thm. 5.19). *Let X be a smooth, quasi-projective variety defined over a field k. Then*

$$H^m_{\text{Zar}}(X, \mathcal{K}_{n,X}) = 0 \qquad (m > n)$$
$$H^n_{\text{Zar}}(X, \mathcal{K}_{n,X}) = CH^n(X)$$

where $CH^n(X)$ is the Chow group of X.

Remark. This formula for $CH^n(X)$ was proved for $n = 2$ by Bloch [B 1] and later by Quillen [Q] for general n.

"Proof":

Since (4) gives a flasque resolution of $\mathcal{K}_{n,X}$ it suffices to take global sections and to take the cohomology for the corresponding complex. This gives the first assertion. For the second (Bloch's formula) first note that $K_0(k(x)) = \mathbb{Z}$, hence

$$\bigoplus_{x \in X^n} \Gamma(X, j_{x*} K_0(k(x))) = \mathcal{Z}^n(X)$$

the group of algebraic cycles on X of codimension n, and similarly $K_1(k(x)) = k(x)^*$, hence

$$\bigoplus_{x \in X^{n-1}} \Gamma(X, j_{x*} k(x)) = \bigoplus_{x \in X^{n-1}} k(x)^*.$$

Now one has to prove that the corresponding map is the natural one, namely taking the divisor of the corresponding rational function on $\{\overline{x}\}$; it then follows that the image of the map is $\mathcal{Z}^n_{\text{rat}}(X)$ (see Chap.I, 1.4.1) and hence the cohomology on the n-th spot is $CH^n(X)$.

3.8. Examples.

For these examples let us adopt the following notations and assumption. Let X be a *smooth, quasi-projective*, irreducible variety defined over a field k, let $k(X)$ be its field of rational functions. Let D (resp. Y) be an irreducible subvariety of codimension one (resp. two) and $j_D : D \hookrightarrow X$ (resp. $j_Y : Y \hookrightarrow X$) the natural inclusion.

3.8.1. Case $n = 0$. $\mathcal{K}_{0,X} = \mathbf{Z}$ the constant sheaf. This we have already seen (see 3.3).

3.8.1. Case $n = 1$. Then $\mathcal{K}_{1,X} = \mathcal{O}^*_X$ (see 3.3) and (4) becomes:

$$1 \to \mathcal{O}^*_X \to k(X)^* \xrightarrow{\text{div}} \bigoplus_D j_{D*}(\mathbf{Z}_D) \to 0$$

is an exact sequence of Zariski sheaves, where $k(X)^*$ is the multiplicative group of the field seen as constant sheaf on X. This is nothing else but the familiar exact sequence for divisors ([H], p.141) provided we keep in mind that since X is smooth the Cartier divisors and the Weil division are the same. Theorem 4 is nothing but the well-known fact that

$$H^1_{\mathrm{Zar}}(X, \mathcal{O}_X^*) = CH^1(X).$$

3.8.2. Case $n = 2$.

Now (4) becomes the following exact sequence of Zariski sheaves

$$0 \to \mathcal{K}_{2,X} \xrightarrow{\alpha} K_2(k(X)) \xrightarrow{T} \bigoplus_D j_{D*}(k(D)^*) \xrightarrow{\mathrm{div}} \bigoplus_Y j_{Y*}(\mathbb{Z}_Y) \to 0$$

where $K_2(k(X))$ is the K_2-group of the function field considered as constant sheaf on X. The map α is the natural one sending in the stalk at $x \in X$ the Steinberg symbol $\{f, g\}$ of $\mathcal{O}_{X,x}$ to the corresponding Steinberg symbol of the function field. The map "div" is: take divisors (as above). Finally, T is the map defined via the so-called "tame symbol" as follows: every D as above defines a valuation $\mathcal{V}_D$ on $k(X)$, then for $\{f, g\} \in K_2(k(X))$ (where $f, g \in k(X)^*$) one takes (at the component D):

$$T\{f, g\} = (-1)^{\mathcal{V}_D(f)\mathcal{V}_D(g)}(f^{\mathcal{V}_D(g)}/g^{\mathcal{V}_D(f)})_{|D}$$

where $(\ldots)_{|D}$ means: take the restriction of the function to D.

Theorem 4 becomes in this case

$$H^2_{\mathrm{Zar}}(X, \mathcal{K}_{2,X}) = CH^2(X)$$

and this is the theorem proved by Bloch in [B 1].

3.9. Remark. The meaning of the groups $H^m_{\mathrm{Zar}}(X, \mathcal{K}_{n,X})$ for $m < n$ is still mysterious.

IIIb Supplement (to be used in Chap. VI).

3.10. Reduction mod m.

Assumptions and notations. In this part let X be a *smooth, projective* variety defined over a field k. We keep the notations and conventions of 3.8. Furthermore, if A is an abelian group (or a sheaf of abelian groups) and $m \in \mathbb{Z}$, then consider the map $\cdot m : A \to A$ defined by $a \mapsto ma$, and put

$$_mA = \mathrm{Ker}(A \xrightarrow{\cdot m} A),$$
$$A/m = \mathrm{Coker}(A \xrightarrow{\cdot m} A).$$

In the following let $(m, p) = 1$ where $p = \mathrm{char}(k)$.

Proposition 1. ([Bl 2], p.79). *There are exact sequences:*

i. $0 \to_m \mathcal{K}_2 \to_m K_2(k(X)) \to \bigoplus_D j_{D*}(\mu_{m,D}) \to 0$

ii. $0 \to \mathcal{K}_2/m \to K_2(k(X))/m \to \bigoplus_D j_{D*}(k(D)^*/m) \to \bigoplus_Y j_{Y*}(\mathbb{Z}/m) \to 0$
where $\mu_{m,D}$ is the group of m–th roots of unity in $k(D)$, considered as constant sheaf on D.

Remark. Grayson has proved ([B 2], p. 81) that, more generally, for every n the Gersten–Quillen sequence (4) remains exact by reducing it mod m (here we restrict to $n = 2$).

For the proof of the proposition we refer to [B 2], p. 80, to [B], p.54 and to [Mur 1], p.241.

Corollary 1.

i) $H^i_{\text{Zar}}(X, {}_m\mathcal{K}_2) = 0$ for $i > 1$
ii) $H^1_{\text{Zar}}(X, \mathcal{K}_2) \xrightarrow{\cdot m} H^1_{\text{Zar}}(X, \mathcal{K}_2) \to H^1_{\text{Zar}}(X, \mathcal{K}_2/m) \to_m CH^2(X) \to 0$ is exact.

Proof.
By the above proposition we have flasque resolutions for the sheaves ${}_m\mathcal{K}_2$ and $\mathcal{K}_2/m$. This gives immediately i. As to ii. we split the map $K_2 \xrightarrow{\cdot m} \mathcal{K}_2$ onto the exact sequences

$$0 \to_m \mathcal{K}_2 \to \mathcal{K}_2 \to Im(m) \to 0,$$
$$0 \to Im(m) \to \mathcal{K}_2 \to \mathcal{K}_2/m \to 0,$$

take cohomology, use part i. and combine.

IIIc. Some final remarks.

3.11. Note that $H^n_{\text{Zar}}(X, \mathcal{K}_{n,X})$ makes sense on any scheme, therefore one can try to use this as a definition for $CH^n(X)$ on an arbitrary scheme (see for instance [Co] and [PW]).

3.12. The usefulness of the formula $CH^1(X) = H^1_{\text{Zar}}(X, \mathcal{O}^*_X)$ comes partially from the fact that one has also another interpretation of $H^1_{\text{Zar}}(X, \mathcal{O}^*_X)$ namely $H^1_{\text{Zar}}(X, \mathcal{O}^*_X) = \text{Pic}(X)$, i.e., the group of isomorphism classes of invertible sheaves (see the lectures of Claire Voisin [Vo] and [H], p. 143 and 224). Such an interpretation is (still) missing for $H^n_{\text{Zar}}(X, \mathcal{K}_{n,X})$ for $n > 1$. Nevertheless the formula $CH^n(X) = H^n_{\text{Zar}}(X, \mathcal{K}_{n,X})$ is significant, not only from an "abstract theoretical" point of view (see 3.11), but also "geometrically" in the sense that it "explains" things and "suggests" things. Let us try to indicate this (at least in a vague and loose way) somewhat further. Namely the formula $CH^2(X) = H^2_{\text{Zar}}(X, \mathcal{K}_2)$ explains Mumford's theorem for a *surface* X with $p_g(X) \neq 0$ (see Chap. I, 1.6.1 and C. Voisin's lecture 5 [Vo]) and it motivated Bloch for his conjecture in the case $p_g(X) = 0$ (Chap. I, 1.6.1 and C. Voisin's lecture 5 [Vo]). For this, as in the case $n = 1$, one should turn $H^n_{\text{Zar}}(X, \mathcal{K}_{n,X})$ first into a functor from preschemes to abelian groups:

$$S \mapsto F(S) = H^n_{\text{Zar}}(X_S, \mathcal{K}_{n,X_S})/p_S^* H^n_{\text{Zar}}(S, \mathcal{K}_{n,S}))$$

(To be precise one should localize this functor with respect to a suitable topology). Next look to the "tangent space" TF of the functor. If $\mathcal{D} = \mathrm{Spec}(k[\epsilon]/\epsilon^2 = 0)$ then by definition

$$TF := \mathrm{Ker}\{F(D) \to F(\mathrm{Spec}\ k)\}$$

which comes via the cartesian square

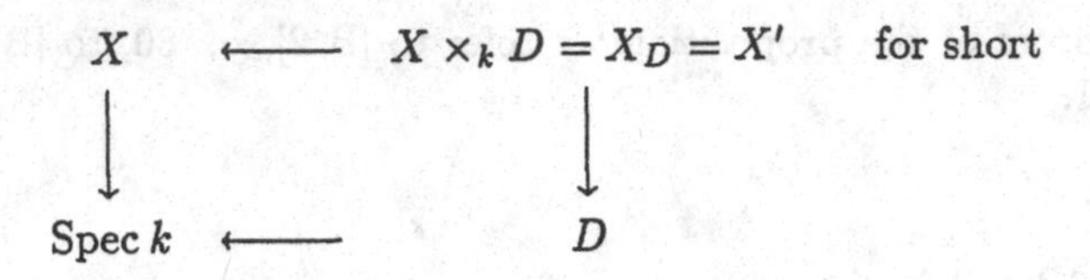

Now first remark that for $S = D$ and $S = \mathrm{Spec}\ k$ we have $H^n_{\mathrm{Zar}}(S, \mathcal{K}_{n,S}) = 0$ for $n > 0$. Also as topological spaces X_D and X are the same. Furthermore putting $\epsilon = 0$ we get sections in the above diagram and a left inverse for the natural homomorphism $\mathcal{K}_{n,X_D} \to \mathcal{K}_{n,X}$. This gives for $n \geq 1$ in our case the formula

$$TF = H^n_{\mathrm{Zar}}(X, \mathrm{Ker}(\mathcal{K}_{n,X_D} \to \mathcal{K}_{n,X}))$$

For $n = 1$ we get the following split exact sequence on X

$$0 \to \mathcal{O}_X \xrightarrow{u} \mathcal{O}^*_{X_D} \to \mathcal{O}^*_X \to 1$$

with $u(f) = 1 + \epsilon f$ and this gives the well–known expression for the tangent space to the Picard functor (cf. lecture 1 of C. Voisin [Vo]):

$$\mathrm{TPic}(X) = H^1(X, \mathcal{O}_X)$$

(always Zariski topology now). For $n = 2$ and assuming, for simplicity, that we are in characteristic zero, there is a result of Van der Kallen ([vdK 2]) saying that $\mathrm{Ker}(\mathcal{K}_{2,X_D} \to \mathcal{K}_{2,X}) = \Omega^1_{X/\mathbb{Q}}$, i.e. the sheaf of *absolute* Kähler differentials on X. Hence

$$TCH^2(X) = H^2(X, \Omega^1_{X/\mathbb{Q}}).$$

There is a short exact sequence

$$0 \to \mathcal{O}_X \otimes_k \Omega^1_{k/\mathbb{Q}} \to \Omega^1_{X/\mathbb{Q}} \to \Omega^1_{X/k} \to 0$$

Let now X be a *surface* and take $k = \mathbb{C}$. From the above short exact sequence we get an exact sequence

$$H^1(X, \Omega^1_{X/\mathbb{C}}) \to H^2(X, \mathcal{O}_X) \otimes_{\mathbb{C}} \Omega^1_{\mathbb{C}/\mathbb{Q}} \to TCH^2(X) \to H^2(X, \Omega^1_{X/\mathbb{C}}) \to 0$$

In case $p_g(X) \neq 0$, hence $H^2(X, \mathcal{O}_X) \neq 0$, we see that since $\dim_{\mathbb{C}} \Omega^1_{\mathbb{C}/\mathbb{Q}} = \infty$ we have $\dim_{\mathbb{C}} TCH^2(X) = \infty$ which "explains" (to some extent at least) Mumford's theorem. On the other hand if $p_g(X) = 0$ then we get $TCH^2(X) \xrightarrow{\sim} H^2(X, \Omega^1_{X/\mathbb{C}})$ which is itself isomorphic to the tangent space to the Albanese variety; this points towards Bloch's conjecture. For further discussions, relating the above to the question of "prorepresentability" of the functor, see Chapter 6 of Bloch's book [B] and the papers [Sti 1] and [Sti 2].

Acknowledgement: I like to thank Jan Stienstra for useful discussions and help for 3.12.

Chapter IV. Introduction to the Deligne–Beilinson cohomology.

In this chapter X denotes a *complex analytic manifold.* We intend to give an *introduction* to Deligne–Beilinson cohomology. For a thorough treatment we refer to [EV]. The reader is also advised to look to [BZ] and to the paper [Be]; good introductions are also [So] and [EZ]. We shall use freely hypercohomology for which we refer to [GH], [Bo] and EGA III, page 32. For the definition of the classical cycle map (to be used in the second part of this chapter) we refer to lecture 2 of Mark Green and for the various versions of this map to [DMOS].

IVa. Deligne–Beilinson cohomology.

4.1. Deligne–Beilinson complex.

Let X be a complex analytic manifold. Recall the holomorphic DeRham complex which gives a resolution for $\mathbb{C}$:

$$\Omega_X^\bullet := 0 \to \Omega_X^0 \to \Omega_X^1 \to \Omega_X^2 \to \dots$$

where Ω_X^i are the *holomorphic* differential forms of degree i. Let $A \subset \mathbb{C}$ be a subring (for instance $A = \mathbb{Z}, \mathbb{Q}$ or $\mathbb{R}$; in our case it usually is $\mathbb{Z}$) and let $A(n)$ denote the subgroup $(2\pi i)^n A \subset \mathbb{C}$. Deligne has introduced the complex:

$$A(n)_{\mathcal{D}} : 0 \to A(n) \to \Omega_X^0 \to \dots \to \Omega_X^{n-1} \to 0 \tag{1}$$

where $A(n)$ is in degree 0 and Ω_X^i in degree $(i+1)$.

Definition 1.

$$H_{\mathcal{D}}^k(X, A(n) := \mathbb{H}^k(X, A(n)_{\mathcal{D}}).$$

This is called the *Deligne–Beilinson cohomology* of X with coefficients $A(n)$; the $\mathbb{H}(X, -)$ is the hypercohomology of the complex. More generally if $Y \hookrightarrow X$ is a closed immersion then one defines the Deligne–Beilinson cohomology with support:

Definition 2.

$$H_{Y,\mathcal{D}}^k(X, A(n) := \mathbb{H}_Y^k(X, A(n)_{\mathcal{D}}).$$

Remark: Note that these are *groups* and not, in general, vector spaces.

4.2. General remark about hypercohomology.

Let $\mathcal{A}$ and $\mathcal{B}$ be two abelian categories. Let $K^\bullet$ be a complex in $\mathcal{A}$ (starting, say, from K^0). Let $F : \mathcal{A} \to \mathcal{B}$ be a left exact functor. Recall ([GH], p. 445 or [EGA] III, p.33) that the hypercohomology is computed via a "double complex", for instance via an injective resolution $K^\bullet \to I^{\bullet\bullet}$ (or via a Čech complex, see [GH]) and then

$$\mathbb{H}^i(F(K^\bullet)) := H^i(\mathrm{tot}F(I^{\bullet\bullet}))$$

where $\mathrm{tot}(F(I^{\bullet\bullet})$ is the total complex of the double complex. There are two spectral sequences both abutting to $\mathbb{H}^i$, namely

$$'E_2^{pq} = H^p(R^qF(K^\bullet)) \Longrightarrow \mathbb{H}^{p+q}(F(K^\bullet))$$
$$''E_2^{pq} = R^pF(H^q(K^\bullet)) \Longrightarrow \mathbb{H}^{p+q}(F(K^\bullet)).$$

Now suppose that we have two complexes and a homomorphism $f : K_1^\bullet \to K_2^\bullet$. Recall that f is called a *quasi-isomorphism* (abbreviated q.i.) if $H(f) : H(K_1^\bullet) \to H(K_2^\bullet)$ is an *isomorphism*. In this case it follows from the second spectral sequence that we have the following (see [GH], p. 447):

Corollary. If $f : K_1^\bullet \to K_2^\bullet$ is a *quasi-isomorphism* then $F(f) : \mathbb{H}^i(F(K_1^\bullet)) \to \mathbb{H}^i(F(K_2^\bullet))$ is an *isomorphism*.

4.3. Examples. Take $A = \mathbb{Z}$

4.3.1. $n = 0$ Then $\mathbb{Z}(0)_\mathcal{D}$ is the constant sheaf $\mathbb{Z}$ and hence $H^k_\mathcal{D}(X, \mathbb{Z}(0)) = H^k(X, \mathbb{Z})$.

4.3.2. $n = 1$. In this case there is a quasi-isomorphism

$$\exp : \mathbb{Z}(1)_\mathcal{D} \to \mathcal{O}_X^*[-1]$$

where $\mathcal{O}_X^*$ means the complex consisting of $\mathcal{O}_X^*$ only, and $[-1]$ means: shifted one place to the right. Explicitly we have

$$\begin{array}{ccccccccc} 0 & \to & 2\pi i\mathbb{Z} & \longrightarrow & \mathcal{O}_X & \longrightarrow & 0 & \to \cdots \\ & & \big\downarrow \exp & & \big\downarrow \exp & & \big\downarrow & \\ 1 & \to & 1 & \longrightarrow & \mathcal{O}_X^* & \longrightarrow & 1 & \to \cdots \end{array}$$

Hence, by the above corollary: $H^k_\mathcal{D}(X, \mathbb{Z}(1)) \simeq H^{k-1}(X, \mathcal{O}_X^*)$ (the change to $k-1$ comes from the shift!). In particular $H^2_\mathcal{D}(X, \mathbb{Z}(1)) \simeq H^1(X, \mathcal{O}_X^*) = \mathrm{Pic}(X)$ (Note: the $H^1(X, \mathcal{O}_X^*)$ and $\mathrm{Pic}(X)$ are in the analytic sense, but if X is a smooth projective variety over $\mathbb{C}$ then the algebraic and analytic versions coincide by the GAGA-theorems; see lecture 1 of C. Voisin).

Exercise: Make the isomorphism $H^2_\mathcal{D}(X, \mathbb{Z}(1)) \simeq H^1(X, \mathcal{O}_X^*)$ explicitly by means of Čech-cocycles (cf. [EV], p. 46).

4.3.3. $n = 2$. In this case there is a q.i. of complexes

$$\mathbb{Z}(2)_{\mathcal{D}} \to \{\mathcal{O}_X^* \xrightarrow{d\log} \Omega_X^1\}[-1].$$

Explicitly:

$$\begin{array}{ccccccc}
0 & \to & (2\pi i)^2\mathbb{Z} & \longrightarrow & \mathcal{O}_X & \xrightarrow{d} & \Omega_X^1 & \to 0 \\
 & & \downarrow & & \downarrow \exp(\frac{1}{2\pi i}-) & & \downarrow \frac{1}{2\pi i} & \\
0 & \to & 1 & \longrightarrow & \mathcal{O}_X^* & \xrightarrow[d\log]{} & \Omega_X^1 & \to 0
\end{array}$$

Hence we have $H_{\mathcal{D}}^k(X, \mathbb{Z}(2)) \xrightarrow{\sim} \mathbb{H}^{k-1}(X, \mathcal{O}_X^* \xrightarrow{d\log} \Omega_X^1)$.

4.4. Product structures.

There is a multiplication of complexes

$$\mu : \mathbb{Z}(n)_{\mathcal{D}} \otimes \mathbb{Z}(m)_{\mathcal{D}} \to \mathbb{Z}(n+m)_{\mathcal{D}}$$

defined (mysteriously!) as follows

$$\mu(x \bullet y) = \begin{cases} x \cdot y & \text{if } \deg x = 0 \\ x \wedge dy & \text{if } \deg x > 0 \text{ and } \deg y = m > 0 \\ 0 & \text{otherwise} \end{cases}.$$

This gives a product structure on the Deligne–Beilinson cohomology as follows

$$\cap : H_{\mathcal{D}}^k(X, \mathbb{Z}(n)) \otimes_{\mathbb{Z}} H_{\mathcal{D}}^{k'}(X, \mathbb{Z}(m)) \to H_{\mathcal{D}}^{k+k'}(X, \mathbb{Z}(n+m))$$

For details see [EV]; also see [BZ] p. 98.

4.5. The cone construction from homological algebra.

Recall from homological algebra that if we have a homomorphism $f : V^\bullet \to W^\bullet$ of complexes (in an abelian category) then we can define the cone complex $C_f^\bullet$ as follows: $C_f^m = V^{m+1} \oplus W^m$ and $d(v, w) = (-dv, fv + dw)$. This gives a short exact sequence of complexes

$$0 \to W^\bullet \to C_f^\bullet \to V^\bullet[1] \to 0 \tag{2}$$

which gives a long exact sequence of homology groups

$$(2') \qquad \begin{array}{ccccccc} \cdots \to H^{k-1}(W^\bullet) & \longrightarrow & H^{k-1}(C_f^\bullet) & \longrightarrow & H^{k-1}(V^\bullet[1]) & \longrightarrow & H^k(W^\bullet) \;\cdots \\ & & & & \| & \nearrow_{f} & \\ & & & & H^k(V^\bullet) & & \end{array}$$

4.6. Deligne–Beilinson cohomology via a cone complex.

Consider again the holomorphic DeRham complex

$$\Omega_X^\bullet : 0 \to \Omega_X^0 \to \Omega_X^1 \to \dots$$

and the two subcomplexes

$$\mathbb{Z}(n)^\bullet : 0 \to \mathbb{Z}(n) \to 0 \to \dots$$
$$F^n\Omega_X^\bullet : 0 \to 0 \to 0 \to \dots \to 0 \to \Omega_X^n \to \Omega_X^{n+1} \to \dots$$

with natural maps $\epsilon(n) : \mathbb{Z}(n) \subset \mathbb{C} \to \Omega_X^\bullet$ and $\sigma(n) : F^n\Omega_X^\bullet \to \Omega_X^\bullet$. Now take as complexes $V^\bullet = \mathbb{Z}(n)^\bullet \oplus F^n\Omega_X^\bullet$ and $W^\bullet = \Omega_X^\bullet$ and as homomorphism $f = \epsilon(n) - \sigma(n)$ and construct the cone complex $C_f^\bullet$ as above. There is a natural inclusion of the Deligne–Beilinson complex (1) into this cone (up to a shift):

$$(3) \qquad i : \mathbb{Z}(n)_\mathcal{D} \to C_f^\bullet[-1] = \{Cone\ (\mathbb{Z}(n) \oplus F^n\Omega_X^\bullet \to \Omega_X^\bullet)\}[-1]$$

and one checks easily that this is a *quasi-isomorphism*. Hence we have a *canonical isomorphism*

$$(4) \qquad H_\mathcal{D}^k(X, \mathbb{Z}(n)) \xrightarrow{\sim} \mathbb{H}^{k-1}(X, C_f^\bullet).$$

The short exact sequence (2) of the cone and the fact that $\mathbb{H}^i(X, \Omega_X^\bullet) = H^i(X\mathbb{C})$ (DeRham) give now the following long exact sequence of (hyper) cohomology groups

$$\dots \to H^{k-1}(X, \mathbb{Z}(n) \oplus \mathbb{H}^{k-1}(X, F^n\Omega_X^\bullet) \to H^{k-1}(X, \mathbb{C}) \to$$

$$(5) \qquad \to H_\mathcal{D}^k(X, \mathbb{Z}(n) \to H^k(X, \mathbb{Z}(n)) \oplus \mathbb{H}^k(X, F^n\Omega_X^\bullet) \to H^k(X, \mathbb{C}) \to \dots.$$

4.7. Assume now that X is compact Kähler
then

$$\sigma(n)\mathbb{H}^i(X, F^n\Omega_X^\bullet) \subset H^i(X, \mathbb{C})$$

gives the *Hodge filtration* on $H^i(X,\mathbb{C})$ (see M. Green, lecture 1 [Gre]). Now take $k = 2p$ and $n = p$ then (5) gives the following exact sequence:

$$0 \to H^{2p-1}(X,\mathbb{C})/\{F^pH^{2p-1}(X,\mathbb{C}) + H^{2p-1}(X,\mathbb{Z}(p))\} \to H^{2p}_{\mathcal{D}}(X,\mathbb{Z}(p) \to$$

$$(6)\quad H^{2p}(X,\mathbb{Z}(p)) \cap \{\epsilon(p)^{-1}(F^pH^{2p}(X,\mathbb{C}))\} \to 0.$$

Since $\epsilon(p)$ is (just) multiplication by $(2\pi i)^p$, we have firstly

$$(7)\qquad H^{2p-1}(X,\mathbb{C})/\{F^pH^{2p-1}(X,\mathbb{C}) + H^{2p-1}(X,\mathbb{Z}(p))\} \xrightarrow{\sim} J^p(X)$$

where $J^p(X)$ is the p–th *Griffiths intermediate Jacobian* of X (see M. Green, lecture 2 [Gre]) and secondly

(8)

$$H^{2p}(X,\mathbb{Z}(p_1)) \cap \{\epsilon(p)^{-1}(F^pH^{2p}(X,\mathbb{C}))\} \xrightarrow{\sim} Hdg^p(X) := H^{2p}(X,\mathbb{Z}) \cap \epsilon^{-1}H^{p,p}(X)$$

where $\epsilon : \mathbb{Z} \to \mathbb{C}$ is the natural inclusion, i.e., $Hdg^p(X)$ is the subgroup of the *integral* (p,p)*–classes* (M. Green, lecture 2 [Gre]). Therefore (6) gives, taking into account (7) and (8) the following exact sequence

$$(9)\qquad 0 \to J^p(X) \to H^{2p}_{\mathcal{D}}(X,\mathbb{Z}(p)) \to Hdg^p(X) \to 0$$

which we have seen already in lecture 2 of Mark Green.

IVb. The Deligne cycle map.

In the following X denotes a *smooth, projective* variety defined over $\mathbb{C}$. Let moreover X be irreducible and $\dim X = d$. Let, as in the previous chapters, $\mathcal{Z}^p(X)$ be the group of algebraic cycles on X of codimension p.

4.8. Prerequisites on the "classical" cycle map.

There is a homomorphism

$$(10)\qquad \gamma_{\text{top}} : \mathcal{Z}^p(X) \to \{ImH^{2p}(X,\mathbb{Z}) \cap H^{p,p}(X)\} \subset H^{2p}(X,\mathbb{Q}) \subset H^{2p}(X,\mathbb{C})$$

where Im means the image in $H^{2p}(X,\mathbb{Q})$ or in $H^{2p}(X,\mathbb{C})$). This homomorphism has been constructed in lecture 2 of M. Green via functionals on $H^{2(d-p)}(X,\mathbb{Z})$, namely if $i : Z \hookrightarrow X$ is a (smooth) subvariety of X of codimension p consider then the functional

$$\omega \mapsto \int_Z i^*\omega$$

where ω is a C^∞-differential form of degree $2d - 2p$ on X (in Mark Green's lecture $\gamma_{\text{top}}(Z)$ is denoted by ψ_Z). Sometimes it is more natural to replace $\mathbb{Z}$ by $\mathbb{Z}(p) = (2\pi i)^p\mathbb{Z} \subset \mathbb{C}$, as is done in Deligne–Beilinson cohomology. (For instance in the exponential sequence $\mathbb{Z}(1)$ is more natural than $\mathbb{Z}$). Then one considers the homomorphism

$$(10')\qquad \gamma : \mathcal{Z}^p(X) \to \{ImH^{2p}(X,\mathbb{Z}(p)) \cap H^{p,p}(X)\} \subset H^{2p}(X,\mathbb{C})$$

defined by $\gamma(\eta) = (2\pi i)^p\gamma_{\text{top}}(\eta)$ for $\eta \in \mathcal{Z}^p(X)$.

4.9. Refinements.

For the following one needs two refinements of the above "classical" cycle map, namely a map into *cohomology with support* both in the integral version and in the DeRham (or Hodge) version. Let $\eta \in \mathcal{Z}^p(X)$, let $|\eta|$ be its support and put $U = X - |\eta|$. Take for simplicity the case that $\eta = Z$ with $i : Z \hookrightarrow X$ a (smooth) subvariety of codimension p.

4.9.1. Integral coefficients. From the Gysin sequence (see lecture 5 of M. Green):

$$\ldots H^{2p-1}(U, \mathbb{Z}(p)) \to H^{2p}_{|Z|}(X, \mathbb{Z}(p)) \to H^{2p}(X, \mathbb{Z}(p)) \to \ldots$$

and using $H^{2p}_{|Z|}(X, \mathbb{Z}(p)) \simeq \mathbb{Z}(p)$ we get an element $c_{\mathbb{Z}}(Z) \in H^{2p}_{|Z|}(X, \mathbb{Z}(p))$ mapping to $\gamma(Z) \in Im H^{2p}(X, \mathbb{Z}(p)) \subset H^{2p}(X, \mathbb{C})$; let its image in $H^{2p}(X, \mathbb{Z}(p)$ itself be denoted by $\gamma_{\mathbb{Z}}(Z)$.

Remark. In case Z is singular one has to work via the desingularisation:

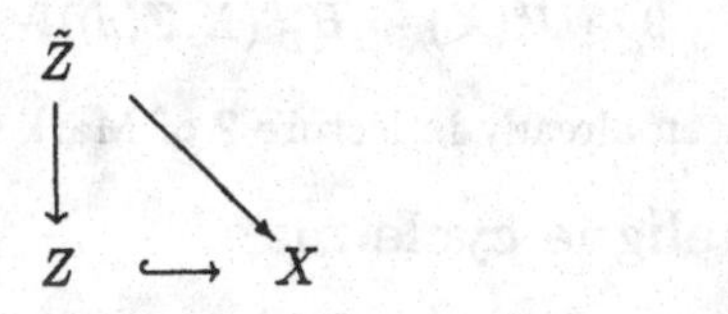

Remark. $\gamma_{\mathbb{Z}}(Z)$ is sometimes called the *Betti class* of Z

4.9.2. DeRham version. It is also possible to construct a class $c_{DR}(Z) \in \mathbb{H}^{2p}_{|Z|}$ $(X, F^p\Omega^\bullet_X)$ mapping under the natural map to a element $\gamma_{DR}(z) \in H^{2p}(X, F^p\Omega^\bullet_X)$, mapping itself to $\gamma(Z) \in H^{p,p}(X) \subset F^pH^{2p}(X, \mathbb{C})$. For this (more subtle) construction we refer to [EZ] and [EV] §6, as well as for the following fact:

4.9.3. Fact: The images of $c_{\mathbb{Z}}(Z)$ and $c_{DR}(Z)$ coincide under the natural maps to $H^{2p}_{|Z|}(X, \mathbb{C})$; let this image be denoted by $c_{\mathbb{C}}(Z)$.

4.9.4. Remark. For the further study of the classical cycle map the reader is advised to look also to [DMOS], p.12-23 and [D] p. 28-29. The compatibility of the different version, both in the classical case and in the refined case, can be reduced to the case of divisors (see [EV], p.79).

4.10. Cycle map in Deligne–Beilinson cohomology.

Always we assume X *smooth, projective* over $\mathbb{C}$, moreover irreducible and of dimension d. Let $Z \in \mathcal{Z}^p(X)$ be a (say smooth) subvariety of X of codimension p.

We use the quasi–isomorphism (3) of 4.6 with $n = p$ and the following piece of the exact sequence (5), but this time for (hyper) cohomology with support on $|Z|$:

$$(11)\quad H^{2p-1}_{|Z|}(X,\mathbb{C}) \longrightarrow H^{2p}_{|Z|,\mathcal{D}}(X,\mathbb{Z}(p)) \longrightarrow H^{2p}_{|Z|}(X,\mathbb{Z}(p)) \oplus \mathbb{H}^{2p}_{|Z|}(X,F^p\Omega^\bullet_X) \xrightarrow{\epsilon(p)-\sigma(p)} H^{2p}_{|Z|}(X,\mathbb{C})$$

Now we use the following facts:
i. $H^{2p-1}_{|Z|}(X,\mathbb{C}) = 0$. This follows from local cohomology theory (see [EV], p. 77 and 83),
ii. the existence of the classes $c_{\mathbb{Z}}(Z) \in H^{2p}_{|Z|}(X,\mathbb{Z}(p))$ and $c_{DR}(Z) \in \mathbb{H}^{2p}_{|Z|}(X,F^p\Omega^\bullet_X)$ which coincide under the natural maps induced by $\epsilon(p) : \mathbb{Z}(p) \to \Omega^\bullet_X$ and $\sigma(p) : F^p\Omega^\bullet_X \to \Omega^\bullet_X$ (see 4.9).
From i) and ii) we get a *unique class*

$$(12)\qquad c_{\mathcal{D}}(Z) \in H^{2p}_{|Z|,\mathcal{D}}(X,\mathbb{Z}(p))$$

mapping to $c_{\mathbb{Z}}(Z)$ and $c_{DR}(Z)$. Moreover from the natural maps from (hyper) cohomology groups with support on $|Z|$ to those without support we get a *unique* class

$$(13)\qquad \gamma_{\mathcal{D}}(Z) \in H^{2p}_{\mathcal{D}}(X,\mathbb{Z}(p))$$

mapping to $\gamma_{\mathbb{Z}}(Z)$ and $\gamma_{DR}(Z)$ (see 4.9.1. and 4.9.2). Extending $\gamma_{\mathcal{D}}$ in the obvious way by linearity we finally get:

Theorem 1. *Let X be smooth, projective and irreducible over $\mathbb{C}$. Then there exists a homomorphism (the Deligne cycle map):*

$$(14)\qquad \gamma^p_{\mathcal{D}} : \mathcal{Z}^p(X) \to H^{2p}_{\mathcal{D}}(X,\mathbb{Z}(p))$$

which, for $\eta \in \mathcal{Z}^p(X)$, maps both to the Betti class $\gamma^p_{\mathbb{Z}}(\eta) \in H^{2p}(X,\mathbb{Z}(p))$ and to the DeRham (or Hodge) class $\gamma^p_{DR}(\eta) \in \mathbb{H}^{2p}(X,F^p\Omega^\bullet_X)$ by the maps from the exact sequence (11).

4.11. Properties of the Deligne cycle map.

We list now some properties of $\gamma_{\mathcal{D}}$; for the proofs we refer to [EV], §7.
a. $\gamma_{\mathcal{D}}$ behaves good for the product structure: $\gamma_{\mathcal{D}}(\eta) \cap \gamma_{\mathcal{D}}(\eta') = \gamma_{\mathcal{D}}(\eta \cdot \eta')$ if the cycles intersect properly (see 4.4 for the cup product).
b. $\gamma_{\mathcal{D}}$ behaves functorially: if $f : X' \to X$ is a proper morphism and $\eta \in \mathcal{Z}^p(X)$ is such that $f^*(\eta)$ is defined, then $f^*(\gamma_{\mathcal{D}}(\eta)) = \gamma_{\mathcal{D}}(f^*(\eta))$. (Remark: there is a homomorphism f^* for the Deligne–Beilinson cohomology, see [EV], p 56)
c. $\gamma_{\mathcal{D}}$ respects rational equivalence: if $\eta_1 \sim \eta_2$ are rational equivalent then $\gamma_{\mathcal{D}}(\eta_1) = \gamma_{\mathcal{D}}(\eta_2)$. The proof (see [EV], p. 85) goes, roughly, as follows: from the rational equivalence we have (Chap.I, 1.4.1) a cycle $\xi \in \mathcal{Z}^p(\mathbb{P}_1 \times X)$ and two

points $a, b \in \mathbb{P}_1$ such that $\xi(a) = \eta_1$ and $\xi(b) = \eta_2$; now take an isomorphism ϕ of $\mathbb{P}_1$ interchanging the points a and b then $\phi \times id_X$ operates as *the identity on the cohomology* of $\mathbb{P}_1 \times X$ and this gives easily the result.

4.12. Combining all these facts one finally obtains the following result:

Theorem 2. *Let X be smooth, projective, irreducible over $\mathbb{C}$. There exists homomorphisms*

$$\gamma_{\mathcal{D}}^p : CH^p(X) \to H_{\mathcal{D}}^{2p}(X, \mathbb{Z}(p))$$

such that

$$\gamma_{\mathcal{D}} = \oplus_{p \geq 0} \gamma_{\mathcal{D}}^p : CH(X) \to H_{\mathcal{D}}(X) := \oplus_{p \geq 0} H_{\mathcal{D}}^{2p}(X, \mathbb{Z}(p))$$

is a ring homomorphism which behaves functorial with respect to morphisms $f : X' \to X$.

4.13. Deligne cycle map and Abel–Jacobi map.

Using the exact sequence (9) and the Deligne cycle map $\gamma_{\mathcal{D}}^p$ we have now the following commutative diagram and the existence of a homomorphism λ^p as indicated:

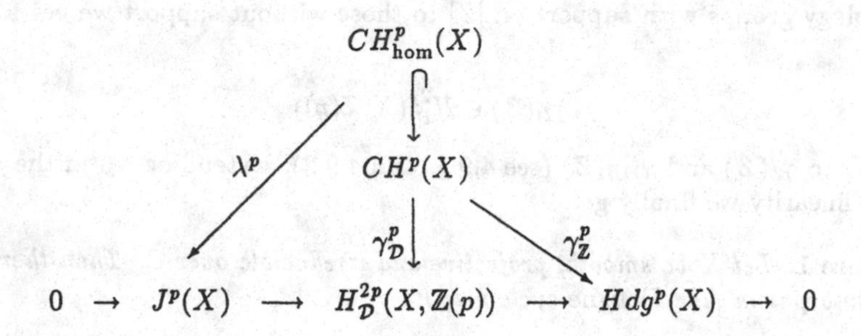

where $CH^p_{\mathrm{hom}}(X) \subset CH^p(X)$ are the cycle classes homological equivalent to zero (Chap. I, 1.4.4).

Theorem 3. *Let X be smooth, projective irreducible over $\mathbb{C}$. there exists a homomorphism*

$$\lambda^p : CH^p_{\mathrm{hom}}(X) \to J^p(X)$$

induced by the Deligne map as indicated. Moreover λ^p coincides with the Abel–Jacobi map AJ (see lecture 2, Mark Green).

The existence of λ^p follows immediately from the above constructions and from the definition of CH^p_{hom}. The fact that λ^p coincide with the Abel–Jacobi map $(AJ)^p$ is very subtle and depends, of course, upon the definition of AJ for which there are the two versions:

a) the classical one via currents (cf. lecture 2, M. Green [Gre]); for the proof that λ coincide with AJ see then [EZ]
b) via extension classes and mixed Hodge structures; for this version see [EV].

Chapter V. The Hodge Conjecture.

In this chapter we consider smooth, projective complex varieties. We discuss the Hodge conjecture, first the "usual"(p,p)–conjecture and next also the general Hodge conjecture as corrected by Grothendieck. It is *not* our intention to give here a survey for all known cases, for this we refer to the excellent paper by Shioda [Sh 3] (of course, in the meantime there are some further examples). What we have in mind is to give some idea of the types of varieties, all rather special, for which the (p,p)–conjecture is known to be true. Concerning the Grothendieck–Hodge conjecture we mention only some examples.

For the origin of the conjecture see first [L] for the Lefschetz $(1,1)$–theorem, next [Ho] for the conjectures itself and [Gr 4] for the corrected formulation of the general conjecture. Except for the above mentioned paper of Shioda we recommend also the paper by Steenbrink [Ste] and the lectures by Lewis [Lew].

Va. The Hodge (p,p)–conjecture.

5.1. Statement of the conjecture.

Let X be a smooth, projective variety defined over $\mathbb{C}$, let moreover X be irreducible and of dimension d. Let $\mathcal{Z}^p(X)$ be the group of algebraic cycles of codimension p on X. In lecture 2 of M. Green [Gre] and in Chapter IV, section 4.8, we have seen that there is a cycle map

$$(1) \qquad \gamma^p_{\mathrm{top}} : \mathcal{Z}^p(X) \otimes \mathbb{Q} \to H^{2p}(X,\mathbb{Q}) \cap H^{p,p}(X)$$

Hodge (p,p)–conjecture: The above map is onto.

The above conjecture is also often denoted by $HC(X,p)$. Let $A^p(X) = Im(\gamma^p_{\mathrm{top}})$ and let $B^p(X)$ be the group on the right hand side; put $A(X) = \oplus_{p\geq 0} A^p(X)$ and $B(X) = \oplus_{p\geq 0} B^p(X)$, $B(X)$ is called the *Hodge ring* of X. Furthermore let $D(X) \subset B(X)$ be the subring obtained by intersection of the divisors. Finally let $A^p_{\mathbb{C}}(X) = A^p(X) \otimes \mathbb{C}$, etc. be the corresponding $\mathbb{C}$–vextorspaces.

5.2. Remarks.

a. The conjecture is true for divisors, even with $\mathbb{Z}$–coefficients, and is known there as the Lefschetz theorem on $(1,1)$–classes (see C. Voisin, lecture 1 [Vo] and [GH], p. 163).
b. The conjecture with $\mathbb{Z}$–coefficients is not true for $p > 1$ [AH].
c. Due to the strong Lefschetz–theorem $L^{d-2p} : H^{2p}(X,\mathbb{Q}) \xrightarrow{\sim} H^{2d-2p}(X,\mathbb{Q})$ (see C. Voisin, lecture 2 [Vo]) and since L^{d-2p} transforms $H^{p,p}(X)$ into $H^{d-p,d-p}(X)$, it

follows that $HC(X,p)$ implies $HC(X,d-p)$. In particular it follows that $HC(X,d-1)$ is true.
d. By the above the Hodge (p,p) conjectures are true for surfaces and threefolds.
e. For varieties where all of the cohomology comes from algebraic cycles (projective spaces, quadrics, Grassmannians, flag varieties) the Hodge (p,p)–conjecture is of course true.

5.3. Further known cases.

The remaining cases where the Hodge (p,p)–conjecture is known to be true fall, roughly speaking, into three types:
a. uniruled fourfolds,
b. large classes of Fermat varieties,
c. some classes of abelian varieties.

5.3.1. Uniruled fourfolds.

Recall that a variety X_d is called *uniruled* if there exists a dominant *rational map* $f: X' = T_{d-1} \times \mathbb{P}_1 \to X$ (note that f is generically finite).

Theorem. ([CM 1]).*The Hodge* $(2,2)$*–conjecture is true for uniruled fourfolds.*

The proof is simple and goes (roughly) as follows. By Hironaka there is a commutative diagram as as follows:

$$
\begin{array}{ccccc}
 & & X'' = B_Y(X') & & \\
 & \swarrow g & & h \searrow & \\
Y & \hookrightarrow & X' & \overset{f}{\dashrightarrow} & X
\end{array}
$$

where $g: X'' \to X'$ consists of successive blowing–ups along smooth centers Y and $h: X'' \to X$ is a morphism which is generically finite. Now one easily sees (for details we refer to [CM 1]) that $HC(X''_d, p) \Longrightarrow HC(X_d, p)$ (note: this argument works for any dimension d), that $HC(X'_4, 2) \Longrightarrow HC(X''_4, 2)$ (this works only for dimension ≤ 4) and finally that $HC(T_3 \times \mathbb{P}_1, 2)$ is true. This proves the theorem.

Applications. The above theorem gives a lot of concrete examples, some of which were proved earlier by more involved arguments. We use the following notations: $V(m) \subset \mathbb{P}_N$ is a hypersurface of degree m in projective N–space, $V(m_1, m_2) \subset \mathbb{P}_N$ a complete intersection of hypersurfaces of degrees m_1 and m_2 etc. Then the Hodge $(2,2)$–conjecture is true for smooth $V(3), V(4)$ and $V(5)$ in $\mathbb{P}_5$ (the case of a cubic fourfold $V(3)$ was proved earlier by Griffiths and Zücker using the theory of normal functions) but also, for instance, for $V(2,2), V(2,3), V(2,4)$ and $V(3,3)$ in $\mathbb{P}_6$. For this, and for a more complete list, see [CM 1] and [CM 2].

5.3.2. Classes of Fermat varieties.

First remark that if we have a hypersurface $X = V(m) \subset \mathbb{P}_{d+1}$ then it follows from the weak Lefschetz theorem (i.e., the theorem relating the cohomology of a variety to that of its hyperplane section; see lecture 2 of C. Voisin [Vo]) that only the $H^d(X)$ is interesting. Therefore we have, in that case, only to look to $HC(V(m), \frac{d}{2})$ with d even and $p = \frac{d}{2}$.

Now consider the *Fermat hypersurface* $F_d(m) \subset \mathbb{P}_{d+1}$ defined by the equation

$$T_0^m + T_1^m + \ldots + T_{d+1}^m = 0.$$

From the point of view of the Hodge conjecture these varieties have been studied by Ran [Ra] and extensively by Shioda and his students (Katsura, Aoki and others) [Sh 1], [Sh 2], [Sh 3]. The Hodge (p,p)–conjecture is, *at least*, true in the following cases (where $p = \frac{d}{2}$) (see [Sh 3]):

a. $m =$ prime, any p (Ran, Shioda),
b. $m \leq 20$, any p (Shioda),
c. $m = 2$ (prime), any p (Shioda, Miyawaki),
d. $m \leq 25, p = 2$ (i.e., $\dim d = 4$) (Shioda and students).

For the proof of these facts we must refer to the original papers (especially to [SK] [Sh 1] and [Sh 4]. In the proofs the following ingredients enter:

i. the operation of the group $G_m^1 = \mu_m \times \ldots \times \mu_m/\Delta$ (with $(d+2)$–factors and Δ is the diagonal) on $F_d(m)$ and especially the operation of this group on the primitive cohomology $H_0^d(F_d(m))$,
ii. the splitting of this cohomology group into the irreducible representations of this group,
iii. the description of the Hodge cycles in terms of these representations,
iv. (a key step!) the inductive building of the Fermat variety $F_{r+s}(m)$ out of the Fermat varieties $F_r(m)$ and $F_s(m)$ by means of blowing ups along $F_{r-1}(m)$ and $F_{s-1}(m)$ and operation of the group μ_m (see [Sh 1]).

5.3.3. Special classes of abelian varieties.

Returning to the notations of 5.1 we have for any (smooth, projective) variety

$$D^p(X) \subseteq A^p(X) \subseteq B^p(X)$$

Now let $X = A$ be an abelian variety. There are two cases:

Case I: $D^p(A) = B^p(A)$. Then clearly $HC(A,p)$ is true. This happens in particular for:

a. generic abelian varieties (Mattuck),
b. $A = E \times \ldots \times E$, E an elliptic curve (Tate),
c. abelian varieties for which the endomorphism algebra satisfies "certain" conditions. Such varieties have been studied by Tankeev, Ribet, Hazama and others. For instance Tankeev has shown that *simple* abelian varieties of *prime dimension* belong to this class.

Case II.: $D^p(A) \neq B^p(A)$. There are two types for which one knows the Hodge–(p,p)–conjecture, namely:

a. "certain" abelian varieties of *dimension 4* which are of so-called "*Weil-type*" ([Sc], [vG]). For this see the nice lecture of Van Geemen.

b. "certain" abelian varieties which are of "Fermat-type". We don't give the precise definition because this is very technical, but roughly speaking A is of Fermat type if A is isogenous to a product of abelian varieties $\prod_j B_j$, with B_j up to isogeny an "admissible" factor of the Jacobian variety $J(F_1(m))$. For instance if C_n is the hyperelliptic curve $y^2 = x^n - 1$ then $J(C_n)$ is of Fermat type. Shioda ([Sh 2]) has shown that for some of such $J(C_n)$ the Hodge-(p,p) conjecture is true.

Examples.

α. $J(C_9)$, this is a variety of dimension 4, Hodge-$(2,2)$ is true, but $D_2 \neq B_2$.

β. $J(C_{15})$, this is a variety of dimension 7, the Hodge-(p,p) are true for all p but again $D \neq B$.

Vb. The general Hodge Conjecture (as corrected by Grothendieck).

5.4. Hodge structure and sub-Hodge structure.

Let us recall (lecture 1 of M. Green [Gre]) that a $\mathbb{Q}$-*Hodge structure* (short:HS) of *weight* n consists of a $\mathbb{Q}$-vector space V such that on $V_{\mathbb{C}} = V \otimes_{\mathbb{Q}} \mathbb{C}$ there is a decomposition $V_{\mathbb{C}} = \oplus_{p+q=n} V^{pq}$, with V^{pq} a $\mathbb{C}$-vector space and such that under complex conjugation we have $V^{pq} = \overline{V^{pq}}$. The *level* of V is $\max\{|p-q|; V^{pq} \neq 0\}$. A $\mathbb{Q}$-*sub-Hodge structure* of V consists of a $\mathbb{Q}$-subvector space W of V such that for $W_{\mathbb{C}} = W \otimes_{\mathbb{Q}} \mathbb{C}$ we have $W_{\mathbb{C}} = \oplus_{p,q}(W_{\mathbb{C}} \cap V^{pq})$.

If X is a smooth, irreducible projective variety over $\mathbb{C}$ then for every n the $H^n(X,\mathbb{C})$ carries a $\mathbb{Q}$-Hodge structure. Moreover if $f: Y \to X$ is a morphism of such varieties, with $\dim X - \dim Y = p$, then $f_*: H^{n-2p}(Y,\mathbb{C}) \to H^n(X,\mathbb{C})$ is a *morphism of Hodge structures*, i.e., it respects the Hodge structures and in particular $Im(f_*)$ gives a $\mathbb{Q}$-sub-Hodge structure of $H^n(X,\mathbb{C})$. Furthermore $f_*(H^{r,s}(X)) \subset H^{r+p,s+p}(X)$ and hence $Im(f_*) \subset H^n(X,\mathbb{C})$ has level $\leq n-p$. Let, as usual, $F^pV = \oplus_{p' \geq p} V^{p'q}$ be the *Hodge filtration* then we see that $Im(f_*) \subset F^pH^n(X,\mathbb{C})$.

5.5. Lemma. *Let $i: Z \hookrightarrow X$ be a subvariety of codimension p, put $j = j(Z): X \backslash Z \hookrightarrow X$ for the inclusion of the complement. Consider*

$$j^*: H^n(X,\mathbb{Q}) \to H^n(X \backslash Z, \mathbb{Q}).$$

Then $\mathrm{Ker}(j^*)$ *is a $\mathbb{Q}$-sub-Hodge structure of level $\leq n-2p$ (i.e., the complexification of* $\mathrm{Ker}(j^*_{\mathbb{C}})$ *is contained in $F^pH^n(X,\mathbb{C})$).*

"Proof"

In case Z is also itself smooth then this follows immediately from the remarks above in 5.4 and from the exactness of the so-called "Gysin-sequence" (lecture 5 of M. Green [Gre]):

$$H^{n-2p}(Z,\mathbb{Q}) \xrightarrow{i_*} H^n(X,\mathbb{Q}) \xrightarrow{j^*} H^n(X \backslash Z, \mathbb{Q})$$

In case Z has singularities, desingularize $\pi : \tilde{Z} \to Z$, consider $\tilde{i} = i \bullet \pi : \tilde{Z} \to X$, then there is a similar exact sequence with $\tilde{Z}$ instead of Z (see [Ste], p. 166).

5.6. Coniveau filtration.

Grothendieck [Gr 4] introduced the following filtration (coniveau filtration) on $H^n(X,\mathbb{Q})$:

$$(2) \qquad F'^p H^n(X,\mathbb{Q}) := \bigcup_{\substack{Z \hookrightarrow X \\ \mathrm{codim}\, p}} \mathrm{Ker} j(Z)^* \subset H^n(X,\mathbb{Q})$$

From the above we have, firstly, that F'^p is a $\mathbb{Q}$–sub–Hodge structure of $H^n(X.\mathbb{Q})$ and, secondly, we have an inclusion:

$$(3) \qquad F'^p H^n(X,\mathbb{Q}) \subset F^p H^n(X,\mathbb{C}) \cap H^n(X,\mathbb{Q}).$$

Now the **general (corrected) Hodge–Conjecture** or also called now the **Grothendieck-Hodge Conjecture** is the following ([Gr 3]):

$GHC(X,n,p)$: $F'^p H^n(X,\mathbb{Q})$ is the largest sub–Hodge–structure of the right hand side of (3).

5.7. Remarks.

a. The original general conjecture made by Hodge [Ho] was that in (3) the left hand side (LHS) equals the right hand side (RHS). However Grothendieck [Gr 4] pointed out that this could not always be true. It would imply in particular that the dimension of the RHS would be even because the LHS, having a Hodge structure, has even dimension. Now if we take $X = E^3$, with E on elliptic curve $E = \mathbb{C}/\mathbb{Z} + \tau\mathbb{Z}$ with $\tau^3 = 2$ (and of course $Im(\tau) \neq 0$) then one can see rather easily that the dimension of the RHS is 17 (see [Lew], 7.15).
b. It is easy to see that $GHC(X,n,p)$ is equivalent with the following statement ([Ste]):
Every $\mathbb{Q}$–sub–Hodge structure W of $H^n(X,\mathbb{Q})$ with level $\leq n - 2p$ is contained in $\mathrm{Ker}\{H^n(X,\mathbb{Q}) \to H^n(X \backslash Z,\mathbb{Q})\}$ for some subvariety Z of X of codimension p.
c. The Hodge–(p,p) conjecture for X is the statement $GHC(X,2p,p)$ which in that case precisely means that $LHS = RHS$ in (3).

5.8. Some known cases.

We mention three types of examples.

5.8.1. Many Fano threefolds. Recall that a threefold X is a Fano threefold if $-K_X$ is ample, where K_X is the canonical class. For such threefolds $H^{3,0}(X) = 0$ and for many of such Fano threefolds it can be shown that the $H^3(X)$ can be obtained via the difference of pairs of lines lying on X ([Tj]). Concrete examples: $GH(X,3,1)$ is true for $X = V(3)$ or $V(4)$ in $\mathbb{P}_4$, $X = V(3,2)$ in $\mathbb{P}_5$ and $X = V(2,2,2)$ in $\mathbb{P}_6$.

5.8.2. Cases via the so–called "induction–principle" of Grothendieck. In [Gr 4] such a principle is indicated, it has been worked out in detail in [Ste]. It covers for instance the following cases: $GHC(V_3(3), 3, 1), GHC(V_4(3), 4, 1)$ and $GHC(V_5(3), 6, 2)$

5.8.3. Bardelli's example [Ba]. An extremely interesting example has been constructed by Bardelli, namely a special threefold $V(2,2,2,2)$ in $\mathbb{P}_7$. Firstly for every such smooth $X = V(2,2,2,2)$ in $\mathbb{P}_7$ one has for the Betti number $b_3(X) = 132$ and since $K_X = 0$ one has $h^{30} = 1$ contrary to the examples in 5.8.1. Now Bardelli takes $X = \mathbb{Q}_0 \cap \mathbb{Q}_1 \cap \mathbb{Q}_2 \cap \mathbb{Q}_3$ in $\mathbb{P}_7$ with special quadrics $\mathbb{Q}_i$ namely, if $(x_0, x_1, x_2, x_3, y_0, y_1, y_2, y_3)$ are the projective coordinates, he takes $\mathbb{Q}_i(x, y) = \mathbb{Q}_i'(x) + \mathbb{Q}_i''(y)$ $(i = o, 1, 2, 3)$. Hence X has an involution $\sigma : X \hookrightarrow X$ obtained by taking $\sigma(x_j) = x_j$, $\sigma(y_j) = -y_j$ $(j = 0, 1, 2, 3)$. This gives a splitting in eigenspaces $H^3(X) = H^3(X)^+ + H^3(X)^-$. In this ways it turns out that the intermediate Jacobian $J(X) = H^{30}(X) + H^{21}(X)/H^3(X, \mathbb{Z})$ contains an *abelian variety* $J(X)^- = H^{21}(X)/H^{21}(X) \cap H^3(X, \mathbb{Z})$ and in fact $\dim J(X)^- = 32$. The tangent space to $J(X)^-$ gives (via complex conjugation) a sub–Hodge structure in $H^{12}(X) + H^{21}(X)$. Now by suitable monodromy theorems and nice geometry on the quadrics Bardelli proves:

Theorem (Bardelli)

i) $J(X)^-$ *is the largest abelian variety in* $J(X)$,
ii) $J(X)^-$ *is parametrized by curves on* X.

This means firstly that the above mentioned Hodge structure is the largest one in $H^{12} + H^{21}$ *and secondly that it is* $F^1H^3(X, \mathbb{Q})$, *i.e., that* $GHC(X, 3, 1)$ *is true.*

5.8.4. Remark. Rossi, a student of Bardelli, has obtained examples of threefolds of general type for which the generalized Hodge conjecture is true (private communication).

Chapter VI. Applications of the theorem of Merkurjev and Suslin to the theory of algebraic cycles of codimension two.

In this chapter let X be a smooth projective variety defined over an algebraically closed field of arbitrary characteristic. We discuss applications of a theorem of Merkurjev and Suslin from algebraic K–theory ([MS]) to the theory of algebraic cycles. More precisely this theorem has applications for cycles of codimension two which are algebraically equivalent to zero.

The chapter is organised as follows. In the first part we discuss the notion of regular homomorphisms of Chow groups into abelian varieties, a concept introduced by Samuel and we state for such maps a result of Hiroshi Saito. Next we formulate the theorem of Merkurjev-Suslin, for this we need the rudiments of étale cohomology. In the third section we discuss another important tool namely the Bloch–Ogus theory. Finally in the last section we combine all these results in order to obtain the applications to algebraic cycles. For a more detailed discussion we refer to our lectures in Sitges in 1983 ([Mur 1]).

We have here discussed only the case of an algebraically closed groundfield. For the interesting applications of the Merkurjev–Suslin theorem to torsion algebraic cycles on varieties over special fields (finite fields, local fields, number fields) we refer to the 1991 CIME lectures by Colliot–Thélène ([CT]).

VIa. Cycles algebraically equivalent to zero and regular homomorphisms into abelian varieties.

6.1. Let X be a smooth, irreducible, projective variety defined over a field k algebraically closed and of arbitrary characteristic. Let $d = \dim X$. Let $0 \leq i \leq d$. Consider $CH^i_{\text{alg}}(X) = Z^i_{\text{alg}}(X)/Z^i_{\text{rat}}(X)$ (see Chapt. I, 1.4.2), i.e. the cycle classes algebraically equivalent to zero.

6.2. Algebraic families of cycle classes.

Let (T, Z) be a couple consisting of a smooth, quasi-projective variety and $Z \in CH^i(T \times X)$. For $t \in T$, put, as usual, $Z(t) = pr_X\{Z \cdot (t \times X)\}$, then $Z(t)$ is called an *algebraic family of cycle classes.* Fix a point $t_0 \in T$, then we have a *map*

$$f_Z : T \to CH^i_{\text{alg}}(X) \tag{1}$$

defined by $f_Z(t) = Z(t) - Z(t_0)$

6.3. Regular homomorphisms into abelian varieties.

Let A be an abelian variety.

Definition. (Samuel). A *homomorphism* $\Phi : CH^i_{\text{alg}}(X) \to A$ is called *regular* if for all algebraic families (T, Z) as above the composition $\Phi \cdot f_Z : T \to CH^i_{\text{alg}}(X) \to A$ is a *morphism* of algebraic varieties.

6.4. Lemma. *Given a regular homomorphisms $\Phi : CH^i_{\text{alg}}(X) \to A$. Then we have:*

1. $Im(\Phi) \subset A$ is an abelian subvariety,
2. $\exists$ algebraic family (B, Z) with B an abelian variety, such that $\Phi \cdot f_Z : B \to Im(\Phi)$ is an isogeny.
3. Let Φ be surjective, then $\exists$ algebraic family (A, Y) and an integer n such that $\Phi \cdot f_Y = n \cdot id_A$.

The proof is standard, we omit it here (see [Mur 1], p. 224).

Remark: It follows from the lemma that we can restrict our attention to surjective regular homomorphisms.

6.5. Definition. (Samuel). Let $\alpha \in CH^i_{\text{alg}}(X)$. Then α is *abelian equivalent to zero* if $\alpha \in \text{Ker}(\Phi)$ for *all regular* homomorphisms $\Phi : CH^i_{\text{alg}}(X) \to A$.

Clearly this defines a subgroup $CH^i_{ab}(X) \subset CH^i_{\text{alg}}(X)$.

6.6. Two natural questions.

I. Does there exists a *universal* regular homomorphism?

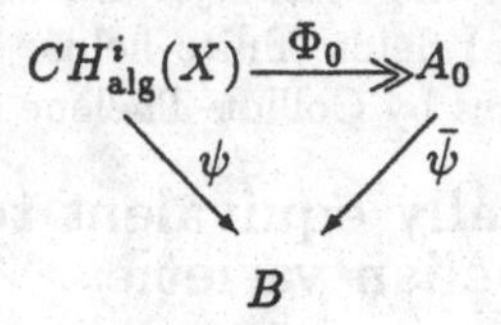

I.e., does there exists such a couple (A_0, Φ_0) with $\Phi_0 : CH^i_{alg}(X) \to A_0$ a regular (and clearly surjective) homomorphism such that every other regular homomorphism factors as indicated. (In fact such a factorization is automatically unique, see [Mur 1] p.225). Clearly if such a couple (A_0, Φ_0) exists it is unique up to a (unique) isomorphism. Sometimes (A_0, Φ_0) is called the "*algebraic representative*" for $CH^i_{alg}(X)$ (Beauville); this should not be confused with the concept of representability of functors which is a much stronger notion. We shall write $A_0 = Ab^i(X)$.

II. In case $k = \mathbb{C}$ we have the Abel–Jacobi map (see lecture 2 of M.Green [Gre]):

$$AJ : CH^i_{hom}(X) \to J^i(X),$$

where $J^i(X)$ is the Griffiths intermediate Jacobian which is only a complex torus in general. However consider now the restrictions of AJ to $CH^i_{alg}(X)$:

$$AJ : CH^i_{alg}(X) \to J^i_a(X) \subset J^i(X).$$

Denote the image of this restriction by $J^i_a(X)$, then it is well–known [Li], p. 826) that $J^i_a(X)$ is an abelian variety and AJ is regular.
Question: Is $(J^i_a(X), AJ)$ universal for $CH^i_{alg}(X)$?

6.7. Examples.

6.7.1. Case $i = 1$. The answer is yes, both for question I and II and $Ab^1(X) = \mathrm{Pic}^0_{red}(X)$, the *Picard variety*; the regular homomorphism comes from the inverse of the Poincaré divisor $G_X : CH^1_{alg}(X) \to \mathrm{Pic}^0(X)$.

6.7.2. Case $i = d = \dim X$. The answer is yes, both for I and II, and clearly $Ab^d(X) = \mathrm{Alb}(X)$, the *Albanese variety*, the regular homomorphism comes from the albanese map.

6.8. Case $i = 2$ ([Mur 1].

Theorem I. ($k = \bar{k}$, arbitrary char.). *For $i = 2$ there exists a universal regular homomorphism*

$$\Phi_0 : CH^2_{alg}(X) \to Ab^2(X).$$

Moreover $\dim Ab^2(X) \leq \frac{1}{2} b_3(X)$ *where* $b_3(X) := \dim H^3_{et}(X, \mathbb{Q}_\ell)$ *with* $\ell \neq p =$ char(k).

Theorem II. ($k = \mathbb{C}$) *For $i = 2$ the Abel–Jacobi map*

$$AJ : CH^2_{\text{alg}}(X) \to J^2_a(X)$$

is the universal regular homomorphism.

6.9. Remarks.

a. For $i = 2$ the proof has the following three ingredients:

i. a result of Hiroshi Saito (algebraic geometry; [SaH],
ii. a theorem of Merkurjev–Suslin (algebraic K–theory; [MS]),
iii. theory of Bloch–Ogus (borderline algebraic K–theory and algebraic geometry [BO]).
b. As far as I know the problem to *both* questions is open for $i \neq 0, 1, 2$ and d.

6.10. Proposition. (Hiroshi Saito). *Let $X = X_d$ be a smooth, irreducible, projective variety defined over an algebraically closed field of arbitrary characteristic. Let $0 \leq i \leq d$. Then the following conditions are equivalent.*

a. $\exists$ universal regular homomorphism $\Phi_0 : CH^i_{\text{alg}}(X) \to A_0$.
b. $\exists$ constant $c(X)$ depending only on X and i such that for every regular surjective homomorphism $\Phi : CH^i_{\text{alg}}(X) \to A$ we have $\dim A \leq c(V)$.

For the proof we refer to [Mur 1] or to the original paper of H. Saito [SaH] (Saito proves in fact a more general result).

VIb. The Merkurjev–Suslin theorem.

6.11. Auxiliary facts from étale topology.

In this part X is a *noetherian scheme.* We will have to consider on X both the *Zariski toplogy* and the *étale topology.* A *topology* on X or a "*site*" is given by giving a category C (to be seen as the system of "open sets") and in this category a family of coverings $\text{Cov}(C)$ in C (cf. [M]). In our case, let X be a noetherian scheme and consider:

a. The *Zariski site* X_{Zar} (the Zariski topology). Here the category C has as objects the usual Zariski open subsets $U \subset X$ and as morphisms the usual inclusions and $\text{Cov}(X_{\text{Zar}})$ are the usual coverings, i.e., families $\{f_i : U_i \to U; \cup f_i(U_i) = U\}_{i \in I}$.
b. The *étale site* $X_{\text{ét}}$ (the étale topology). Here the category consists of X–schemes $\varphi : U \to X$ with φ étale (recall:étale = flat + non–ramified and locally of finite type) and morphisms are morphism $f : U_1 \to U_2$ in this category with f étale. Finally $\text{Cov}(X_{\text{ét}})$ consists of families in this category as follows:

$$\{f_i : U_i \to U; \cup f_i(U_i) = U\}_{i \in I}.$$

In a site C we have the notion of sheaf and of cohomology theory.

A presheaf is a contravariant functor $F : \mathcal{C} \to$ (Sets) (resp. (Groups) or (Abelian groups)) and this presheaf is a sheaf if for every $\{f_i : U_i \to U\}_{i \in I} \in \mathrm{Cov}(\mathcal{C})$ the following sequence is exact

$$F(U) \underset{f_i}{\longrightarrow} \prod F(U_i) \overset{p_1}{\underset{p_2}{\rightrightarrows}} \prod_{i,j} F(U_i \times_U U_j)$$

(i.e. the first arrow is injective and its image equals the set of coincidences of p_1 and p_2).

6.12. Now on the étale site we we consider in particular the sheaves $\mathcal{G}_m$ and μ_n defined as follows: for $\varphi : U \to X$ étale, put

$$\Gamma(U, \mathcal{G}_m) = \Gamma(U, \mathcal{O}_U^*)$$
$$\Gamma(U, \mu_n) = \{\xi \in \Gamma(U, \mathcal{O}_U); \xi^n = 1\}.$$

Now "recall" the following facts:

Theorem. (Kummer sequence).

Let X be over a field $k, p = \mathrm{char}(k)$ and $(p, n) = 1$. Then in $X_{\text{ét}}$ the following sequence is exact:

$$1 \to \mu_n \to \mathcal{G}_m \overset{n}{\longrightarrow} \mathcal{G}_m \to 1$$
$$\xi \longmapsto \xi^n$$

Theorem. (Hilbert 90)

$$H^i_{\text{ét}}(X, \mathcal{G}_m) = \mathrm{Pic}(X)$$

Corollary. *The following sequence is exact in $X_{\text{ét}}$ if $(p, n) = 1$:*

$$\ldots \to \Gamma(X, \mathcal{O}_X^*) \overset{\cdot n}{\longrightarrow} \Gamma(X, \mathcal{O}_X^*) \to H^1_{\text{ét}}(X, \mu_n) \to \mathrm{Pic}(X) \overset{n}{\longrightarrow} \mathrm{Pic}(X) \to \ldots$$

Now let $X = \mathrm{Spec}(F)$ with F a field. Write $H^i_{\text{ét}}(F, -)$ instead of $H^i_{\text{ét}}(\mathrm{Spec} F, -)$. Since $\mathrm{Pic}(\mathrm{Spec} F) = 0$ we get from the above corollary:

Corollary. *If $(n, p) = 1$ with $p = \mathrm{char}(F)$ then we have an exact sequence*

$$F^* \overset{\cdot}{\longrightarrow} F^* \overset{\partial_n}{\longrightarrow} H^1_{\text{ét}}(F, \mu_n) \to 0$$

i.e., (with the notation of Chap.III, 3.10)

$$F^*/n \overset{\sim}{\longrightarrow} H^1_{\text{ét}}(F, \mu_n).$$

6.13. The theorem of Merkurjev and Suslin

Using the cup product in étale cohomology we get from the above the following homomorphism

$$F^* \otimes_{\mathbb{Z}} F^* \to H^1_{\text{ét}}(F, \mu_n) \otimes H^1_{\text{ét}}(F, \mu_n) \xrightarrow{\cup} H^2_{\text{ét}}(F, \mu_n^{\otimes 2})$$

explicitly as follows

$$a \otimes b \longmapsto \partial_n a \otimes \partial_n b \longmapsto \partial_n a \cup \partial_n b$$

Write $\partial_n a \cup \partial_n b = (a, b)_n$. There is the following result by Tate and Bass.

Theorem. *For* $(n, p) = 1$ *we have* $(a, 1 - a)_n = 1$

Since clearly also $(a_1 a_2, b)_n = (a_1, b)_n + (a_2, b)_n$, and similarly for b, it follows by the theorem of Matsumoto (Chap. III, 3.2.3, thm 2) that we have a homomorphism $\tilde{R}_n$

$$\tilde{R}_n : K_2(F) \to H^2_{\text{ét}}(F, \mu_n^{\otimes 2})$$
$$\{a, b\} \longmapsto (a, b)_n$$

and then clearly also a homomorphism

$$R_n : K_2(F)/n \to H^2_{\text{ét}}(F, \mu_n^{\otimes 2})$$

the so-called *norm residue map.*

Theorem. (Merkurjev–Suslin; [MS]). *With the above assumptions, i.e.* $(n, p) = 1$ *for* $p = \text{char}(F)$, *the* R_n *is an isomorphism.*

VIc. Bloch–Ogus theory.

In this section X is an algebraic variety (or if you want, a scheme) defined over a field k.

6.14. Relation between the Zariski site and the étale site.

There is a morphism of sites ([M], p.56)

$$\pi : X_{\text{ét}} \to X_{\text{Zar}}$$

obtained by remarking that an open immersion $U \hookrightarrow X$ is also an étale morphism. Such a morphism of sites is the generalization of a continuous map of topological spaces. If $\mathcal{F}$ is a sheaf on $X_{\text{ét}}$ then we can construct its direct higher images $R^i\pi_*\mathcal{F}$; these are sheaves on X_{Zar}, obtained by sheafifying the Zariski presheaves $U \mapsto H^i_{\text{ét}}(U, \mathcal{F})$ for $U \subset X$ Zariski–open.

Theorem. (Leray spectral sequence, [M], p.89).

There is a spectral sequence

$$E_2^{rs} = H^r_{\text{Zar}}(X, R^s\pi_*\mathcal{F}) \Longrightarrow H^{r+s}_{\text{ét}}(X, \mathcal{F})$$

6.15. The sheaves $\mathcal{H}^i(\mu_n^{\otimes q})$

Apply the above to the sheaves $\mathcal{F} = \mu_n^{\otimes q}$ (always $(n,p) = 1, p = \text{char}(k), q \in \mathbb{Z}$) on $X_{\text{ét}}$. Then we get Zariski sheaves as follows

$$\mathcal{H}^i(\mu_n^{\otimes q}) := R^i\pi_*(\mu_n^{\otimes q}).$$

These sheaves have been studied by Bloch–Ogus and the basic fact is ([BO], [B] p.4.12):

Theorem. *Let X be a smooth (irreducible) variety. Let $(n,p) = 1$. Then on X_{Zar} we have a flasque resolution*

$$0 \to \mathcal{H}^i(\mu_n^{\otimes q}) \to H^i_{\text{ét}}(k(X), \mu_n^{\otimes q}) \to \bigoplus_{x\in X^1} j_{x*}H^{i-1}_{\text{ét}}(k(x), \mu_n^{\otimes(q-1)}) \to$$
$$\cdots \to \bigoplus_{x\in X^{i-1}} j_{x*}H^1_{\text{ét}}(k(x), \mu_n^{\otimes(q-i+1)}) \to \bigoplus_{x\in X^i} j_{x*}H^0(k(x), \mu_n^{\otimes(q-i)}) \to 0$$

Remarks.

1. Compare this with the Gersten–Quillen resolution of the sheaves $\mathcal{K}_{i,X}$ in Chap.III, 3.6.
2. The conventions are the same as in Chap.III, 3.6. Namely $X^r = \{x \in X;$ codim $\{\overline{x}\} = r\}$. For such a point x we take the étale cohomology group $H^j_{\text{ét}}(k(x), -)$ and consider it first as a constant sheaf on $\{\overline{x}\}$ and next extend it by zero outside of $\{\overline{x}\}$.
3. For the proof and the maps we must refer to the original paper [BO] and to [B].

Corollary. *With the same assumptions:*

a. $E_2^{r,s} = H^r_{\text{Zar}}(X, \mathcal{H}^s(\mu_n^{\otimes q})) = 0$ *for* $r > s$.
b. There exists a homomorphism

$$\gamma_n : E_2^{i-1,i} = H^{i-1}_{\text{Zar}}(X, \mathcal{H}^i(\mu_n^{\otimes q})) \to H^{2i-1}_{\text{ét}}(X, \mu_n^{\otimes q}).$$

c. For $i = 2$ this homomorphism is injective

$$\gamma_n : E_2^{1,2} = H^1_{\text{Zar}}(X, \mathcal{H}^2(\mu_n^{\otimes q})) \to H^3_{\text{ét}}(X, \mu_n^{\otimes q}).$$

Proof.

For a. this follows now immediately since we have a flasque resolution of $\mathcal{H}^s(\mu_n^{\otimes q}))$ and then b. and c. follow by inspection of the Leray spectral sequence.

VId. Application to algebraic cycles of codimension 2.

6.16. Proposition. *Let X be a smooth, projective algebraic variety defined over an algebraically closed field. Let $n \in \mathbb{Z}$ with $(n,p) = 1$ with $p = \mathrm{char}(k)$. Then there exists (natural) homomorphisms $\alpha_n, \beta_n, \gamma_n$ as follows:*

$$\begin{array}{ccc} H^1_{\mathrm{Zar}}(X, \mathcal{K}_{2,X}/n) & \xrightarrow[\alpha_n]{} & {}_nCH^2(X) \\ \downarrow \cong & & \\ H^1_{\mathrm{Zar}}(X, \mathcal{H}^2(\mu_n^{\otimes 2})) & \xhookrightarrow[\gamma_n]{} & H^3_{\mathrm{et}}(X, \mu_n^{\otimes 2}) \end{array}$$

with α_n surjective, β_n an isomorphism and γ_n injective. (We follow the conventions of Chap. III, 3.10, ${}_nCH^2(X)$ is the kernel of multiplication by n).

Proof.

The existence and injectivity of γ_n is shown in the corollary of 6.15. The existence and surjectivity of α_n we have seen in Chap. III, 3.10, cor. 1. Finally the existence of the isomorphism β_n follows from the following commutative diagram (see Chap.III, Prop. 1 in 3.10, and the theorem in 6.15 and with the conventions of Chap. III, 3.8):

$$\begin{array}{ccccccccc} 0 \to & \mathcal{K}_{2,X}/n & \to & K_2(k(x))/n & \to & \bigoplus_D j_{D*}(k(D)^*/n) & \to & \bigoplus_Y j_{Y*}(\mathbb{Z}/n) & \to 0 \\ & \downarrow \beta_n & & \downarrow R_n & & \downarrow \cong & & \downarrow \cong & \\ 0 \to & \mathcal{H}^2(\mu^{\otimes 2}) & \to & H^2_{\mathrm{et}}(k(X), \mu_n^{\otimes 2}) & \to & \bigoplus_D j_{D*}H^1(k(D), \mu_{n,D}) & \to & \bigoplus_Y j_{Y*}(\mathbb{Z}/n) & \to 0 \end{array}$$

As to the vertical maps, the right one is obvious, the next one is the isomorphism from the corollary in 6.12 and R_n is the norm residue map of 6.13. Since by Merkurjev–Suslin the R_n is an isomorphism we get the existence of the isomorphism β_n.

6.17. Fix a prime $\ell \neq p$. Recall the definition of the *Tate group* $T_\ell(G)$ of an abelian group G:

$$T_\ell(G) := \varprojlim_{\nu}({}_{\ell^\nu}G)$$

By taking $n = \ell^\nu$ and passing to the limit for $\nu \to \infty$ we get from 6.10 the following:

Corollary. *There exist homomorphism α, β and γ as follows*

$$\begin{array}{ccc} \varprojlim_\nu H^1_{\mathrm{Zar}}(X, \mathcal{K}_{2,X}/l^\nu) & \xrightarrow[\alpha_n]{} & T_l CH^2(X) \\ \beta \downarrow \cong & & \\ \varprojlim_\nu H^1_{\mathrm{Zar}}(X, \mathcal{H}^2(\mu_{l^\nu} \otimes 2)) & \xhookrightarrow[\gamma_n]{} & H^3_{\mathrm{et}}(X, Z_l^{\otimes 2}) \end{array}$$

with α surjective, β an isomorphism and γ injective. (Remark: note that indeed the surjectivity of the α_{ℓ^ν} implies the surjectivity of α since the α_{ℓ^ν} is a map to a finite group).

6.18. Proof of theorem I of 6.8.

Let X be as in that theorem, A an abelian variety, and let $\Phi : CH^2_{\mathrm{alg}}(X) \to A$ be a surjective regular homomorphism. By the result 6.10 of Hiroshi Saito the theorem will follow from the following:

Claim. $2 \dim A \leq b_3(X) := \dim H^3_{\mathrm{\acute{e}t}}(X, \mathbb{Q}_\ell)$.

Proof of the claim:
By Lemma 6.4 there exists an algebraic family (B, Z), with B an abelian variety such that $\Phi \cdot f_Z : B \to A$ is an isogeny. Now fix a prime ℓ such that $(\ell, \deg \Phi \cdot f_Z) = 1$ and $(\ell, p) = 1$. Then we have isomorphisms ${}_{\ell^\nu}B \cong_{\ell^\nu} A$. Passing to the projective limit for $\nu \to \infty$ we get for the Tate groups $T_\ell(B) \cong T_\ell(A)$ hence from

$$B \xrightarrow{f_Z} CH^2(X) \to A$$

we get $T_\ell(A) \subseteq T_\ell(CH^2(X))$. Then finally using the corollary of 6.17 we get $2 \dim A = rkT_\ell(A) \leq b_3(X)$.

6.19. Concerning the proof of theorem II

For that proof we need one extra tool namely the *Bloch map*. Let X be a smooth, projective variety over an algebraically closed field k. Let $CH^i(X)_{\ell-\mathrm{tors}} = \varinjlim_\nu \{{}_{\ell^\nu}CH^i(X)\}$ be the ℓ-torsion in $CH^i(X)$ and let $H^{2i-1}_{\mathrm{\acute{e}t}}(X, \mathbb{Q}_\ell/\mathbb{Z}_\ell(i)) = \varinjlim_\nu H^{2i-1}_{\mathrm{\acute{e}t}}(X, \mu_{\ell^\nu}^{\otimes i})$, then the Bloch map is a homomorphism

$$\lambda^i_\ell : CH^i(X)_{\ell-\mathrm{tors}} \to H^{2i-1}_{\mathrm{\acute{e}t}}(X, \mathbb{Q}_\ell/\mathbb{Z}_\ell(i)).$$

For the construction of the Bloch map see the original paper [B 3], and also [CT].
Assume now that $k = \mathbb{C}$. For the proof of theorem II we refer to [Mur 1], p. 255-258. An essential point in the proof is that the above Bloch map coincides on the torsion points with the Abel–Jacobi map. The proof gives also the following *corollary*:

$$AJ : CH^2(X)_{\mathrm{alg,tors}} \to J^2_a(X)_{\mathrm{tors}}$$

is an *isomorphism*. For the case of a surface this gives back (for $k = \mathbb{C}$) the theorem of Roitman (see lecture 5 of C. Voisin [Vo]).

Chapter VII Grothendieck's theory of motives.

The theory of motives has been created in the sixties by Grothendieck. One of Grothendieck's aims was to understand and explain the striking similarities between the various cohomology theories of algebraic varieties. Going beyond that Grothendieck states: "the theory of motives is a systematic theory of arithmetic properties of algebraic varieties, as embodied in their groups of classes of cycles for numerical equivalence" ([Gr 3], p.198).

In this lecture we discuss part of the theory of motives, or to be precise, of *pure* motives. The category of pure motives (or better: the various categories of pure motives, see 7.5) is well established, contrary to the category of *mixed* motives for which we even don't have a conjectural construction. In the first part of the lecture we discuss Grothendieck's construction. We put emphasis on the so–called category of Chow motives which is the most precise one from the point of view of algebraic cycles. In the presentation we follow largely the elegant treatment of Scholl [Sch]. In the second part we discuss the surprising proof of U. Jannsen of the semi–simplicity of the category of motives via numerical equivalence [Ja 2]. Finally in the last part we formulate some conjectures and present some evidence for them [Mur 3].

As general references we recommend for pure motives the papers [Kl 1], [Ma 1], [Mur 2]and [Sch], for mixed motives [Ja 1] and Ja 3].

In this chapter k is an arbitrary field, $\mathcal{V} = \mathcal{V}(k)$ is the category of *smooth, projective* varieties defined over k. For $X \in \mathcal{V}$ let, as before, $CH(X) = \oplus_i CH^i(X)$ be the Chow ring of algebraic cycles modulo rational equivalence and $CH_{\mathbb{Q}}(X) = CH(X) \otimes \mathbb{Q}$.

VIIa. The construction of pure motives.

7.1. Correspondences. Let $X, Y \in \mathcal{V}$, put $\mathrm{Corr}(X,Y) := CH_{\mathbb{Q}}(X \times Y)$. $\mathrm{Corr}(X,Y)$ is called the group of *correspondences* between X and Y. For $f \in \mathrm{Corr}(X,Y)$ let ${}^tf \in \mathrm{Corr}(Y,X)$ be its transpose. There is *composition* of correspondences, for X, Y and $Z \in \mathcal{V}$ one has a bilinear map

$$\mathrm{Corr}(X,Y) \times \mathrm{Corr}(Y,Z) \to \mathrm{Corr}(X,Z)$$

defined for $f \in \mathrm{Corr}(X,Y), g \in \mathrm{Corr}(Y,Z)$ by $g \bullet f = (pr_{13})_*\{p_{12}^*(f) \cdot p_{23}^*(g)\}$, where the intersection product is on $X \times Y \times Z$. The *degree* of a correspondence is defined as follows: let X be irreducible and of dimension d then $\mathrm{Corr}^s(X_d, Y) := CH_{\mathbb{Q}}^{d+s}(X \times Y)$. For instance if $\varphi : X_d \to Y_e$ is a usual morphism of algebraic irreducible varieties of dimension d and e respectively, then we have for its graph Γ_φ that $\Gamma_\varphi \in \mathrm{Corr}^{e-d}(X,Y)$ and ${}^t\Gamma_\varphi \in \mathrm{Corr}^0(X,Y)$. Note that if $f \in \mathrm{Corr}^0(X,Y)$ then we have a homomorphism $f : CH_{\mathbb{Q}}^i(X) \to CH_{\mathbb{Q}}^i(Y)$ defined by $a \mapsto f(a) = (p_Y)_*\{f \cdot (a \times Y)\}$ for $a \in CH_{\mathbb{Q}}^i(X)$. Finally there is the notion of projector: a *projector* of X is a correspondence $p \in \mathrm{Corr}^0(X,X)$ such that $p \bullet p = p$.

7.2. Category of Chow motives $\mathcal{M}_r$. First one defines the category $\mathcal{M}_r^+(k) = \mathcal{M}_r^+$ of *effective Chow motives.* Objects are pairs $M = (X,p)$ with $X \in \mathcal{V}$ and p a projector of X. Next if $M = (X,p)$ and $N = (Y,q)$ then

$$\mathrm{Hom}_{\mathcal{M}_r^+}(M,N) := q \bullet \mathrm{Corr}^0(X,Y) \bullet p$$

and composition of morphisms in $\mathcal{M}_r^+$ is defined via composition of correspondences.

Next one enlarges $\mathcal{M}_r^+$ to the full category of Chow motives $\mathcal{M}_r(k) = \mathcal{M}_r$ as follows. Objects of $\mathcal{M}_r$ are triples $M = (X,p,m)$ with X and p as above and $m \in \mathbb{Z}$. Let $N = (Y,q,n)$, put

$$\mathrm{Hom}_{\mathcal{M}_r}(M,N) := q \bullet \mathrm{Corr}^{n-m}(X,Y) \bullet p$$

Note that $\mathcal{M}_r^+ \subset \mathcal{M}_r$ is a full subcategory. The extension from $\mathcal{M}_r^+$ to $\mathcal{M}_r$ is rather formal, objects in $\mathcal{M}_r$ should be considered as kind of "twists" (by the "Tate" motive, see 7.3) of objects of $\mathcal{M}_r^+$.

Properties of $\mathcal{M}_r$.

1. $\mathcal{M}_r$ is an additive category,
2. $(X,p,m) \oplus (Y,q,m) = (X \amalg Y, p \amalg q, m)$,
3. There is a tensor product: $(X,p,m) \otimes (Y,q,n) = (X \times Y, p \times q, m+n)$,
4. $\mathcal{M}_r$ is pseudo abelian, i.e. every projector p of M has an image and $M = Im(p) \oplus Im(1-p)$ (however $\mathcal{M}_r$ is *not* abelian! [Sch]).
5. $\exists$ a contravariant functor $h_r : \mathcal{V} \to \mathcal{M}_r^+ \subset \mathcal{M}_r$ defined by $h_r(X) = (X, id, 0)$ and if $\varphi : X \to Y$ is a morphism then $h_r(\varphi) = {}^t\Gamma_{\mathbb{Q}} : h_r(Y) \to h_r(X)$.

7.3. Examples.

In the following examples we restrict mostly our attention to $\mathcal{M}_r^+$.

1. For $X \in \mathcal{V}$ we have $h_r(X) = (X, id)$
2. Fix a point $e \in X$, consider $\pi_0 = e \times X$, this is a projector; put $h_r^0(X) = (X, \pi_0)$
3. Similarly, if for simplicity X is irreducible of dim d, let $\pi_{2d} = X \times e$; this is a projector and write $h_r^{2d}(X) = (X, \pi_{2d})$.
4. Let $\pi^+ = id - \pi_0 - \pi_{2d}$, write $h_r^+(X) = (X, \pi^+)$; now we have $h_r(X) = h_r^0(X) \oplus h_r^+(X) \oplus h_r^{2d}(X)$.
5. Let $\mathbf{1} = (\mathrm{Spec}, id, 0)$; this is the identity for the tensor product
6. Put $L = (\mathbb{P}_1, \pi_2)$; this is the so-called Lefschetz motive In the category $\mathcal{M}_r$ we have $L \simeq (\mathrm{Speck}, id, -1)$ and $T = (\mathrm{Spec}k, id, 1)$ is the so-called Tate motive, which is "inverse" to L.
7. Let $\varphi : X'_d \to X_d$ be a morphism of smooth, projective, irreducible varieties both of dimension d and let φ be generically finite (i.e. φ dominant) of degree m. Let $p = \frac{1}{m}{}^t\Gamma_\Phi \bullet \Gamma_\Phi$, then p is a projector of X' and hence we have $h_r(X') = (X',p) \oplus (X', 1-p)$; moreover in $\mathcal{M}_r^+$ we have $(X',p) \simeq h_r(X)$.

7.4. Motives of curves.

The previous examples (with the exception of 7) are trivial, however the fact that the notion of motive is an interesting one is shown by the following example. Let X be a smooth, projective curve C (say over $k = \overline{k}$); in this case write also

$h_r^+(C) = h_r^1(C)$ and now we have a decomposition $h_r(C) = h_r^0(C) \oplus h_r^1(C) \oplus h_r^2(C)$. The motive $h_r^1(C)$ is closely related to the Jacobian variety of C as is shown by the following theorem.

Theorem. *Let C and C' be smooth, projective curves. Then*

$$\mathrm{Hom}_{\mathcal{M}_r}(h_r^1(C), h_r^1(C')) = \mathrm{Hom}_{ab-var}(J(C), J(C')) \otimes \mathbb{Q}$$

"Proof".

The theorem is only a translation of the following:

Theorem (Weil).

$$CH^1(C \times C')/(\text{horizontal} + \text{vertical components}) \simeq \mathrm{Hom}_{ab-var}(J(C), J(C')).$$

Corollary. *Let $\mathcal{M}' \subset \mathcal{M}_r^+$ be the full subcategory defined by $\mathcal{M}' = \{M; \exists$ curve C such that $h_r^1(C) = M \oplus M'\}$. Then $\mathcal{M}'$ is equivalent to the category of abelian varieties up to isogeny.*

"Proof"

This is a consequence of the above theorem and the fact that up to isogeny every abelian variety is a direct summand of the Jacobian of a curve.

7.5. Variants by using other equivalence relations.

In the construction above we took as intersection ring the Chow ring $CH_{\mathbb{Q}}(-)$, the motives obtained in this way are therefore called *Chow motives.* However Grothendieck's construction works for any intersection ring, therefore we can replace rational equivalence by any "adequate" equivalence relation (see Chap. I, 1.3), using then as intersection ring the ring $CH_{\mathbb{Q}}(-)/\sim$, i.e., taking as correspondences $CH_{\mathbb{Q}}(X \times Y)/\sim$ and this gives *another category* of motives which we denote by $\mathcal{M}_\sim$. Clearly there is a natural functor $\mathcal{M}_r \to \mathcal{M}_\sim$. Of particular interest are homological equivalence and numerical equivalence, the corresponding categories of motives are denoted by $\mathcal{M}_h$ (the "hom-motives") and by $\mathcal{M}_n$ (the "neq-motives"). Note that for homological equivalence we first have to choose a Weil cohomology theory (see Chap. I, 1.4.4), but, at least in characteristic zero, due to the comparison theorems, the result does not depend on that choice. In case our field k is embeddable in $\mathbb{C}$ there is still "another" variant, suggested and used by Deligne, namely instead of working with algebraic cycles work with absolute Hodge cycles (cf [DMOS]), the corresponding category of motives is denoted by $\mathcal{M}_{AH}$.

The relations between the various categories can be summarised by the following diagram of categories and natural functors between them (keeping in mind the relations in Chap. I, 1.4.5):

$$\begin{array}{ccccccc} \mathcal{V} & \xrightarrow[h_r]{} & \mathcal{M}_r & \longrightarrow & \mathcal{M}_h & \longrightarrow & \mathcal{M}_n \\ & & & & & \searrow & \\ & & & & & & \mathcal{M}_{\mathrm{AH}} \end{array}$$

Conjectures and their consequences: The conjecture that homological and numerical equivalence coincide (see Chap. I, 1.4.6) would imply that $\mathcal{M}_h$ and $\mathcal{M}_n$ coincide. The (usual) Hodge conjecture (see Chap. V) would imply that $\mathcal{M}_h$ and $\mathcal{M}_{AH}$ coincide and – as is well known (see for instance [Kl 3], cor. 5.3) – it would also imply that homological and numerical equivalence coincide in characteristic zero.

Remark. Grothendieck has put special emphasis on the category $\mathcal{M}_n$ constructed by means of numerical equivalence (see [Gr 3], p.198 and also the introduction of [Kl 3]), therefore this category is sometimes called "the" category of (pure) motives. The importance of this category is stressed by Jannsen's theorem below (see VIIB).

7.6. Realizations.

7.6.1. Take the category of Chow motives $\mathcal{M}_r$. A correspondence $f \in \mathrm{Corr}^0(X,Y)$ induces homomorphisms $f : CH^i_{\mathbb{Q}}(X) \to CH^i_{\mathbb{Q}}(Y)$ and $f : H^i(X) \to H^i(Y)$ for every suitable cohomology theory. Therefore for $M = (X,p) \in \mathcal{M}_r^+$ we can define

$$CH^i_{\mathbb{Q}}(M) := Im(p) \subset CH^i_{\mathbb{Q}}(X)$$
$$H^i(M) := Im(p) \subset H^i(X)$$

These Chow groups and cohomology groups (in all versions: DeRham, Betti, étale, crystalline) are called *realizations* of M. Note that we can transport to M "every natural structure"like for instance groups $H^i_{\mathrm{Zar}}(M, \mathcal{K}_{j,X})$ and, if the field is embeddable in $\mathbb{C}$, Hodge structures on $H^i(M,\mathbb{Q})$, etc.

7.6.2. Next take the category $\mathcal{M}_h$. The motives $M \in \mathcal{M}_h^+$ carry, in the same way as above, cohomology groups (in all versions) as realizations and also if k is embeddable in $\mathbb{C}$ Hodge structures. However we have lost the Chow groups (and $H^i(X, \mathcal{K}_{j,X})$).

7.6.2. bis. The same remark holds for $\mathcal{M}_{AH}$.

7.6.3. Next consider $\mathcal{M}_n$. We can only give *conjecturally* realizations to $M \in \mathcal{M}_n^+$, namely if we assume the conjecture that homological equivalence and numerical equivalence coincide we can attach cohomology groups to M.

7.7. Grothendieck's standard conjectures (principal version).

Fix a *Weil cohomology theory* $H(X)$ (cf Chap. I, 1.4.4; for instance take $H(X) = H_{\text{ét}}(X_{\bar{k}}, \mathbb{Q}_\ell)$). Let $X \in \mathcal{V}$ irreducible and $d = \dim X$. By the hard

Lefschetz theorem (see lecture 1 of Green [Gre] and lecture 2 of Voisin [Vo]) there is a unique cohomological operator Λ "quasi-inverse" to the Lefschetz operator L. More precisely, if $0 \le r \le d$, then Λ is the unique linear map making the following diagram commutative

$$\begin{array}{ccc} H^{d-r}(X) & \xrightarrow{L^r} & H^{d+r}(X) \\ \Lambda \downarrow & & \downarrow L \\ H^{d-r-2} & \xrightarrow{L^{r+2}} & H^{d+r+2}(X) \end{array}$$

where the two horizontal maps are isomorphisms by the hard Lefschetz theorem. The first standard conjecture (the *Lefschetz standard conjecture*) states:

(LStC): The operator Λ is algebraic, i.e., is induced by an algebraic cycle on $X \times X$

Next recall that the Lefschetz operator L gives on the cohomology groups a decomposition into primitive elements. For $0 \le j \le d$ the primitive part $P^j(X) \subset H^j(X)$ is defined as $P^j(X) := \mathrm{Ker}(L^{d-j+1})$. It is well-known that the cup product pairing, provided with a suitable factor, is positive definite on the primitive parts (the Hodge–Riemann bilinear relations, see lecture 2, section 2.4 of Claire Voisin [Vo]). Now for $0 \le i \le d$ consider the cycle map $\gamma^i : \mathcal{Z}^i(X) \to H^{2i}(X)$ and put $A^i(X) = Im(\gamma^i) \otimes \mathbb{Q}$. Writing $< -, - >$ as notation for the cup product pairing in complementary dimensions, the second standard conjecture (the *Hodge standard conjecture*) is as follows:

(HStC): For all $i \le \frac{d}{2}$ the $\mathbb{Q}$-valued pairing on $A^i(X) \cap P^{2i}(X)$,

$$x, y \mapsto (-1)^i \langle L^{d-2i}x, y \rangle$$

is positive definite.

Variations and implications:

For the original statements see [Gr 3]. There are various equivalent, or almost equivalent, statements for the standard conjectures: for these as well as for the consequences of (LStC) and (HStC) we refer to the excellent papers [Kl 1] and [Kl 3]. Here we mention only that the key conjecture "homological equivalence = numerical equivalence" would imply (LStC) (see [Kl 3], prop.5.1); conversely (LStC) together with (HStC) imply "homological = numerical equivalence" (ibid, prop. 5.1). Also (LStC) implies that the Künneth components are algebraic (ibid, section 4). Also the two standard conjectures together would imply that the category $\mathcal{M}_n$ is an abelian, semi-simple category (ibid; thm.5.6); this is now a fact by Jannsen's theorem (see VIIB).

Current state of the standard conjectures.

(LStC) is true for curves, for surfaces (Grothendieck) and for abelian varieties (Kleiman, Lieberman). Also in characteristic zero (HStC) is true by Hodge theory for the classical cohomology, but then by the comparison theorems, also for the étale cohomology. In arbitrary characteristic (HStC) is true for surfaces (B. Segre, Grothendieck and others). For all of this see the papers [Kl 1] and [Kl 3].

VIIb. Jannsen's theorem on the semi–simplicity of $\mathcal{M}_n$.

In 1991 U. Jannsen proved, without assuming the standard conjectures, that the category $\mathcal{M}_n$, i.e., the category based upon numerical equivalence, is abelian and semi–simple. In this section we reproduce part of Jannsen's ingenious proof; the proof is the more surprising since it is based upon facts known already in the sixties. On the other hand unfortunately the proof does not settle the crucial question: do homological and numerical equivalence coincide?

7.8. Theorem (U. Jannsen) [Ja 2].

Let $\mathcal{M}_\sim$ be the category of motives constructed via algebraic cycles and via an adequate equivalence relation (as described in section 7.5). Then the following properties are equivalent:

1. $\mathcal{M}_\sim$ is an abelian, semi–simple category.
2. The equivalence relation $\sim$ is numerical equivalence.
3. For $X \in \mathcal{V}$, irreducibel and of dimension d the $CH^d_{\mathbb{Q}}(X \times X)/\sim$ is a finite dimensional, semi–simple $\mathbb{Q}$–algebra (note: the multiplicative structure comes from composition of correspondences).

7.9. Proof of $1 \Rightarrow 2$.

First "recall" that for every adequate equivalence relation $\sim$ we have $\mathcal{Z}^i_\sim(X) \subseteq \mathcal{Z}^i_n(X)$ (cf. Chap. I, 1.4.5). In $\mathcal{M}_\sim$ we have as unit element $\mathbf{1} = (\mathrm{Spec}\ k, id, 0)$. Assume now that $Z \in \mathcal{Z}^i_n(X)$, but $Z \notin \mathcal{Z}^i_\sim(X)$. Then Z determines a non–trivial correspondence $\mathrm{Corr}^i(\mathrm{Spec}\ k, X) = CH^i(\mathrm{Spec}\ k \times X)/\sim$, hence an element $f \in \mathrm{Hom}_{\mathcal{M}_\sim}(\mathrm{Spec}\ k, X)$ with $f \neq 0$. Now $\mathbf{1}$ being irreducible in $\mathcal{M}_\sim$ there must be, by our assumptions on $\mathcal{M}_\sim$, an element $g \in \mathrm{Hom}_{\mathcal{M}_\sim}(X, \mathrm{Spec}\ k)$ such that $g \bullet f = id$. Such g is given by an algebraic cycle $W \in CH^{d-i}_{\mathbb{Q}}(X)$ if $d = \dim X$. The composition $g \bullet f$ is given by the cycle $(pr_{13})_*\{(Z \times \mathrm{Spec}\ k) \cdot (\mathrm{Spec}\ k \times W)\}$ on $\mathrm{Spec}\ k \times \mathrm{Spec}\ k$ where the intersection is on $\mathrm{Spec}\ k \times X \times \mathrm{Spec}\ k = X$, i.e., is determined by the intersection number of Z and W on X. Since $Z \in \mathcal{Z}^i_n(X)$ this intersection number is zero, and hence this cycle on $\mathrm{Spec}\ k \times \mathrm{Spec}\ k$ is zero, i.e., $g \bullet f = 0$ which contradicts $g \bullet f = id$, hence $Z \in \mathcal{Z}^i_\sim(X)$.

7.10. Proof $2 \Rightarrow 3$

Key Lemma. *Let $X \in \mathcal{V}$, irreducible and* $\dim X = d$*; let $0 \leq i \leq d$. Put $A^i(X) = CH^i_{\mathbb{Q}}(X)$ / (numerical equivalence).*
Then:
a. $\dim_{\mathbb{Q}} A^i(X)$ *is finite.*
b. $A^d(X \times X)$ is a finite dimensional, semi–simple $\mathbb{Q}$–algebra.

Proof of a. (see [Kl 1], Thm. 3.5)

In fact we claim: $\dim_{\mathbb{Q}} A^i(X) \leq b_{2i} := \dim_{\mathbb{Q}_\ell} H^{2i}_{ét}(X_{\bar{k}}, \mathbb{Q}_\ell)$. In order to see this let $\gamma^i : \mathcal{Z}^i(X) \to H^{2i}(X)$ be the cycle map, where $H^{2i}(X) = \dim H^{2i}_{ét}(X_{\bar{k}}, \mathbb{Q}_\ell)$. Let $\alpha_1, \alpha_2, \ldots, \alpha_m \in \mathcal{Z}^{d-i}(X)$ be such that $\gamma(\alpha_1), \ldots, \gamma(\alpha_m)$ is a basis of the $\mathbb{Q}_\ell$-subvectorspace of $H^{2(d-i)}(X)$ generated by $Im(\gamma^{d-i})$. Consider now a map $\lambda : \mathcal{Z}^i(X) \to \mathbb{Z}^m$ defined, for $\beta \in \mathcal{Z}^i(X)$, by $\beta \mapsto (\#(\beta, \alpha_1) \ldots, \#(\beta, \alpha_m))$ where $\#(\beta, \alpha)$ is the intersection number. Clearly $\mathcal{Z}^i_n(X) \subset \mathrm{Ker}(\lambda)$; conversely if $\beta \in \mathrm{Ker}(\lambda)$ and $\alpha \in \mathcal{Z}^{d-i}(X)$ then $\alpha = \sum_j q_j \gamma(\alpha_j)$ in $H^{2(d-i)}(X)$ and hence using the fact that under the cycle map intersection corresponds with cup product we get $\#(\beta, \alpha) = \gamma(\beta) \cap \gamma(\alpha) = \sum_j q_j \#(\beta, \alpha_j) = 0$, hence $\mathrm{Ker}(\lambda) \subset \mathcal{Z}^i_n(X)$. Hence $\mathrm{Ker}(\lambda) = \mathcal{Z}^i_n(X)$, hence $\mathcal{Z}^i(X)/\mathcal{Z}^i_n(X) = Im(\lambda) \subset \mathbb{Z}^m$, which proves the claim.

7.11. Proof of b. (this is the heart of Jannsen's proof).

Put $A = A^d(X \times X) \otimes \mathbb{Q}_\ell$ and $B = \{CH^d_{\mathbb{Q}}(X \times X)/\text{(homological equivalence)}\} \otimes \mathbb{Q}_\ell$; note $B \subset H^{2d}(X \times X) = H^{2d}_{ét}(X_{\bar{k}} \times X_{\bar{k}}, \mathbb{Q}_\ell)$. Both A and B are finite dimensional $\mathbb{Q}_\ell$-algebras (the multiplicative structure comes from composition of correspondences). In order to prove b. of the key lemma it suffices to prove that A is semi-simple, i.e., that A has no Jacobian radical.

Now recall that if F is a field (in our case $F = \mathbb{Q}_\ell$) and R is a finite dimensional F-algebra then the Jacobian radical $J(R)$ of R is the union of all nilpotent two-sided ideals of R and every element of $J(R)$ is nilpotent ([P], p.61) and R is semi-simple if and only if $J(R) = (0)$ (ibid, p.41).

Consider now the natural surjective homomorphism $\Phi : B \to A$ (recall $\mathcal{Z}^i_{\mathrm{hom}} \subseteq \mathcal{Z}^i_n$; Chap. I, 1.4.5). Write $J = J(A)$ and $J' = J(B)$; now we claim $\Phi(J') = J$. Firstly since $\Phi(J')$ is a nilpotent two-sided ideal in A we have $\Phi(J') \subset J$. Conversely, since B/J' is semi-simple, also $A/\Phi(J')$ is semi-simple ([P], cor. on p.42), but then since $J/\Phi(J')$ is two-sided nilpotent in $A/\Phi(J')$ we must have $J = \Phi(J')$.

Now take $f \in J(A^d(X \times X)) \subset J$; we have to prove $f = 0$. Lift f to $f' \in J'$, hence f' is nilpotent in B. Now for any $g \in B$ we have that $f' \bullet g$ (and also $g \bullet f'$) is nilpotent since they are in the two-sided ideal J' of B. Now comes the key to the proof:

Lefschetz Trace formula ([Kl 1], 1.3.6):

Let X and B be as above. Let $f', g \in B$. Then one has for the $\mathbb{Q}_\ell$-linear pairing $\langle f', {}^t g \rangle$ of the cup product in $X \times X$ the formula

$$\langle f', {}^t g \rangle = \sum_{i=0}^{2d} (-1)^i Tr_i(f' \bullet g)$$

where, for $h \in B, Tr_i(h)$ is the trace of h acting on $H^i(X) = H^i_{ét}(X_{\bar{k}}, \mathbb{Q}_\ell)$ (with $\ell \neq \mathrm{char}(p)$).

7.12. Continuation of the proof. Since $f' \bullet g \in J'$ it is nilpotent and hence $Tr_i(f' \bullet g) = 0$ for all i and any $g \in B$. This implies that for every $g \in CH^d_{\mathbb{Q}}(X \times X)$ the cup product $\langle f', g \rangle = 0$, hence $f = 0$ in $A^d(X \times X)$.

7.13. About the proof 3 ⇒ 1.

That part is rather formal, we do not repeat it here. It depends on two lemma's.

Lemma 1. *Condition 3 implies that for every $M \in \mathcal{M}_\sim$ we have:*

a. $\dim_{\mathbb{Q}} End_{\mathcal{M}_\sim}(M)$ *is finite*
b. $End_{\mathcal{M}_\sim}(M)$ *is a semi-simple* $\mathbb{Q}$*-algebra.*

Lemma 2. *If $\mathcal{M}$ is a $\mathbb{Q}$-linear, pseudo-abelian category such that $End_{\mathcal{M}}(M)$ is a finite dimensional, semi-simple $\mathbb{Q}$-algebra for every $M \in \mathcal{M}$, then $\mathcal{M}$ is an abelian, semi-simple category.*

For the proofs see [Ja 2].

VIIc. Conjectures and facts on Chow motives.

Although the category $\mathcal{M}_r$ of Chow motives is not abelian ([Sch]) we still think that it is an important and interesting category. We think that motives, being of algebraic and arithmetical nature, should give insight not only in cohomology, but also in the full Chow group of algebraic cycles provided we work with Chow motives.

7.14. Chow–Künneth decomposition.

Definition. [Mur 3] Let $X \in \mathcal{V}$, let X be irreducible and $\dim X = d$. We say that X has a *Chow–Künneth decompostion* if there exist $\pi_i \in CH^d_{\mathbb{Q}}(X \times X)$ for $0 \leq i \leq 2d$, such that:

1. $\sum_i \pi_i = \Delta(X)$ (the diagonal of X)
2. $\pi_i \bullet \pi_j = \begin{pmatrix} 0 & j \neq i \\ \pi_i & j = i \end{pmatrix}$
(in particular the π_i are projectors)
3. (over $\bar{k}$)π_i modulo homological equivalence is the usual Künneth component $\Delta(2d-i, i)$.

If such π_i exists put $h^i_r(X) = (X, \pi_i)$; clearly we have then $h_r(X) = \sum_i h^i_r(X)$.

Conjecture A. Every $X \in \mathcal{V}$ has a Chow–Künneth decomposition.

Remarks.

1. Note that conjecture A implies in particular that the Künneth components are algebraic; however the conjecture is more subtle.
2. Note that, as is already seen in the easy example of curves, the π_i are *not* unique *as cycle classes.* What we do expect however, is that the $h^i_r(X) = (X, \pi_i)$ are unique, up to isomorphism, *as motives.*

7.15. Facts. Chow–Künneth decomposition exists at least in the following cases:

1. Curves.
2. Surfaces. ([Mur 2])
3. Abelian varieties ([Sh]). In fact, using the so–called Fourier transform, it can be shown [DM]) that in that case there are canonical π_i determined by the condition

that ${}^t\Gamma_n \bullet \pi_i = \pi_i \bullet {}^t\Gamma_n = n^i \pi_i$, where $n : A \to A$ is the multiplication by n on the abelian variety A. Moreover Künnemann has given explicit formulas for such π_i ([Kü-1]).
4. For any $X_d \in \mathcal{V}$ there exists, except the trivial components $h_r^0(X)$ and $h_r^{2d}(X)$, also Chow motives $h_r^1(X)$ (the "Picard-motive") and $h_r^{2d-1}(X)$ (the "Albanese-motive") which are potential components for the Chow–Künneth decomposition (and which carry as Chow groups precisely the Picard, resp. the Albanese variety of X) ([Mur 2]).

7.16. Conjectural filtration on the Chow groups.

Let $X_d \in \mathcal{V}$. In [Mur 3] we have "described" a conjectural filtration of the Chow groups based upon such a Chow–Künneth decomposition. This goes via the following list of conjectures:

Conjecture A: see 7.14.

Now the π_i operate on the Chow groups $CH^j_{\mathbb{Q}}(X)$. In the following fix j now $(0 \le j \le d)$.

Conjecture B: $\pi_{2d}, \pi_{2d-1}, \ldots, \pi_{2j+1}$ on the one hand, and $\pi_0, \pi_1, \ldots, \pi_{j-1}$ on the other hand, operate as zero on $CH^j_{\mathbb{Q}}(X)$.

Assuming conjectures A and B, *we define by induction* a decreasing filtration on $CH^j_{\mathbb{Q}}(X)$ as follows:

a. $F^0 = CH^j_{\mathbb{Q}}(X)$
b. $F^\nu = \mathrm{Ker}(\pi_{2j-\nu+1}|F^{\nu-1})$ for $\nu = 1, 2, \ldots, j+1$ (where, as usual, $\pi_r|F^s$ means the restriction of π_r to F^s).
Then we have $F^{j+1}CH^j_{\mathbb{Q}}(X) = 0$ (see [Mur 3], remark 1.4.2.).

Conjecture C: the filtration is independent of the ambiguity in the choices of the cycle classes π_i.

One sees easily now that $F^1 \subset CH^j_{\mathbb{Q},\mathrm{hom}}(X)$, where $CH^j_{\mathbb{Q},\mathrm{hom}}(X)$ are the classes homologically equivalent to zero ([Mur 3], 1.4.4). Finally one makes the crucial

Conjecture D: $F^1 = CH^j_{\mathbb{Q},\mathrm{hom}}(X)$.

7.17. Status of conjectures.

The above conjectures are trivially true for curves, they are true for surfaces ([Mur 2]). Furthermore conjectures A, B and D are true for threefolds of type $X = C_1 \times C_2 \times C_3$ or $X = S \times C$, with C a curve and S a surface ([Mur 3], Part II). For abelian varieties A is true (see above) and part of B and C is true and the remaining part is equivalent with a conjecture of Beauville [Be]; for details see [Mur 3], Part I.

7.18. Final Remarks.

1. Relation with Bloch and Beilinson conjectures. U. Jannsen has shown ([Ja 3]) that the above conjectures are equivalent with Beilonson's conjectures on a filtration on the Chow groups as stated in [Ja 1], section 11.1 and that both conjectural filtrations coincide. One should also compare the above conjectures with the recent works of M. Saito [SaM], S.Saito [SaS] and W. Raskind [Ras]..

2. Filtration by coniveau. The Chow groups have in addition to the filtration discussed above (coming from motives) also another filtration coming from the level or coniveau (also sometimes called arithmetical filtration; see [Gr 4]). For instance algebraic equivalence somes into this setting. The "motive" and the "coniveau" filtration should be combined in order to get the complete picture (see [Mur 3], example 3.2.1).

3. As further evidence for the importance of Chow motives we mention Künnemann's recent work on the Lefschetz decomposition of Chow motives on abelian varieties ([Kü 2]).

References

[EGA] Grothendieck, A. and Dieudonné, J.: Éléments de Géométrie Algébrique, Chap III, Publ. Math. IHES, Vol 11 (1961).

[SGA6] Grothendieck, A., Berthelot, P. and Illusie, L.: Séminaire de Géométrie Algébrique, Théorie des Intersections et Théorème de Rieman-Roch, LNM 225 (1971), Springer Verlag, Berlin, etc.

[Ba] Bardelli,F.: On Grothendieck's generalized Hodge conjecture for a family of threefolds with trivial canonical bundle, Journ.reine und ang.Math. 422 (1991), 165-200.

[Be] Beauville, A.; Sur l'anneau de Chow d'une variété abélienne, Math.Ann. 273 (1986), 647-651.

[Bei] Beilinson, A.; Higher regulators and values of L-functions, Journ. Soviet Math. 30 (1985), 2036-2070.

[B] Bloch, S.: Lectures on algebraic cycles, Duke Univ. Math. Series IV, 1980.

[B 1] Bloch, S.: K_2 and algebraic cycles, Ann. of Math. 99 (1974), 349-379.

[B 2] Bloch, S.: Torsion algebraic cycles, K_2, and the Brauer group of function fields, LNM 844 (1981), 75-102, Springer verlag, Berlin, etc.

[B 3] Bloch, S.: Torsion algebraic cycles and a theorem of Roitman, Compos. Math. 39 (1979), 108-127.

[Bo] Bloch, S. and Ogus, A.: Gersten's conjecture and the homology of schemes, Ann. Scient. Éc.Norm. Sup. 7 (1974), 183-202.

[Bo] Borel, A. et al: Intersection cohomology, Progress in Math. 50 (1984), Birkhäuser Verlag, Basel, etc.

[BS] Borel, A. and Serre, J.-P.: Le théorème de Riemann–Roch (d'après Grothendieck), Bull. Soc. Math. de France 86 (1958), 97-136.

[BZ] Brylinski, J.-L. and Zucker, S.: An overview of recent advances in Hodge theory, Encycl. of Math. Sciences, Vol 69 (1990), 39-142, Springer-Verlag, Berlin, etc.

[Che] Chevalley, C.; Anneaux de Chow et applications, Séminaire Chevalley 1958, Secr. Math. Paris.

[Cho] Chow, W.L.: On the equivalence classes of cycles in an algebraic variety, Ann. of Math. 64 (1956), 450-479.

[Cl] Clemens, H.: Homological equivalence modulo algebraic equivalence is not finitely generated, Publ.Math.IHES 58 (1983), 231-258.

[Co] Collino, A.: Quillen's K-theory and algebraic cycles on singular varieties. In: Geometry of Today, Rome 1984, Progress in Math., Vol 60, 75-85 (1985) Birkhäuser Verlag, Basel, etc.

[CM 1] Conte, A. and Murre, J.P.: The Hodge conjecture for fourfolds admitting a covering by rational curves, Math.Ann. 238 (1978), 461-513.

[CM 2] Conte, A. and Murre J.P.: The Hodge conjecture for Fano complete intersections of dimension four, Proc. Conf. Algebraic Geometry, Angers 1979, 129-142.

[CT] Colliot-Thélène, J.-L.: Cycles algébriques de torsion et K-théorie algébrique, Cours donné au CIME, Juin 91, Prepublic. 92-14, Univ. Paris-Sud, Orsay.

[De] Deligne, P.: Théorie de Hodge II, Publ. Math. IHES 40, (1972), 5-57.

[DMOS] Deligne, P., Milne, J.S., Ogus, A. and Shih, K.: Hodge cycles, motives and Shimura varieties, LNM 900, (1982), Spinger Verlag, Berlin, etc.

[DM] Deninger, C. and Murre, J.P.: Motivic decomposition of abelian schemes and the Fourier transform, Journal reine und ang. Math. 422 (1991), 201-219.

[El] Elkik, R.: L'équivalence rationelle, Exp II Sém. Douady-Verdier, Astérisque 36-37, Paris 1976.

[EV] Esnault, H. and Viehweg, E.: Deligne-Beilinson cohomology. In: Beilinson's conjectures on special values of L-functions, Perspect. in Math. Vol.4 (1988), 43-91, Academic Press, Boston,etc.

[EZ] El Zein, P. and Zucker, S.: Extendability of normal functions associated to algebraic cycles. In: Topics in Transc.Alg.Geom., Annals of Math.Studies 106 (1984), 269-288, Princeton Univ. Press.

[F] Fulton, W.: Intersection Theory, Ergeb.der Math. 3 Folge, Band 2, 1984, Springer Verlag, Berlin, etc.

[FL] Fulton, W. and Lang, S.: Riemann-Roch Algebra, Grund.der Math. Wiss. 277, 1985, Springer Verlag, Berlin, etc.

[vG] Geemen, B. van: An introduction to the Hodge conjecture for abelian varieties, this volume.

[GH] Griffiths, P. and Harris, J.: Principles of Algebraic Geometry, 1978, Wiley-Inter-
science, New York, etc.

[Gre] Green, M.L.: Infinitesimal methods in Hodge theory, CIME notes Torino 93, this volume.

[Gri] Griffiths, P.A.: On the periods of certain rational integrals, I, II, Annals of Math. 90 (1969), 460-541.

[Gr 1] Grothendieck, A.: La théorie des classes de Chern, Bull. Soc. Math. France 86 (1958), 137-154.

[Gr 2] Grothendieck, A.: Technique de descente et théorèmes d'existence en géométrie algébrique, V. Les schémas de Picard. Théorémes d'existence. Séminaire Bourbaki, t. 14, 1961-62, No 232, Secretar. Math. Paris, 1962.

[Gr 3] Grothendieck, A.: Standard conjectures on algebraic cycles. In: Algebraic Geometry, Bombay, 1968, Oxford Univ. Press, (1969), 193-199, Oxford.

[Gr 4] Grothendieck, A.: Hodge's general conjecture is false for trivial reasons, Topology 8 (1969), 299-303.

[H] Hartshorne, R.; Algebraic Geometry, Grad. Texts in Math. 52, 1977, Spinger Verlag, Berlin, etc.

[H 1] Hartshorne, R.: Equivalence relations on algebraic cycles and subvarieties of small codimension. In: Algebraic Geometry, Arcata 1974. AMS Proc. Symp. Pure Math. 29, 129-164, (1975).

[Hi] Hironaka, H.: Smoothing of algebraic cycles of small dimension, Amer. J. Math. 90, 1-54 (1968).

[Ho] Hodge, W.V.D.: The topological invariants of algebraic varieties, Proc. Int. Congr. Math, Cambridge 1950, 182-192 (1950).

[Ja 1] Jannsen, U.: Mixed motives and algebraic K-theory, LNM 1400,(1990), Springer Verlag, Berlin, etc.

[Ja 2] Jannsen, U.: Motives, numerical equivalence and semi-simplicity, Invent. Math. 107 (1992), 447-452.

[Ja 3] Jannsen, U.: Motivic sheaves and filtrations on Chow groups, Forthcoming in AMS conference on motives in Seattle 1991.

[vdK] Kallen, W. van der: Generators and relations in algebraic K-theory, Proc. Int. Congress Math., Helsinki 1978, 305-310 (1978).

[vdK 2] Kallen, W. van der: Le K_2 des nombres duaux, C.R. Ac. Sci. Paris 273, 1209-1207 (1971).

[Kl 1] Kleiman, S.L.: Algebraic cycles and the Weil conjectures. In: Dix exposés sur la chohomologie des schémas, 359-389, North Holland, Amsterdam 1968.

[Kl 2] Kleiman, S.L.: Motives. In: Algebraic Geometry Oslo Conf. 1970, 53-82, Wolters-Noordhoff, Groningen 1972.

[Kl 3] Kleiman, S.L.: The standard conjectures, Forthcoming in AMS conference on motives in Seattle, 1991.

[Kü 1] Künnemann, K.: On the motive of an abelian scheme. Preprint 1991 forthcoming in AMS conference on motives in Seattle, 1991.

[Kü 2] Künnemann, K.: A Lefschetz decomposition for Chow motives of abelian schemes, Invent.Math. 113, 85-103 (1993).

[L] Lefschetz, S.: L'analysis situs et la geométrie algébrique (1924), see: Selected Papers, Chelsea, 283-439 (1971).

[Le] Lewis, J.D.: A survey of the Hodge conjecture, lectures C.R.M., Montreal, 1990.

[Li] Lieberman, D.: Intermediate Jacobians. In: algebraic Geometry Oslo Conf. 1970, 125-139, Wolters-Noordhoff, Groningen 1972.

[LP et al] Lluis-Puebla, E., Loday, J.L., Gillet, H., Soulé, C. and Snaith, V. : Higher algebraic K-theory, an overview, LNM 1491 (1992), Springer-Verlag, Berlin, etc.

[Ma 1] Manin, Y.I.: Correspondences, motifs and monoidal transformations, Math. USSR Sb. 6, (1968) 439-470.

[Ma 2] Manin, Y.I.: Lectures on the K-functor in algebraic geometry, Russ. Math. Surveys 24, No 5 (1969), 1-89.

[M] Milne, J.S.: Étale cohomology, Princ. Math. Series 33, Princeton Univ. Press, Princeton, 1980.

[Mi] Milnor, J.: Introduction to algebraic k-theory, Ann. of Math. Studies 72, Princeton Univ. Press, Princeton 1970.

[MS] Merkurjev, A.S. and Suslin, A.A.: K–cohomology of Severi–Brauer varieties and norm residue homomorphisms, Math. USSR Izv. 21 (1983), 307-340.

[Mum] Mumford, D.: Rational equivalence of zero–cycles on surfaces, J. Math. Kyoto Univ., 9 (1969), 195-204.

[Mur 1] Murre, J.P.: Applications of algebraic K–theory to the theory of algebraic cycles, In: Proc. Conf. Algebraic Geometry, Sitges 1983, LNM 1124 (1985), 216-261, Springer Verlag, Berlin, etc.

[Mur 2] Murre, J.P.: On the motive of an algebraic surface, Journal reine und angew. Math. 409 (1990), 190-204.

[Mur 3] Murre, J.P.: On a conjectural filtration on the Chow groups of an algebraic variety, Indag. Math, New Series, 4 (1993), 177-201.

[P] Pierce, R.S.: Associative Algebras, Grad. Texts in Math. 88, Springer Verlag, Berlin, etc., 1982.

[PW] Pedrini, C. and Weibel, C.: K–theory and Chow groups of singular varieties. In: AMS Contemp. Math. 55 (1986), 339-370.

[Q] Quillen, D.: Higher algebraic K–theory, LNM 341, 85-147, Springer Verlag, Berlin, etc., 1973.

[Ra] Ran, Z.: Cycles on Fermat hypersurfaces, Comp.Math. 42 (1980), 121-142.

[Ras] Raskind, W.: Higher ℓ–adic Abel–Jacobi mappings and filtrations on Chow groups, Preprint 1993.

[Sa 1] Samuel, P.: Rational equivalence on algebraic cycles, Am.J. of Math. 78 (1956), 383-400.

[Sa 2] Samuel, P.: Rélations d'équivalence en géométrie algébrique, Proc. Int. Cong. Math., Edinburgh 1958, 470-487.

[SaH] Saito, H.: Abelian varieties attached to cycles of intermediate dimension, Nagoya Math.J. 75 (1979) 95-119.

[SaM] Saito, M.: On the bijectivity of some cycle maps, Forthcoming in AMS conference on motives Seattle 1991.

[SaS] Saito, S.: Motives and filtrations on Chow groups, preprint 1992.

[Sc] Schoen, C.: Hodge classes on self–products of a variety with an automorphism, Comp.Math. 65 (1988), 3-32.

[Sch] Scholl, A.J.: Classical motives, Forthcoming in AMS conference on motives Seattle 1991.

[She] Shermenev, A.M.: The motif of an abelian variety, Funkt.Anal. 8 (1974), 47-53.

[Sh 1] Shioda, T.: The Hodge conjecture for Fermat varieties, Math. Ann. 245 (1979), 175-184.

[Sh 2] Shioda,T.: Algebraic cycles on abelian varieties of Fermat type, Math.Ann. 258 (1981), 65-80.

[Sh 3] Shioda, T.: What is known about the Hodge conjecture? In: advanced studies in pure Math. 1, 55-68 (1983), Kinokuniya Comp. Tokyo and North Holland Publ. Comp. Amsterdam.

[SK] Shioda, T. and Katsura, T.: On Fermat varieties, Tohoku Math. J. 31, 97-115 (1979).

[So] Soulé, C.: Régulateurs. Sémin. Bourbaki 1984-85, exp. 644, Astérisque 133-134 (1986), 237-253.

[Sr] Srivinas, V.: Algebraic K–theory, Progress in Math. Vol 90, 1991, Birkhäuser Verlag, Boston, etc.

[Ste] Steenbrink, J.H.M.: Some remarks on the Hodge conjecture, LNM 1246, 165-175 (1987), Springer Verlag, Berlin, etc.

[Sti 1] Stienstra, J.: On the formal completion of the Chow group $CH^2(X)$ for a smooth projective surface in characteristic zero, Indag. Math. 45, 361-382 (1983).

[Sti 2] Stienstra, J.: Cartier–Dieudonné theory for Chow groups, Journ.reine und angew. Math. 355, 1-66 (1985).

[Tj] Tjurin, A.N.: Five lectures on three dimensional varieties, Russ.Math.Surveys 27 (1972), 1-53.

[Vo] Voisin, C.: Transcendental methods in the study of algebraic cycles, CIME notes Torino 1993, this volume.

[We] Weil, A.: Sur les critères d'équivalence en géométrie algébrique, Math. Ann. 128, 95-127 (1954).

TRANSCENDENTAL METHODS IN THE STUDY OF ALGEBRAIC CYCLES

Claire VOISIN

Université d'Orsay et IHES

In these lectures, in contrast with the orientation of M. Green's lectures, we put the accent on the global arguments used when working with families of algebraic varieties (monodromy arguments), and also on the relations between non representability of Chow groups and the "transcendental part of Hodge structures", which explains the title, excepted for the fact that there are very few things that cannot be done within algebraic geometry.

Each lecture is introduced, and they are independent, excepted for lecture 5 - lecture 6, and lecture 7 - lecture 8. The lectures should be read in parallel with those of M. Green and J.P. Murre which they are supposed to complete. They are organized as follows :

1) Divisors

2) Topology and Hodge theory

3) Noether-Lefschetz locus

4) Monodromy

5) 0-cycles I

6) 0-cycles II

7) Griffiths groups

8) Application of the Noether-Lefschetz locus to threefolds.

I am very grateful to the C.I.M.E. foundation and to Fabio Bardelli for giving me the opportunity to deliver these lectures and to collaborate with M. Green and J.P. Murre. I also thank the IHES for its hospitality during the academic year of 1992/93.

M. Green et al.: LNM 1594, A. Albano and F. Bardelli (Eds.), pp. 153–222, 1994.

Lecture one : Divisors

0. This lecture is devoted to the classical and well understood subject of divisors on an algebraic variety. What we want to do is to review the main features of the theory of divisors, the codimension one cycles, from the point of view of algebraic geometry and Hodge theory, which fit very well in this case. In particular, we want to insist on the finiteness statements which have been shown by Clemens and Mumford to be specific of the codimension one cycles.

In contrast, the remaining lectures will illustrate the fact that when the objects constructed from the Hodge theory become more transcendental, the corresponding theory of algebraic cycles is less harmonious.

1. Recall that on a smooth algebraic variety X, one can identify the Weil divisors, which are formal sums $\Sigma n_i D_i$ of codimension one irreducible subvarieties affected with integral coefficients $n_i \in \mathbf{Z}$, and Cartier divisors, which are sections of the Zariski sheaf $\mathcal{K}_X^*/\mathcal{O}_X^*$ that is are described in a covering (U_i) of X by Zariski open sets by a collection of rational functions $\psi_i \in \mathcal{K}_X^*$ such that ψ_i/ψ_j is invertible in $U_i \cap U_j$. To the Cartier divisor ψ_i corresponds its local factorization into product of prime, locally principal, ideals. To such a Cartier divisor is also associated an algebraic line bundle, the rank one $\mathcal{O}_X$-submodule of the constant sheaf $\mathcal{K}_X$ generated by ψ_i^{-1} on U_i. The isomorphism class of this line bundle is determined by the class in $H^1_{\text{zar}}(\mathcal{O}^*_{X,\text{alg}})$ of the Čech cocycle $g_{ij} = \psi_i/\psi_j \in \mathcal{O}^*_{U_{ij}}$.

The line bundles associated to two Cartier divisors (ψ_i), (ψ_i') are isomorphic if and only if there is a global rational function ψ on X such that $\psi_i' = \psi\psi_i$ modulo an invertible function on U_i, and then the corresponding Weil divisors D and D' differ by the principal divisor $\operatorname{div}\psi$. One obtains this way a bijection between the group $CH^1(X) :=$ {Weil divisors modulo principal Weil divisors} and the group Pic X of algebraic line bundles modulo linear equivalence (i.e. isomorphisms). The abelian group structure of Pic X is given by the tensor product of line bundles.

The study of Pic X splits into two parts : One introduces the notion of algebraic equivalence between line bundles. A line bundle L on X is algebraically equivalent to zero if there is an irreducible curve C and a line bundle $\mathcal{L}$ on $C \times X$, two points c and $c' \in C$ such that $\mathcal{L}_{|c\times X} \simeq \mathcal{O}_X$ and $\mathcal{L}_{|c'\times X} \simeq L$. These line bundles form a subgroup $\operatorname{Pic}^0 X \subset \operatorname{Pic} X$.

In [5] one can find, for a complete variety X, a purely algebraic construction of $\operatorname{Pic}^0 X$ as an abelian variety, that is a complete commutative algebraic group. One of the main tools is the identification of the space of the first order deformations of a line bundle to the space $H^1(\mathcal{O}_X)$ via the correspondence: {multiplicative one cocycle $g_{ij} + \epsilon f_{ij}$ on $X \times \operatorname{Spec} k[\epsilon]/\epsilon^2 \rightarrow$ additive one cocycle $f_{ij}/g_{ij} \in \mathcal{O}_{v_{ij}}$}.

The quotient $\operatorname{Pic} X/\operatorname{Pic}^0 X$ can also be studied by purely algebraic methods. Using vanishing theorems one can prove that if $S \subset X$ is a smooth surface, complete intersection of ample divisors, one has an isomorphism $\operatorname{Pic}^0 X \simeq \operatorname{Pic}^0 S$ and an injection $\operatorname{Pic} X \subset \operatorname{Pic} S$. So $\operatorname{Pic} X/\operatorname{Pic}^0 X \hookrightarrow \operatorname{Pic} S/Pic^0 S$. Now on the surface S, the Grothendieck-Hirzebruch-Riemann-Roch formula is a powerful tool to study line bundles. The results that one can obtain this way are the following :

a) A line bundle is said to be numerically equivalent to zero if it is of degree zero on any curve $C \subset X$. This determines a subgroup $\operatorname{Num} X \subset \operatorname{Pic} X/\operatorname{Pic}^0 X$ and one has : $\operatorname{Num} X$ is a finite group.

b) The quotient $NS(X) = \operatorname{Pic} X/\operatorname{Num} X$ is a finitely generated group of rank $\rho(X)$, called the Néron-Severi group.

c) Hodge index theorem : If X is a surface, the intersection theory puts an intersection form on $NS(X)$, [1], which by definition of numerical equivalence is non degenerate. Then its index is $\rho(X) - 1$.

2. Now we assume that X is a complex projective algebraic variety and we turn to the Hodge theoretic description of line bundles. The starting point is the theorem of Serre [6], also called "GAGA principle" :

2.1. Theorem : The functor $E \rightarrow E^{an}$ which to a coherent sheaf of $\mathcal{O}_X$-modules associates the corresponding coherent sheaf of $\mathcal{O}_X^{an}$-modules, for the usual topology, via the continuous map of schemes $X^{an} \rightarrow X^{zar}$ is an equivalence of category.

It follows that for algebraic X we can identify $\operatorname{Pic} X$ with the group of analytic line bundles, which is itself isomorphic to $H^1(X, \mathcal{O}_X^{an*})$. (As before to a line bundle corresponds the class of the Čech cocycle of its transition functions, constructed from a set of trivializations). The cohomology is now understood in the usual topology. In the sequel, we will use $\mathcal{O}_X, \mathcal{O}_X^*$ for $\mathcal{O}_X^{an}, \mathcal{O}_X^{an*}$ and work with the usual topology. For the cohomology of coherent sheaves, the confusion is allowed by Serre's theorem.

The second main tool is the exponential exact sequence.

2.2. $0 \rightarrow \mathbb{Z} \stackrel{2i\pi}{\rightarrow} \mathcal{O}_X \stackrel{\exp}{\rightarrow} \mathcal{O}_X^* \rightarrow 0$, which shows that $\mathcal{O}_X^*$ is quasi isomorphic up to a shift to the complex: $\mathbb{Z}(1) : 0 \rightarrow \mathbb{Z} \stackrel{2i\pi}{\rightarrow} \mathcal{O}_X \rightarrow 0$, where $\mathcal{O}_X$ is put in degree one. This complex is the first Deligne complex and its cohomology $H^2_D(X,\mathbf{Z}(1)) \simeq H^1(\mathcal{O}_X^*)$ is the first Deligne cohomology group. The generalization of it will be explained by Jacob Murre. There is an obvious map of complexes : $\mathbf{Z}(1) \to \mathbf{Z}$, which has for kernel the sheaf $\mathcal{O}_X$ concentrated in degree one. So one can associate to a line bundle L its Deligne-Chern class $C_1^D(L) \in H^1(\mathcal{O}_X^*) \simeq H^2_D(X,\mathbf{Z}(1))$ and its topological Chern class $c_1(L) \in H^2(X,\mathbf{Z})$. Using the exponential exact sequence, $c_1(L)$ can also be represented as a Čech cocycle $\alpha_{ijk} = \frac{1}{2i\pi}(\log g_{ij} + \log g_{jk} + \log g_{ki}) \in \mathbf{Z}$, and it follows that its image in $H^2(X,\mathbf{C})$ identifies, via the De Rham isomorphisms, to the class of the closed two form :

2.3. $\omega_L = \frac{1}{2i\pi}\partial\overline{\partial}\log h(\sigma)$, where h is a hermitian metric on L and σ is a local nowhere zero holomorphic section of L on which ω_L actually does not depend. This is the Chern-Weil construction of the first Chern class.

From the formula 2.3 it is not difficult to show that $c_1(L)$ is also Poincaré dual of the homology class $\Sigma n_i[D_i] \in H_{2n-2}(X,\mathbf{Z})$ of the divisor $\Sigma n_i D_i$ of any meromorphic section of L (Lelong formula).

If one considers the long exact sequence associated to 2.2, one finds that the following is exact at the middle :

2.4. $\quad H^1(\mathcal{O}_X^*) \xrightarrow{c_1} H^2(X,\mathbf{Z}) \to H^2(X,\mathcal{O}_X)$, which gives :

2.5. The Lefschetz theorem on $(1,1)$ classes :

An integral class $\alpha \in H^2(X,\mathbf{Z})$ is the first Chern class of a Line bundle (or the Poincaré dual of the homology class of a Weil divisor) if and only if its $(0,2)$ component vanishes.

Here we have identified the map $H^2(X,\mathbf{Z}) \to H^2(X,\mathcal{O}_X)$ to the composite map $H^2(X,\mathbf{Z}) \to H^2(X,\mathbf{C}) \to H^2(X,\mathcal{O}_X)$, and the last map is described in the DeRham/Dolbeault cohomology as the map $\{d\text{-closed 2-form } \omega\} \to \{(0,2)$ component of ω, which is $\overline{\partial}$-closed$\}$. Using the Hodge decomposition $H^2(X,\mathbf{C}) = H^{2,0} \oplus H^{1,1} \oplus H^{0,2}$, this map is also the projection on the last factor, and its kernel is $F^1H^2(X,\mathbf{C}) := H^{2,0}(X) \oplus H^{1,1}(X)$. Finally using the fact that under complex conjugation $\overline{H^{2,0}(X)} = H^{0,2}(X)$ one sees that a class $\alpha \in H^2(X,\mathbf{Z}) \cap F^1H^2(X)$ is in fact in $H^{1,1}_{\mathbf{R}}(X) = \{$ classes of real closed 2-forms of type $(1,1)\}$, as is expected from formula 2.3.

Returning to the beginning of the long exact sequence associated to 2.2, one sees that the kernel of the map c_1 is the quotient $H^1(\mathcal{O}_X)/H^1(X,\mathbf{Z})$. What one learns then from the Hodge theory is the following :

One has a decomposition $H^1(X,\mathbb{C}) = H^{1,0} \oplus H^{0,1}$, where $H^{0,1} \simeq H^1(\mathcal{O}_X) \simeq \overline{H^{1,0}(X)} = \overline{H^0(\Omega_X)}$. It follows that the composite $H^1(X,\mathbb{R}) \to H^1(X,\mathbb{C}) \to H^{0,1}(X)$ is an isomorphism of real vector spaces, and that $H^1(X,\mathbb{Z})$ projects to a lattice in $H^1(\mathcal{O}_X)$. The quotient $H^1(\mathcal{O}_X)/H^1(X,\mathbb{Z})$ is then a complex torus, with the complex structure given by the complex structure on its tangent space $H^1(\mathcal{O}_X)$. It is in fact an algebraic torus (or an abelian variety) : As explained in [4], a complex torus $T = \mathbb{C}^m/\Gamma$ is algebraic if one has a skew symmetric form $\omega_{\mathbb{Z}}$ on Γ, whose extension $\omega_{\mathbb{R}}$ to $\mathbb{C}^m = \Gamma \otimes \mathbb{R}$ is of type $(1,1)$ for the complex structure on $\mathbb{C}^m$, and whose associated sesquilinear form h is positive definite on $\mathbb{C}^n$, where $h(X,X) = -\omega(X,iX)$.

Equivalently, one can consider the complexification $\omega_{\mathbb{C}}$ of ω to a form on $\Gamma \otimes \mathbb{C}$, use the splitting $\Gamma \otimes \mathbb{C} \simeq \Gamma^{1,0} \oplus \Gamma^{0,1}$ given by the complex structure on T and define h on $\Gamma^{1,0}$ by

2.6. $h_\omega(X,X) = \frac{i}{2}\, \omega_{\mathbb{C}}(X,\overline{X}).$

By 2.5, $\omega \in \Lambda^2 H_1(T,\mathbb{Z})^* \simeq H^2(T,\mathbb{Z})$ will then be the class of a line bundle, which is ample by the positivity of h.

To construct such a polarization on $T = H^1(\mathcal{O}_X)/H^1(X,\mathbb{Z})$, one fixes $\alpha \in H^2(X,\mathbb{Z})$ the Chern class of an ample line bundle on X, and one defines :

2.7. $\omega_{\mathbb{Z}}(\varphi,\psi) = -\int_X \alpha^{n-1}\varphi \wedge \psi$

Using the previous notations, we have $\Gamma \otimes \mathbb{C} = H^1(X,\mathbb{C})$ with $\Gamma^{1,0} \simeq H^1(\mathcal{O}_X)$.

The fact that $\omega_{\mathbb{R}}$ is of type $(1,1)$ on T is equivalent to the vanishing of $\omega_{\mathbb{C}|H^1(\mathcal{O}_X)}$, which is clear by $\omega_{\mathbb{C}}(\varphi^{0,1},\psi^{0,1}) = -\int_X \alpha^{n-1} \wedge \varphi^{0,1} \wedge \psi^{0,1}$ and $\alpha^{n-1} \wedge \varphi^{0,1} \wedge \psi^{0,1} = 0$ on X.

The positivity condition reduces then to the statement that $h_{\mathbb{C}}(\varphi^{0,1},\varphi^{0,1}) = -\frac{i}{2}\int_X \alpha^{n-1} \wedge \varphi^{0,1} \wedge \overline{\varphi^{0,1}} = \frac{i}{2}\int_X \alpha^{n-1} \wedge \varphi^{1,0} \wedge \overline{\varphi^{1,0}}$ is strictly positive when $\varphi^{0,1} = \overline{\varphi^{1,0}}$ is a non-zero element in $H^1(\mathcal{O}_X)$. This is a particular case of the "Hodge-Riemann bilinear relations" which will be explained in the next lecture. In fact the positivity of $h_\omega(\varphi^{0,1},\varphi^{0,1})$ can be checked directly by a reduction to the case of a curve : α being the class of an ample line bundle on X, there is a smooth curve $C \subset X$, such that $N\alpha^{n-1}$ is the Poincaré dual of the homology class of C, for some integer $N > 0$. Then :

2.8. $\frac{i}{2}\int_X \alpha^{n-1}\overline{\varphi^{0,1}} \wedge \varphi^{0,1} = \frac{1}{N}\frac{i}{2}\int_C \overline{\varphi^{0,1}} \wedge \varphi^{0,1}$ and it is easy to check that $i\overline{\varphi^{0,1}} \wedge \varphi^{0,1}$ is a positive 2-form on C (with respect to the orientation given by its complex structure).

3. The fact that T is an abelian variety will now imply that it is isomorphic to $\mathrm{Pic}^0 X$. It is clear that algebraically equivalent line bundles are topologically equivalent, hence have the same c_1. So points of $\mathrm{Pic}^0 X$ correspond to points in T.

On the other hand one can construct a line bundle on $T \times X$ called the Poincaré line bundle, using the Lefschetz theorem on $(1,1)$ classes :

Consider $H^1(T,\mathbf{Z}) \otimes H^1(X,\mathbf{Z}) \subset H^2(T \times X, \mathbf{Z})$ in the Künneth decomposition. By definition of T, $H^1(T,\mathbf{Z})$ is canonically the dual of $H^1(X,\mathbf{Z})$ so we have a natural class $e \in H^2(T \times X, \mathbf{Z})$ corresponding to $\mathrm{Id} \in H^1(X,\mathbf{Z})^* \otimes H^1(X,\mathbf{Z})$.

Next we show that e is of type $(1,1)$ in the Hodge decomposition of $H^2(T \times X)$. That is we have to check that $e \in H^{1,0}(T) \otimes H^{0,1}(X) \oplus H^{0,1}(T) \otimes H^{1,0}(X)$. But from $T = H^1(\mathcal{O}_X)/H^1(X,\mathbf{Z})$, we deduce that $(1,0)$-forms on T identify to linear forms on $H^1(X,\mathbf{C})$ which vanish on $H^{1,0}(X)$ and $(0,1)$-forms on T identify to linear forms on $H^1(X,\mathbf{C})$ which vanish on $H^{0,1}(X)$. If (ω_i) is a basis for $H^{1,0}(X)$, $(\overline{\omega}_i)$ the conjugate basis of $H^{0,1}(X)$, and $(\omega_i^*, \overline{\omega}_i^*)$ the dual basis of $H^1(X,\mathbf{C})^* = H^1(T,\mathbf{C})$ one finds that $e = \Sigma\, \omega_i^* \otimes \omega_i + \Sigma\, \overline{\omega}_i^* \otimes \overline{\omega}_i$, where the ω_i^*'s vanish on $H^{0,1}(X)$ hence are in $H^{0,1}(T)$ and the $\overline{\omega}_i^*$'s vanish on $H^{1,0}(X)$ hence are in $H^{1,0}(T)$. So e is of type $(1,1)$.

It follows from 2.5 that $e = c_1(\mathcal{L})$ for some line bundle $\mathcal{L}$ on $T \times X$, which is uniquely defined up to line bundles with vanishing Chern classes, coming from T and X. $\mathcal{L}$ is uniquely defined if we impose $\mathcal{L}_{|0 \times X} = \mathcal{O}_X$ and $\mathcal{L}_{|T \times x} = \mathcal{O}_T$ for some fixed point $x \in X$.

T is now a smooth connected algebraic variety which parametrizes line bundles on X via $t \mapsto \mathcal{L}_t := \mathcal{L}_{|T \times X}$. By definition of algebraic equivalence $\mathcal{L}_t$ is then algebraically equivalent to $\mathcal{O}_X = \mathcal{L}_0$. To conclude that $T = \mathrm{Pic}^0 X$ it suffices now to check:

3.1. The map $E : T \to T$ defined by $E(t) = \mathcal{L}_t \in \mathrm{Ker}(c_1 : \mathrm{Pic}\, X \to H^2(X,\mathbf{Z})) \simeq T$ is the identity.

3.1. is a very general fact concerning the Abel-Jacobi map and has a generalization given by the theorem on normal functions. So it may be useful to see the topological meaning of this statement.

First of all, let us describe in more geometric terms the map $\mathrm{Ker}\, c_1 \to H^1(\mathcal{O}_X)/H^1(X,\mathbf{Z})$: Coming back to Weil divisors, an element of $\mathrm{Ker}\, c_1$ is represented by $D = \Sigma n_i D_i$ such that $\Sigma n_i D_i$ is homologous to zero on X. Then if Γ is a $2n-1$ real chain such that $\partial\Gamma = \Sigma n_i D_i$, $\int_\Gamma$ acts on $H^{n,n-1}(X)$ and is well defined up to periods $\int_T$ for $T \in H_{2n-1}(X,\mathbf{Z})$. So one obtains a point $\Phi(D)$ in $H^{n,n-1}(X)^*/H_{2n-1}(X,\mathbf{Z})$. ($\Phi$ is the

Abel-Jacobi map for divisors; its careful definition, in particular the fact that $\int_\Gamma$ is well defined, will be given by M. Green in a more general context).

Now by the Poincaré duality $H^{n,n-1}(X)^*/H_{2n-1}(X,\mathbb{Z}) \simeq H^1(\mathcal{O}_X)/H^1(X,\mathbb{Z}) = T$. One version of Abel's theorem is the statement that $\Phi(D)$ is equal to $c_1^D(D) \in T$, via the last isomorphism. The proof of 3.1 is then immediate. A meromorphic section of $\mathcal{L}$ gives $\mathcal{D} \subset T \times X$ such that the homology class of $\mathcal{D}$ is the Poincaré dual of $c_1(\mathcal{L})$. For $t \in T$, we choose a path γ_t from 0 to t in T. Then $\bigcup_{s\in\gamma_t} D_s$ gives a $2n-1$ chain Γ_t in X with boundary $D_t - D_0$, and we have the identification $E(t) = \int_{\Gamma_t} \in H^{n,n-1}(X)^*/H_{2n-1}(X,\mathbb{Z}) \simeq T$.

It follows that the map $E_* : H_1(T,\mathbb{Z}) \to H_1(T,\mathbb{Z}) \simeq H_{2n-1}(X,\mathbb{Z})$ induced in homology by E is described geometrically as $\gamma \mapsto \mathcal{D}_\gamma = \bigcup_{t\in\gamma} D_t$, for γ a loop in T. In other words one has $E_*(\gamma) = p_{2*}(p_1^*\gamma \cap [\mathcal{D}])$ where p_1, p_2 are the projections from $T \times X$ to T and X, and $[\mathcal{D}]$ is the homology class of $\mathcal{D}$. Now $[\mathcal{D}]$ is Poincaré dual of $c_1(\mathcal{L})$, and the identification of $c_1(\mathcal{L})$ to $\mathrm{Id} \in H^1(T,\mathbb{Z}) \otimes H^1(X,\mathbb{Z})$ gives immediately $E_* = \mathrm{Id}_{H_1(T,\mathbb{Z})}$. Hence $E = \mathrm{Id}_T$. (It is not difficult to show that E is additive).

It follows now that the map c_1 embeds $\operatorname{Pic} X/\operatorname{Pic}^0 X$ in the finitely generated group $H^2(X,\mathbb{Z})$. The group $\operatorname{Num} X$ identifies to the torsion of $H^2(X,\mathbb{Z})$ for the following reason: by 2.5, torsion classes are first Chern classes of line bundles, and if $\alpha = c_1(L)$ one has for $C \subset X, d^0 L_{|C} = \int_c \alpha = 0$ for $\alpha \in \operatorname{Tors} H^2(X,\mathbb{Z})$. So we have the inclusion $\operatorname{Tors} H^2(X,\mathbb{Z}) \subset \operatorname{Num} X$. The converse uses the weak Lefschetz theorem and the Hodge index theorem, which will be explained in the next lecture : the first one will imply that if $\alpha = c_1(L)$ is a non-torsion class in $H^2(X,\mathbb{Z})$, for a surface $S \subset X$, a complete intersection of ample divisors, $\alpha_{|S} = c_1(L_{|S})$ is a non-torsion class in $H^2(S,\mathbb{Z})$. The second one will imply that the intersection form of $H^2(S,\mathbb{Z})/\text{Torsion}$ restricted to the set of divisor classes is non degenerate. Hence there is a curve $C \subset S$ such that $d^0 L_{|C} \neq 0$.

References :

[1] A. Beauville, Surfaces algébriques complexes, chapitre 1, Astérisque n° 54, Société Mathématique de France (1978).

[2] P. Griffiths, J. Harris, Principles of algebraic Geometry, pp 128-148, Pure and Applied Mathematics, John Wiley and Sons (1978).

[3] A. Grothendieck, Techniques de construction et théories d'existence en géométrie algébrique IV : Les schémas de Hilbert, Séminaire Bourbaki 1960/61, exposé n° 221.

[4] D. Mumford, Abelian varieties, chapter one, Tata Institute for fundamental Research, studies in Mathematics, Oxford University Press 1970.

[5] C.S. Seshadri, La variété de Picard d'une variété complète, Séminaire Chevalley 1958/59, exposé n°8.

[6] J.-P. Serre, Géométrie algébrique et géométrie analytique, Ann. de l'Inst. Fourier 6 (1956) 1-42.

[7] A. Weil, Variétés kählériennes, chapitres 5 et 6, Actualités Scientifiques et Industrielles 1267, Hermann, seconde édition 1971.

Lecture two : Topology and Hodge theory

0. In this lecture, we describe the consequences of the Hodge theory on the topology of algebraic varieties. The most important are the Lefschetz decomposition and the Hodge index theorem, or more generally, the Hodge Riemann bilinear relations, which lead to the notion of polarized Hodge structure.

From the point of view of topology, the weak Lefschetz theorem shows that the cohomology of a complex projective variety X of dimension n differs from the cohomology of its hyperplane section Y only in degrees n and $n-1$.

This suggests that working inductively on the dimension, one has only to study the primitive middle dimensional cohomology of X and Y to understand its relationship with the theory of algebraic cycles. However, excepted for the case of line bundles, the weak Lefschetz theorem has no direct analogue at the level of algebraic cycles.

The weak Lefschetz theorem has a more precise version given by the Lefschetz decomposition, which is proved by the Hodge theory, despite of its algebraic flavour. Interpretation of it in terms of motives is the contents of one of the standard conjectures [3]. We begin with the Morse theoretic proof of the weak Lefschetz theorem [4], and explain how the hard Lefschetz theorem follows then from the Hodge-Riemann bilinear relations. For the Lefschetz decomposition and polarizations on Hodge structures we follow [5].

1. Let $U \subset \mathbf{C}^N$ be a smooth complex analytic subvariety of dimension n. Let $<,>$ be the standard hermitian inner product on $\mathbf{C}^N$. Then for a general point $0 \in \mathbf{C}^N$, the function $f_0 : U \to \mathbf{R}$ defined by $f_0(X) = <\vec{0x}, \vec{0x}>$ is a Morse function on U, that is an exhaustion function with only one non degenerate critical point for each critical value. The Morse theory says that for each critical point x_0 of such a function f, such that $\mathrm{Hess}_{x_0} f$ has index k, the set $U_{f(x_0)+\epsilon} := \{x \in U / f(x) \leq f(x_0) + \epsilon\}$, for small ϵ, is obtained from $U_{f(x_0)-\epsilon}$ by glueing a k-disk on a $k-1$ sphere contained in $U_{f(x_0)-\epsilon}$.

Now we have the following (cf. [4]) :

1.1. Lemma : If $f = f_0$ is the squared distance function with respect to a hermitian metric on $\mathbf{C}^N$, the index k of any non degenerate critical point of f_0 satisfies :

$$k \leq n = \dim_{\mathbf{C}} U.$$

From this one deduces :

1.2. Theorem : An affine variety of complex dimension n has the homotopy type of a CW-complex of real dimension $\leq n$.

As an immediate consequence, one obtains the following vanishing theorems :

1.3. Corollary : Let U be an affine variety of dimension n. Then $H_k(U, \mathbf{Z}) = 0 = H^k(U, \mathbf{Z})$ for $k > n$.

Now, if $X \subset \mathbf{P}^N$ is a smooth projective variety of dimension n and $Y = \mathbf{P}^{N-1} \cap X$ is a smooth hyperplane section, $U := X \backslash Y$ is an affine variety of dimension n, $U \subset \mathbf{P}^N \backslash \mathbf{P}^{N-1} = \mathbf{C}^N$, and the Poincaré duality says that $H_k(U, \mathbf{Z}) = H^{2n-k}(X, Y, \mathbf{Z})$. The long exact sequence of relative cohomology for the pair (X, Y) and the corollary 1.3 now imply :

1.4. The weak Lefschetz theorem : The inclusion $j : Y \hookrightarrow X$ induces isomorphisms $j^* : H^k(X, \mathbf{Z}) \to H^k(Y, \mathbf{Z})$ for $k < n-1$ and an injection $j^* : H^{n-1}(X, \mathbf{Z}) \to H^{n-1}(Y, \mathbf{Z})$.

Let $\omega \in H^2(X, \mathbf{Z})$ be the Poincaré dual of the homology class of $Y \subset X$. Using Poincaré duality on X and Y, one obtains $j_* : H^k(Y) \to H^{k+2}(X)$, dual of $j^* : H^{2n-k-2}(X) \to H^{2n-k-2}(Y)$.

By the definition of ω it is then immediate that $L = L_X := j_* j^* : H^k(X) \to H^{k+2}(X)$ is equal to the cup-product with ω.

Identifying a neighbourhood of Y in X with a neighbourhood of the zero section of its normal bundle N in X, ω is also identified with the Thom class of N, and it is then a standard fact that $L_Y := j^* j_* : H^k(Y) \to H^{k+2}(Y)$ is equal to the cup-product with $\omega_{|Y}$. Now we have :

1.5. The hard Lefschetz theorem : For $k \leq n$, $L^{n-k} : H^k(X, \mathbf{Q}) \to H^{2n-k}(X, \mathbf{Q})$ is an isomorphism.

By the weak Lefschetz theorem and induction on the dimension it suffices to check it for $k = n-1$, as shown by the following diagram where both vertical maps are isomorphisms for $k \leq n-2$:

$$\begin{array}{ccc} H^k(X,\mathbb{Q}) & \xrightarrow{L^{n-k}} & H^{2n-k}(X,\mathbb{Q}) \\ j^* \downarrow & & \uparrow j_* \\ H^k(Y,\mathbb{Q}) & \xrightarrow[L^{n-k-1}]{} & H^{2n-k-2}(Y,\mathbb{Q}) \end{array}$$

For $k = n-1$, we still have the commutative diagram :

$$\begin{array}{ccc} H^{n-1}(X,\mathbb{Q}) & \xrightarrow{L} & H^{n+1}(X,\mathbb{Q}) \\ j^* \downarrow & & \uparrow j_* \\ H^{n-1}(Y,\mathbb{Q}) & \simeq & H^{n-1}(Y,\mathbb{Q}) \end{array}$$

where j^* is injective and its dual j_* is surjective. So via j^*, the kernel of L identifies with the kernel of the intersection form $<>_Y$ on $H^{n-1}(Y)$, restricted to $\operatorname{Im} j^* = (\operatorname{Ker} j_*)^\perp$. We have $\operatorname{Ker} j_* \subset \operatorname{Ker}(L_Y : H^{n-1}(Y,\mathbb{Q}) \to H^{n+1}(Y,\mathbb{Q}))$ and $\operatorname{Ker} j_*$ is a sub-Hodge structure of $\operatorname{Ker} L_Y$. The Hodge-Riemann bilinear relations to be explained below imply :

1.6. On $\operatorname{Ker} L_Y \cap H^{p,q}(Y)$, $p+q = n-1$, the hermitian form $<\varphi,\psi>_H = i^{p-q}\int_Y \varphi \wedge \overline{\psi}$ is non degenerate of a definite sign.

In particular it remains non degenerate of a definite sign on $\operatorname{Ker} j_* \cap H^{p,q}(Y)$, and it follows that $<>_Y$ is non degenerate on $\operatorname{Ker} j_*$, hence on $(\operatorname{Ker} j_*)^\perp$. So $L : H^{n-1}(X,\mathbb{Q}) \to H^{n+1}(X,\mathbb{Q})$ is injective, hence an isomorphism.

If one defines the primitive cohomology $H^k(X)^0$, for $k \leq n$ as the kernel of $L^{n-k+1} : H^k(X,\mathbb{Q}) \to H^{2n-k+2}(X,\mathbb{Q})$ the hard Lefschetz theorem implies now :

1.7. Lefschetz decomposition : The natural map $\varphi := \oplus L^k : \bigoplus_{m-k\leq n} H^{m-2k}(X)^0 \to H^m(X)$ is an isomorphism.

Notice that $L^k : H^{m-2k}(X)^0 \to H^m(X)$ is a morphism of Hodge structure, so the Lefschetz decomposition is in fact a decomposition of $H^m(X,\mathbb{Q})$ into a direct sum of primitive sub-Hodge structures.

Sketch of proof of 1.7. By the hard Lefschetz theorem, one may assume that $m \leq n$. Now one works by induction on m. If 1.7 is known for $m-2$, one considers $L^{n-m+1}: H^m(X) \to H^{2n-m+2}(X)$. Its kernel is by definition $H^m(X)^0$. Now $L^{n-m+2}: H^{m-2}(X) \to H^{2n-m+2}(X)$ is an isomorphism, which implies that L^{n-m+1} is surjective and that $H^m(X) \simeq H^m(X)^0 \oplus LH^{m-2}(X)$, so 1.7 is also true for m, by injectivity of $L: H^{m-2}(X) \to H^m(X)$.

The projection from $H^m(X)$ to $H^{m-2}(X)$ given by the isomorphism $H^m(X) \simeq H^m(X)^0 \oplus LH^{m-2}(X)$ is induced up to coefficient by the operator Λ of Hodge theory, which acts on forms (see §2). It is not known if it can be represented by an algebraic cycle of codimension $n-1$ in $X \times X$ (acting on $H^\bullet(X)$ via Poincaré duality and Künneth decomposition), although this is obviously the case for the operator L.

2. We turn now to the Hodge theoretic approach to the Lefschetz theorems. The weak and hard Lefschetz theorems will be obtained as a consequence of the Lefschetz decomposition, which in Kähler geometry exists at the level of forms.

Let X be a Kähler manifold of dimension n. Let ω be the Kähler form, and let $L: A^k(X) \to A^{k+2}(X)$ be the pointwise operator of multiplication by ω on complex k-forms on X. One defines the Hodge operator $*: A^k(X) \to A^{2n-k}(X)$ by $(\varphi, \psi)\,\mathrm{Vol} = \varphi \wedge *\psi$, where $\mathrm{Vol} = \frac{\omega^n}{n!}$ and $(,)$ is the induced pointwise hermitian metric on $A^k(X)$.

Let $\Lambda = *^{-1}L*$. Then obviously Λ is the formal adjoint of L for the hermitian metric $<,>$ defined on $A^k(X)$ by $<\varphi, \psi> = \int_M (\varphi, \psi)\,\mathrm{Vol}$.

One says that a form $\varphi \in A^k(X)$ is primitive if it satisfies $\Lambda\varphi = 0$.

One has then the following result of hermitian geometry, which comes from the computation of the commutator $[L, \Lambda]$ as a number operator ($[L, \Lambda]$ acts as multiplication by $p-n$ on p-forms).

2.1. Proposition : Any k-form $\varphi \in A^k(X)$ can be written uniquely as $\varphi = \sum_{r \geq \mathrm{Sup}(k-n,0)} L^r \varphi_{k-2r}$ where φ_{k-2r} is primitive of degree $k-2r$. There exists non commutative polynomials $P_{k,r}(L, \Lambda)$ with rational coefficients such that $\varphi_{k-2r} = P_{k,r}(L, \Lambda)(\varphi)$.

The existence of the Lefschetz decomposition 1.7 follows now from the following facts:

2.2. i) Let $\varphi \in A^k(X)$, $k \leq n$; then $\Lambda\varphi = 0$ is equivalent to $L^{n-k+1}\varphi = 0$.

ii) The operators L and Λ commute with the Laplacian operator Δ.

ii) implies that if φ is harmonic its primitive components φ_{k-2r} are also harmonic, so 2.1 and 2.2 i) imply the surjectivity in 1.7. The injectivity follows also from 2.1 : Suppose $0 = \Sigma L^r\varphi_{k-2r}$ in $H^k(X)$, with $L^{n-k+2r+1}\varphi_{k-2r} = 0$ in $H^{2n-k+2r+2}(X)$. One may assume that φ_{k-2r} is harmonic so $L^r\varphi_{k-2r}$ is also harmonic, and the equalities $0 = \Sigma L^r\varphi_{k-2r}$, $0 = L^{n-k+2r+1}\varphi_{k-2r}$ hold at the level of forms. By 2.2 i) and unicity in 2.1, one deduces $\varphi_{k-2r} = 0$.

2.3. On the forms $\varphi \in A^k(X)$, we have defined the hermitian inner product $< \varphi, \psi >= \int_X(\varphi,\psi)\,\mathrm{Vol} = \int_X \varphi \wedge *\psi$. For a non zero form φ, one has $< \varphi, \varphi > > 0$. One deduces from this :

2.4. Hodge-Riemann bilinear relations : Let $Q_k(\varphi,\psi) = (-1)^{k(k-1/2)} \int_X \varphi\wedge\psi\wedge\omega^{n-k}$, for $\varphi, \psi \in H^k(X)$, $k \leq n$. Then :

i) $Q_k(\varphi,\psi) = 0$ for $\varphi \in H^{p,q}(X), \psi \in H^{p',q'}(X)$ and $(p',q') \neq (q,p)$.

ii) $i^{(p-q)}Q_k(\varphi,\overline{\varphi}) = (-1)^{k(k-1)/2} i^{(p-q)} \int_X \varphi \wedge \overline{\varphi} \wedge \omega^{n-k} > 0$, for $\varphi \in H^{p,q}(X)$, $\varphi \neq 0$, and φ primitive.

This follows from the following fact of hermitian geometry :

2.5. If $\varphi \in A^{p,q}(X)$ is primitive $(p+q = k \leq n)$, one has the equality $*\varphi = (-1)^{k(k-1)/2}$ $i^{(p-q)}\frac{1}{(n-k)!}L^{n-k}\overline{\varphi}$.

Of course, to deduce 2.4 of 2.5, one uses the fact that if φ is a primitive cohomology class, its harmonic representative is a primitive form, by 2.2 ii).

The consequences of 2.4 depend on the parity of k :

A) If k is even, 2.4 describes completely the indices of the real symmetric intersection forms Q_k on $H^k(X)$. In particular if n is even one can deduce from 2.4 and 1.7 (cf. [5], p. 78) the following :

2.6. Hodge index theorem : The topological intersection form $Q(\varphi,\psi) = \int_X \varphi \wedge \psi$ on $H^n(X)$ for n even has for index $i(X) = \sum_{(a,b)}(-1)^a h^{a,b}$.

Let us also mention the following consequence of 2.4, which is very important for the consistency of the Hodge conjecture, when applied to the monodromy groups.

2.7. Let X be an algebraic variety and ω the class of an ample line bundle on X. For each integer p consider $Hdg^{2p}(X) := H^{2p}(X,\mathbf{Z}) \cap H^{p,p}(X)$. Then the group of automorphisms of $H^{2\cdot}(X,\mathbb{Z})$ preserving the forms Q_k, the operator L and the subspace $Hdg^{2\cdot}(X) \subset H^{2\cdot}(X,\mathbb{Z})$ acts as a finite group on $Hdg^{2\cdot}(X)$.

B) When k is odd, as explained in the first lecture, one can interpret 2.4 as the positivity condition necessary to polarize certain complex tori constructed from the Hodge decomposition on $H^k(X)^0$. These tori are defined as the quotients $F/H^k(X,\mathbb{Z})^0$ where $F = \sum_{\substack{p+q=k \\ p=q+1(4)}} H^{p,q}(X)^0$ is considered as a quotient of $H^k(X,\mathbb{C})^0$. They are the Weil intermediate jacobians, which unfortunately do not in general vary holomorphically with X ([2]).

In contrast, the Griffiths intermediate jacobians (see M. Green's lectures) vary holomorphically with X, but are only complex tori, having a line bundle with non degenerate indefinite curvature, the signs of which are determined by 2.4. These two tori contain the "algebraic part of the intermediate jacobian" which is the complex torus $L^{\ell,\ell+1}/L_{\mathbb{Z}}^{2\ell+1}$, where $L \subset H^{2\ell+1}(X,\mathbb{Z})^0$ is maximal to satisfy : $L_{\mathbb{C}} = L_{\mathbb{C}} \cap F^{l+1}H^{2\ell+1}(X,\mathbb{C})^0 \oplus \overline{L_{\mathbb{C}} \cap F^{l+1}H^{2\ell+1}(X,\mathbb{C})^0}$. This torus is an abelian variety polarized by $Q_{2\ell+1}$ according to 2.4 and (Lecture one, 2.6 - 2.7).

3. To conclude this lecture we mention the following applications of the Lefschetz decomposition and the Hodge-Riemann bilinear relations :

The first one is due to Deligne [1] :

3.1. Theorem : Let $f : \mathcal{X} \to B$ be a smooth and projective morphism; then the Leray spectral sequence

$$H^p(R^q f_*\mathbb{C}) \Rightarrow H^{p+q}(X,\mathbb{C}) \text{ degenerates at } E_2.$$

The point is that a relatively ample line bundle will give a flat Lefschetz decomposition on the flat bundle $R^q f_*\mathbb{C}$. The operators $L^k : R^q f_*\mathbb{C} \to R^{q+2k} f_*\mathbb{C}$ are compatible with d_2, and it suffices to check the vanishing of d_2 on the primitive cohomology : but if n is the dimension of the fibers, one has the following diagram :

$$\begin{array}{ccc}
H^p((R^q f_*\mathbb{C})^0) & \xrightarrow{L^{n-q+1}} & H^p(R^{2n-q+2} f_*\mathbb{C}) \\
\downarrow d_2 & & \downarrow d_2 \\
H^{p+2}(R^{q-1} f_*\mathbb{C}) & \xrightarrow{L^{n-q+1}} & H^{p+2}(R^{2n-q+1} f_*\mathbb{C})
\end{array}$$

Now the first L^{n-q+1} is zero and the second one is an isomorphism, which implies that the first d_2 is zero. The vanishing of the other d_r's is proved in the same way.

For the second application, let us say that a Hodge structure over $\mathbf{Q}(H^k_{\mathbf{Q}}, H^k_{\mathbf{C}} = \bigoplus_{p+q=k} H^{p,q})$ is polarized if there is a bilinear form Q on $H^k_{\mathbf{Q}}$ symmetric for k even, skew for k odd, defined over $\mathbf{Q}$ and satisfying the conditions 2.4.

Then one has :

3.2. Let H be a polarized Hodge structure and let $L \subset H$ be a sub-Hodge structure. Then L is a direct factor in H, i.e. there exists a sub-Hodge structure $L' \subset H$ such that $L \oplus L' = H$.

It suffices to note that Q is non degenerate on $L_{\mathbf{Q}}$ because of the definiteness of $i^{(p-q)}Q(\varphi, \overline{\varphi})$ on $H^{p,q}$ and that $L^{\perp} := L'$ is also a sub-Hodge structure of H by the first condition in 2.4.

References :

[1] P. Deligne, Théorèmes de Lefschetz et critères de dégénérescence de suites spectrales, Publ. Math. IHES, 35 (1968) 107-126.

[2] P. Griffiths, Periods of integrals on algebraic manifolds I, Amer. J. Math. 90 (1968), 568-626.

[3] S.L. Kleiman, Algebric cycles and the Weil conjectures. Dans : Dix exposés sur la cohomologie des schémas, North-Holland, Amsterdam (1963) 359-386.

[4] J. Milnor, Morse theory, Part I, Princeton University Press (1963), Annals of Mathematics Studies, number 51.

[5] A. Weil, Variétés kählériennes, chapitres I et II, Actualités scientifiques et industrielles 1267, Hermann, seconde édition 1971.

Lecture 3 : Noether-Lefschetz loci

0. This lecture is devoted to the Noether-Lefschetz loci associated with a variation of Hodge structure, that is the loci where an integral class is a Hodge class. The name is borrowed from the case of a variation of Hodge structure given by the H^2 of a family of varieties, where by the Lefschetz theorem on $(1,1)$ classes these loci can also be defined as the sets of points in the family corresponding to varieties having a line bundle with prescribed first Chern class. It also refers to the celebrated Noether-Lefschetz theorem, proved rigorously by Lefschetz, which says that a general surface in $\mathbf{P}^3$ of degree at least four carries no line bundle other than the multiples of $\mathcal{O}(1)$.

For Hodge structures of higher degree, such interpretation depends on the Hodge conjecture, and conversely one could hope that the compared study of the deformation of subschemes and the deformation of the associated Hodge classes would lead to a proof of the "variational Hodge conjecture" [7]. In fact, excepted for particular situations (special cycles in hypersurfaces, Lagrangian subvarieties in symplectic manifolds) this study is very difficult to carry out, and the only general criterion, the notion of semi-regularity due to Bloch [1], subsequently refined by Ran [6], cannot be checked in general. (That is, general principles do not give the existence of semi-regular representative of a Hodge class, even modulo an ample class, excepted in the case of divisors). We start with the Hodge theoretic local description of the Noether-Lefschetz loci, and continue with the geometric description of the deformation theory of the Hodge class associated to a subscheme [1]. We also introduce some infinitesimal invariants of a Hodge class and explain its relations with geometry, following [2]. Finally, we turn to the case of divisors, explain shortly some recent results in Noether-Lefschetz theory for surfaces in $\mathbf{P}^3$, and conclude with the proof of the infinitesimal criterion of M. Green ([3], [4]) which will be used as an existence criterion for one cycles in threefolds in lecture 8.

1. Let U be a connected and simply connected complex space and let $H^{2k}_{\mathbf{Z}}$, $\mathcal{H}^{2k}$, $F^\ell\mathcal{H}^{2k}$ be a variation of Hodge structure on U. So we have $\mathcal{H}^{2k} = H^{2k}_{\mathbf{Z}} \otimes \mathcal{O}_U$, a flat bundle with the Gauss-Manin connection $\nabla : \mathcal{H}^{2k} \to \mathcal{H}^{2k} \otimes \Omega_U$, such that $\nabla H^{2k}_{\mathbf{Z}} = 0$, and $F^\ell\mathcal{H}^{2k}$ is a decreasing filtration by holomorphic subbundles satisfying $\nabla F^\ell\mathcal{H}^{2k} \subset F^{\ell-1}\mathcal{H}^{2k} \otimes \Omega_U$. Now the local system $H^{2k}_{\mathbf{Z}}$ is trivial on U, so one can identify it with the group of its global sections and with its fiber at any point (notation $\lambda \to \lambda_t$, $t \in U$).

One says that $\lambda_t \in H^{2k}_{\mathbf{Z}(t)}$ is a Hodge class if it belongs to $F^k H^{2k}_{(t)}$.

For any $\lambda \in H^{2k}_{\mathbf{C}} := H^{2k}_{\mathbf{Z}} \otimes \mathbf{C}$, one defines $U_\lambda \subset U$ by $U_\lambda = \{t \in U / \lambda_t \in F^k\mathcal{H}^{2k}_{(t)}\}$. One has then :

1.1. Lemma : U_λ is analytic and can be defined locally by $h^{k-1,k+1}$ holomorphic equations, where $h^{k-1,k+1} = \operatorname{rank} \mathcal{H}^{k-1,k+1}$, $\mathcal{H}^{k-1,k+1} = F^{k-1}/F^k\mathcal{H}^{2k}$.

The fact that U_λ is analytic is clear, because U_λ is defined by the vanishing of the projection of the flat, hence holomorphic section $\lambda \in \mathcal{H}^{2k}$ in the quotient $\mathcal{H}^{2k}/F^k\mathcal{H}^{2k}$.

To prove the second statement, one uses transversality : One chooses $\mathcal{G} \subset \mathcal{H}^{2k}$ such that $\mathcal{G} \oplus F^{k-1}\mathcal{H}^{2k} = \mathcal{H}^{2k}$. One has then a holomorphic projection $\pi : \mathcal{H}^{2k} \to F^{k-1}\mathcal{H}^{2k}$. Let $0 \in U_\lambda$ and let U'_λ be defined in a neighbourhood of 0 by the vanishing of $\pi(\lambda)$ in the quotient $F^{k-1}/F^k\mathcal{H}^{2k}$. In U'_λ, λ belongs to $\mathcal{G} \oplus F^k\mathcal{H}^{2k}$, so $\lambda = \lambda_1 + \lambda_2$, $\lambda_1 \in F^k\mathcal{H}^{2k}, \lambda_2 \in \mathcal{G}$.

Now $\mathcal{G}$ has an induced connection $\nabla_\mathcal{G} = \pi_\mathcal{G} \circ \nabla$, and by transversality, one finds on U'_λ :

$$\nabla\lambda_1 \in F^{k-1}\mathcal{H}^{2k} \otimes \Omega_{U'_\lambda},\ \nabla\lambda = 0 \Rightarrow \nabla_\mathcal{G}\lambda_2 = 0.$$

Because $\lambda_2 = 0$ at 0, it follows that $\lambda_2 = 0$ on a neighbourhood of 0 in U'_λ, hence that $U'_\lambda = U_\lambda$ in this neighbourhood .

Now, if one wants to extend these definitions to the case of a non simply connected U and preserve the fact that the U_λ are analytic, one has to work with polarized Hodge structures and restrict the definition to the case of an integral class λ. Otherwise, it could happen that the monodromy on λ along U_λ produces infinitely many new classes λ' in such a way that the analytic space U_λ has in fact infinitely many branches U'_λ at one point. This situation is excluded for λ integral, and for polarized Hodge structures, by the finiteness of the monodromy action on the set of Hodge classes, mentioned in lecture 2.

1.2. Now we want to describe at finite orders near 0 the subscheme U_λ. Suppose we have a map $S_m \underset{j_m}{\mapsto} U_\lambda$, where $S_m = \operatorname{Spec} \mathbb{C}[\epsilon]/\epsilon^m$, such that $j_m(0) = 0$, and let $S_m \subset S_{m+1} \underset{j_{m+1}}{\mapsto} U$ be an extension of j_m to a map with value in U. Then $j_{m+1}(S_{m+1}) \subset U_\lambda$ if and only if the flat class λ remains in $F^k\mathcal{H}^{2k}$ on S^{m+1}. Now let $\tilde\lambda$ be a holomorphic section of $j^*_{m+1}F^k\mathcal{H}^{2k}$ extending λ (by assumption, on S_m, λ is a section of $F^k\mathcal{H}^{2k}$). Then because λ is flat on S_m, $\nabla\tilde\lambda = \epsilon^{m-1}d\epsilon\, \varphi$ where φ is in $F^{k-1}\mathcal{H}^{2k}_{(0)}$ by transversality. $\tilde\lambda$ can be modified by an arbitrary section of the form $\epsilon^m\psi$, hence $\nabla\tilde\lambda$ can be modified by an arbitrary element of the form $\epsilon^{m-1}d\epsilon\,\psi$, where $\psi \in F^k\mathcal{H}^{2k}_{(0)}$. So the image $0_{m+1}(\lambda)$ of φ in $F^{k-1}\mathcal{H}^{2k}_{(0)}/F^k\mathcal{H}^{2k}_{(0)} = \mathcal{H}^{k-1,k+1}_0$ does not depend on $\tilde\lambda$, and one has : $j_{m+1}(S_{m+1}) \subset U_\lambda \iff$ there exists an extension $\tilde\lambda \in j^*_{m+1}F^k\mathcal{H}^{2k}$ which is flat $\iff 0_{m+1}(\lambda) = 0$ in $\mathcal{H}^{k-1,k+1}_{(0)}$.

In particular, for $m = 1$, we have described the Zariski tangent space of U_λ at 0 :

1.3. Lemma : $TU_{\lambda(0)} = \operatorname{Ker}\overline{\nabla}(\lambda^{k,k})$, where $\overline{\nabla} : \mathcal{H}^{k,k}_{(0)} \to \operatorname{Hom}(TU_{(0)}, \mathcal{H}^{k-1,k+1}_{(0)})$ is the linear map constructed by transversality from the Gauss-Manin connection :

$$\begin{array}{cccc}
\nabla : & F^{k+1}\mathcal{H}^{2k} & \to & F^k\mathcal{H}^{2k}\otimes\Omega_U \\
 & \downarrow & & \downarrow \\
\nabla : & F^k\mathcal{H}^{2k} & \to & F^{k-1}\mathcal{H}^{2k}\otimes\Omega_U \\
 & \downarrow & & \downarrow \\
\overline{\nabla} : & \mathcal{H}^{k,k} & \to & \mathcal{H}^{k-1,k+1}\otimes\Omega_U
\end{array}$$

and $\lambda^{k,k} \in \mathcal{H}^{k,k}_{(0)}$ is the projection of the class λ, which is by assumption in $F^k\mathcal{H}^{2k}$ at 0.

The lemma follows from the equality $\overline{\nabla}(\lambda)(j_{2*}\partial/\partial\epsilon) = 0_2(\lambda)$, which is clear by the description of $0_2(\lambda)$ given above. Finally, from lemmas 1.1 and 1.3, we deduce :

1.4. Lemma : If $\overline{\nabla}(\lambda^{k,k})$ is surjective at $0 \in U_\lambda$, U_λ is smooth of codimension $h^{k-1,k+1}$ at 0.

2. The case where λ is the Hodge class of subvariety :

2.1. We assume now that we have a family of smooth complex varieties $\mathcal{X} \xrightarrow{\pi} U$, and we consider the associated variation of Hodge structures : $H^{2k}_{\mathbf{Z}} = R^{2k}\pi_*\mathbf{Z}$, $F^\ell/F^{\ell+1}\mathcal{H}^{2k} = R^{2k-\ell}\pi_*(\Omega^\ell_{\mathcal{X}/U})$. Let $Z_0 \subset X_0$ be a smooth subvariety (Bloch works with $\ell.c.i$ subschemes). Let $\lambda \in H^{2k}(X_0,\mathbf{Z}) \cap H^{k,k}(X_0,\mathbf{Z})$ be the associated Hodge class. We want to compare the deformation theory of the pair (Z_0, X_0) and the subscheme U_λ defined in 1). Let us assume that for $S_m \xrightarrow{j_m} U$, as in 1.2, we could extend Z_0 to a subvariety

$$\begin{array}{ccc}
Z_m & \hookrightarrow & \mathcal{X}_m \\
{\scriptstyle\pi_2}\searrow & & \swarrow{\scriptstyle\pi} \\
 & S_m &
\end{array}$$

then this implies that $j_m(S_m) \subset U_\lambda$; to see this we have to show that the flat extension λ_m of λ is in $F^k\mathcal{H}^{2k}{}_{|S_m}$. But if n is the dimension of the fibers X_t, $\mathcal{H}^{2k}$ is dual of $\mathcal{H}^{2n-2k}$ and it is easy to see that the flat extension λ_m of λ, as an element of $\mathcal{H}^{2n-2k*}$ is given by the restriction map $\mathcal{R}^{2n-2k}\pi_*(\Omega^\bullet_{\mathcal{X}_m/S_m}) \to \mathcal{R}^{2n-2k}\pi_*(\Omega^\bullet_{Z_m/S_m}) = \mathcal{R}^{n-k}\pi_*(K_{Z_m/S_m}) = \mathcal{O}_{S_m}$, which implies clearly that it annihilates $F^{n-k+1}\mathcal{H}^{2n-2k}{}_{|S_m} = \mathcal{R}^{2n-2k}\pi_*F^{n-k+1}(\Omega^\bullet_{\mathcal{X}_m/S_m})$.

2.2. If we have an extension $j_{m+1} : S^{m+1} \to U$ of j_m as in 1) we have now the obstruction $0_{m+1}(\lambda) \in \mathcal{H}^{k-1,k+1}_{(0)}$ constructed in 1.2, and the obstruction $0'_{m+1}(Z_m) \in H^1(N_{Z_0})$ to extend Z_m to a subvariety $Z_{m+1} \subset \mathcal{X}_{m+1}$, where N_{Z_0} is the normal bundle of Z_0 in X_0. The obstruction $0'_{m+1}(Z_m)$ can be defined as the extension class of the exact sequence :

$$0 \to \mathcal{O}_{Z_0} \to J^{\mathcal{X}_{m+1}}_m \otimes \mathcal{O}_{Z_0} \to \underbrace{J^{\mathcal{X}_m}_m \otimes \mathcal{O}_{Z_0}}_{= N^*_Z} \to 0$$

where $J^{\mathcal{X}_{m+1}}_m, J^{\mathcal{X}_m}_m$ are ideal sheaves of Z_m in $\mathcal{X}_{m+1}$ and $\mathcal{X}_m$ respectively. More analytically the deformation $\mathcal{X}_{m+1}$ can be represented by a form $\alpha = \sum_0^m \epsilon^i \alpha_i$ where $\alpha_i \in A^{0,1}(T^{1,0}_{X_0})$, (that is α_i is a $(0,1)$ form on X_0 with value in the bundle $T^{1,0}_{X_0}$ of $(1,0)$ vector fields), and α has to satisfy up to order m, the integrability condition : $(*)$ $\bar{\partial}\alpha - [\alpha, \alpha] = 0$. Because $Z_m \subset \mathcal{X}_m$ is analytic, we may assume that up to order $m-1, \alpha_{|Z_0}$ is in $A^{0,1}(Z_0)(T^{1,0}_{Z_0})$. Then $(*)$ together with the fact that $\alpha_0 = 0$ shows that the image of $\alpha_{m|Z_0}$ in $A^{0,1}(Z_0)(N_{Z_0})$ is $\bar{\partial}$ closed, hence has a class in $H^1(N_{Z_0})$, which gives the obstruction $0'_{m+1}(Z_m)$.

To compare $0'_{m+1}(Z_m)$ and $0_{m+1}(\lambda)$, Bloch introduces the semi-regularity map γ : $H^1(N_{Z_0}) \to H^{k+1}(\Omega^{k-1}_{X_0}) = \mathcal{H}^{k-1,k+1}_{(0)}$ which is defined as the dual of the composite :

2.2.1. $\quad H^{n-k-1}(\Omega^{n-k+1}_{X_0}) \to H^{n-k-1}(Z_0, \Omega^{n-k+1}_{X_0}{}_{|Z_0}) \to H^{n-k-1}(Z_0, K_{Z_0} \otimes N^*_{Z_0})$

where the last map is induced by the exact sequence : $0 \to N^*_{Z_0} \to \Omega_{X_0|Z_0} \to \Omega_{Z_0} \to 0$, which gives to the $n-k+1^{th}$ power :

$$\Omega^{n-k+1}_{X_0} \to K_{Z_0} \otimes N^*_{Z_0}.$$

Bloch shows : [1].

2.3. Theorem : One has the relation $0_{m+1}(\lambda) = \gamma(0'_{m+1}(Z_m))$ in $H^{k-1,k+1}(X_0)$.

The proof of Bloch is purely algebraic : an analytic proof for smooth Z_0 can be sketched as follows : as an element of $H^{n-k+1,n-k-1}(X_0)^*$ the obstruction $0_{m+1}(\lambda)$ has the following description : Let $\varphi \in F^{n-k+1}H^{2n-2k}(X_0)$ and let $\tilde{\varphi} = \sum_0^m \epsilon^i \varphi_i$ represents a holomorphic section of $j^*_{m+1}(F^{n-k+1}\mathcal{H}^{2n-2k})$ extending φ, where φ_k are $2n-2k$ closed forms on X_0, such that $\Sigma \epsilon^i \varphi_i$ is in $F^{n-k+1}A^{2n-2k}$ up to order m for the complex structure defined by α (cf. 2.2); then

2.3.1. $\quad O_{m+1}(\lambda)(\varphi) = \int_{Z_0} \varphi_m.$

Now using the fact that $\alpha = \sum_0^m \epsilon^i \alpha_i$ with $\alpha_{i|Z_0} \in A^{0,1}(Z_0)(T^{1,0}_{Z_0})$ for $i < m$, one finds that $\varphi_{i|Z_0} = 0$ for $i < m$ and that :

2.3.2. $\quad \varphi_{m|Z_0} = \int \alpha_m(\varphi_0)_{|Z_0}.$

It is then clear by 2.3.1, 2.3.2 , 2.2.1 and the description of $0'_{m+1}(Z_m)$ as the projection of $\dot{\alpha}_{m|Z'_0}$ that $(0_{m+1}(\lambda)(\varphi) = \int_{Z_0} \varphi_m = \int_{Z_0} \operatorname{Int} \alpha_m(\varphi_m) = < \gamma^*(\varphi),\ 0'_{m+1}(Z_m) >$, which is the contents of the theorem.

As a consequence of theorem 2.3 one has now :

2.4. Theorem : If γ is injective, the locus U_λ coincide schematically with the image in U of the space of deformations of the pair (Z_0, X_0).

One says that Z is semi-regular in this case, because this is the generalization of the notion of regularity $(H^1(N_{Z_0}) = 0)$ which implies the stability of Z_0 under small deformations of X_0 [5].

2.5. We want to explain now the construction of invariants of Hodge classes [2], which are an attempt to recover information on the ideal of a subscheme from the infinitesimal behaviour of its Hodge class.

Consider $Z_0 \subset X_0$, a subvariety of codimension k. Associated to a (maybe infinitesimal) deformation $\mathcal{X} \to U$ of X_0, we have the map $\varphi = \oplus \varphi_p : TU_{(0)} \to \oplus \operatorname{Hom}(H^{p,q}(X_0), H^{p-1,q+1}(X_0))$ describing the infinitesimal variation of Hodge structure of X_0. These maps can be iterated and by the flatness of the Gauss-Manin connection, we obtain that the ℓth iterations $\varphi_p(U_1) \circ \ldots \circ \varphi_{p+\ell-1}(U_\ell) : H^{p+\ell-1,2n-2k-p-\ell+1}(X_0) \to H^{p-1,2n-2k-p+1}(X_0)$ is symmetric in $U_1, \ldots, U_\ell$. So we obtain in particular a map :

$$\psi^\ell : H^{n-k+\ell,n-k-\ell}(X_0) \to \operatorname{Hom}(S^\ell TU_{(0)}, H^{n-k,n-k}(X_0)).$$

The invariants that Griffiths and Harris associate to $\lambda = [Z] \in H^{k,k}(X_0)$ is then :

2.5.1. $H^{n-k+\ell,n-k-\ell}(-\lambda) = \{\omega \in H^{n-k+\ell,n-k-\ell}(X_0) / \forall V \in S^\ell TU_{(0)}, \psi^\ell(\omega)(V).\lambda = 0\}.$

Now the relation with the geometry is given by :

2.6. Lemma : The image of $H^{n-k-\ell}(\Omega_{X_0}^{n-k+\ell} \otimes I_{Z_0})$ in $H^{n-k-\ell}(\Omega_{X_0}^{n-k+\ell})$ is contained in $H^{n-k+\ell,n-k-\ell}(-\lambda)$.

The lemma follows from the description of $\psi^\ell(\omega)(V)$ as the interior product of ω by the image of $V = U_1 \otimes \ldots \otimes U_\ell$ in $H^\ell(\Lambda^\ell T_{X_0})$, where we use the Kodaira-Spencer map

$TU_{(0)} \to H^1(T_{X_0})$: It is clear then that for $\omega \in H^{n-k-\ell}(\Omega_{X_0}^{n-k+\ell} \otimes I_{Z_0}), \psi_\ell(\omega)(V) \in H^{n-k}(\Omega_{X_0}^{n-k} \otimes I_{Z_0})$ which integrates to zero on Z_0.

One can find examples of cycles in hypersurfaces for which equality holds in 2.6.

In the case of a class $\lambda \in H^2(S, \mathbf{Z}) \cap H^{1,1}(S)$ on a surface S, the space $H^{2,0}(-\lambda) \subset H^0(K_S)$ is made of sections of the canonical bundle and it was believed for a time that for surfaces in $\mathbf{P}^3$, if one has a reduced component S_λ of the Noether-Lefschetz locus, along which the space $H^{2,0}(-\lambda)$ is non zero, the class λ is supported on the divisor of a form ω in $H^{2,0}(-\lambda)$. This was proved in [8] for degree less than seven, and disproved in general in [9].

3. We finish this lecture with the proof of a very important lemma, due to M. Green, which gives an infinitesimal and purely algebraic criterion for the existence and density of the Noether-Lefschetz locus for divisors. We will work in the geometric setting although the proof works for any VHS of weight two.

So let $\mathcal{S} \xrightarrow{\pi} U$ be a family of smooth projective varieties parametrized by a smooth and connected basis U, and consider the associated variation of Hodge structure : $H^2_{\mathbf{Z}} = R^2\pi_*\mathbf{Z}, F^2\mathcal{H}^2 = \mathcal{H}^{2,0} \subset F^1\mathcal{H}^2 \subset \mathcal{H}^2 = H^2_{\mathbf{Z}} \otimes \mathcal{O}_U$, with $\mathcal{H}^{1,1} = F^1\mathcal{H}^2/F^2\mathcal{H}^2, \mathcal{H}^{0,2} = \mathcal{H}^2/F^1\mathcal{H}^2$, and the corresponding infinitesimal variation of Hodge structure :

3.1.
$$\begin{array}{cccc} \nabla: & F^2\mathcal{H}^2 & \to & F^1\mathcal{H}^2 \otimes \Omega_U \\ & \downarrow & & \downarrow \\ \nabla: & F^1\mathcal{H}^2 & \to & \mathcal{H}^2 \otimes \Omega_U \\ & \downarrow & & \downarrow \\ \overline{\nabla}: & \mathcal{H}^{1,1} & \to & \mathcal{H}^{0,2} \otimes \Omega_U \\ & \downarrow & & \downarrow \\ & 0 & & 0 \end{array}$$

For $\lambda \in \mathcal{H}^{1,1}_{(0)} = H^1(\Omega_{S_0})$, we have the map :

$$\overline{\nabla}(\lambda) : TU_{(0)} \to \mathcal{H}^{0,2}_{(0)} = H^2(\mathcal{O}_{S_0}).$$

We define the C^∞ bundle $\mathcal{H}^{1,1}_{\mathbf{R}}$ as the real part of the bundle $\mathcal{H}^{1,1}$; more precisely, let $\mathcal{H}^2_{\mathbf{R}} = H^2_{\mathbf{R}} \otimes C^\infty_{\mathbf{R}}(U)$ and $\mathcal{H}^{1,1}_{\mathbf{R}} = C^\infty F^1\mathcal{H}^2 \cap \mathcal{H}^2_{\mathbf{R}}$. Then $\mathcal{H}^{1,1}_{\mathbf{R}}$ is the sheaf of sections of the real vector bundle on U with fiber $H^{1,1}_{\mathbf{R}}(S_t) = H^{1,1}(S_t) \cap H^2(S_t, \mathbf{R})$. We shall use the notation $H^{1,1}_{\mathbf{R}}$ for the total space of this vector bundle.

We have then :

3.2. Lemma (M. Green) : Assume that there exists $\lambda \in H^1(\Omega_{S_0})$ such that $\overline{\nabla}(\lambda) : TU_{(0)} \to H^2(\mathcal{O}_{S_0})$ is surjective. Then the set $\{(t,\lambda)/\lambda \in H^2(S_t,\mathbf{Q}) \cap H^{1,1}(S_t)\}$ is dense in $H^{1,1}_{\mathbf{R}}$.

Proof : The assumption is a Zariski open property on the algebraic vector bundle $\mathcal{H}^{1,1}$ on U so if it satisfied at 0, it will also be satisfied in a Zariski open set of U; furthermore if it is satisfied by $\lambda \in H^1(\Omega_{S_0})$ it is also satisfied by λ in a Zariski open dense subset of $H^{1,1}_{\mathbf{R}}(S_0)$, because $H^{1,1}(S_0) = H^{1,1}_{\mathbf{R}}(S_0) \otimes \mathbf{C}$. So we have only to prove :

3.2.1. Let $\lambda \in H^{1,1}_{\mathbf{R}}(S_0)$, satisfying the condition : $\overline{\nabla}(\lambda) : TU_{(0)} \to H^2(\mathcal{O}_{S_0})$ is surjective; then there exists a sequence (t_n, λ_n) with $\lambda_n \in H^2(S_n, \mathbf{Q}) \cap H^{1,1}(S_{t_n})$, and (t_n, λ_n) converges to $(0, \lambda)$ in $H^{1,1}_{\mathbf{R}}$.

In a neighbourhood of 0, the local system $H^2_{\mathbf{Z}}$ is trivial, hence we have a flat trivialization of the bundle $\mathcal{H}^2$. this gives a diagram of holomorphic maps :

$$\begin{array}{ccc} F^1H^2 & \xrightarrow{P_1} & H^2(S_0,\mathbf{C}) \\ & \searrow \quad \nearrow P & ; \\ & H^2 & \end{array}$$

where F^1H^2 is the total space of $F^1\mathcal{H}^2$, and H^2 is the total space of $\mathcal{H}^2$. P_t gives an isomorphism $H^2(S_t,\mathbf{C}) \simeq H^2(S_0,\mathbf{C})$ which preserves the rational structure, and by definition of a flat trivialization the Gauss-Manin connection is described by $P(\nabla\sigma) = dP(\sigma)$ for a section $\sigma : U \to H^2$ of the bundle $\mathcal{H}^2$.

It follows from this and the definition of $\overline{\nabla}$, that the condition "$\overline{\nabla}(\lambda)$ surjective " is equivalent to : The map P_1 is a submersion at $\lambda \in F^1H^2_{(0)} = F^1H^2(S_0)$. Now λ being real, $P_1(\lambda) \in H^2(S_0,\mathbf{R})$, so it can be approximated by $\lambda_n \in H^2(S_0,\mathbf{Q})$. P_1 being submersive at λ, one can find t_n, such that $\lim_{n\infty} t_n = 0$, and $\widetilde{\lambda}_n \in F^1H^2_{(t_n)}, \lim_{n\infty} \widetilde{\lambda}_n = \lambda$, such that $P_1(\widetilde{\lambda}_n) = \lambda_n$ Then $\widetilde{\lambda}_n \in F^1H^2(S_{t_n}) \cap H^2(S_{t_n},\mathbf{Q})$, because the flat trivialization preserves the rational structure. So the lemma is proved.

3.3. One should notice that this criterion does not work for higher weight variation of Hodge structures $H^{2k}_{\mathbf{Z}}$, with $F^{k+2}\mathcal{H}^{2k} \neq 0$. The point is that one can analogously construct the holomorphic map $P_k : F^k H^{2k} \to H^{2n}(X_0, \mathbf{C})$ but by transversality it will not be submersive at any point when $F^{k+2}\mathcal{H}^{2k} \neq 0$.

This suggests that the Noether-Lefschetz locus is not dense in general, and may be not dense, for example, for four dimensional hypersurfaces of $\mathbf{P}^5$ of degree at least six ($h^{4,0} \neq 0$), but I don't know how to prove this.

References :

[1] S. Bloch : Semi-regularity and De Rham cohomology, Inventiones Math. 17, S1. 66 (1972).

[2] P. Griffiths-J. Harris : Infinitesimal variations of Hodge structure II : An infinitesimal invariant of Hodge classes, Compositio Math. Vol 50 (1985) 207-265.

[3] C. Giliberto, J. Harris, R. Miranda : General components of the Noether-Lefschetz locus and their density in the space of all surfaces, Math. Ann. 282 (1988) 666-680.

[4] S.0. Kim : Noether-Lefschetz locus for surfaces, Trans. Amer. Math. Soc. 1 (1991).

[5] K. Kodaira : On stability of compact submanifolds of complex manifolds, Amer. J. Math. Vol. 85 (1963) 79-94.

[6] Z. Ran : Hodge theory and the Hilbert scheme, J. of Differential Geom. 37 (1993) 191-198.

[7] J. Steenbrink : Some remarks about the Hodge conjecture, in *Hodge theory* proceedings Sant Cugat, Lecture Notes in Math. n° 1246, 165-175, Springer Verlag.

[8] C. Voisin : Sur le lieu de Noether-Lefschetz en degré 6 et 7, Compositio Mathematica 75 (1990) 47-68.

[9] C. Voisin : Contrexemple à une conjecture de J. Harris, C.R. Acad; Sci. Paris t. 313, Série I, 685-687 (1991).

Lecture 4. Monodromy

In this lecture we describe the theory of Lefschetz degenerations from the point of view of topology and Hodge theory. This theory is fully understood and modern developments of (mixed) Hodge theory have produced more general results concerning the degenerations of algebraic varieties. In fact the degenerations that have been considered to give a unified treatment of monodromy and asymptotic Hodge theory are the degenerations to a normal crossing variety, which are, by the semi-stable reduction theorem of Mumford, the most general ones, up to base change. We refer to [10] for an efficient survey of the general results.

However, for the applications we have in mind, the Lefschetz theory has the advantage of giving an explicit description of the middle homology of a variety and of the monodromy action on the middle homology of its hyperplane sections. Furthermore, normal crossing varieties which have been used to give supplementary results in the theory of algebraic cycles ([4], [5], [12]) don't have triple intersections, so behave locally as a Lefschetz degeneration with parameters. We refer to [4] for a full treatment of the generalized Picard-Lefschetz formula and of the degeneration of Hodge structure in this case.

We begin with the description of the vanishing cycle associated to a node, and sketch the proof of the Picard-Lefschetz formula. We apply this to the Lefschetz description of the vanishing homology of X and Y, where $Y \subset X$ is a hyperplane section. We give then several applications of the Picard-Lefschetz formula. We conclude with a concrete (but non rigorous) description of the behaviour of the Hodge filtration near a Lefschetz degeneration, in the spirit of [7], [9], and its relations with algebraic cycles : The limit Hodge class of the vanishing cycle in even dimension, the generalized intermediate jacobian and its extension class in odd dimension. We refer to [2], [4] for a rigorous treatment, and for applications of this last point.

1. Let $V \subset \mathbf{C}^n$ be a neighbourhood of 0 and let $f : U \to \mathbf{C}$ be a holomorphic map, such that $df(x) \neq 0$ for $x \neq 0$, and $df(0) = 0$, but $\mathrm{Hess}_0 f := \Sigma \frac{\partial^2 f}{\partial z_i \partial z_j} dz_i dz_j$ is a non degenerate symmetric bilinear form. Then the holomorphic Morse lemma says that shrinking U if necessary, there exists holomorphic coordinates u_i centered at zero such that :

1.1.1. $\quad f = \Sigma u_i^2.$

Now we assume that U is a closed ball of radius 1 and that f is given by (1.1.1). Consider for $|t| < 1$ and for a choice of $\sqrt{t}$ the sphere $S^{n-1}_{\sqrt{t}} = \{(u_1, \ldots, u_n) \ / \ u_i = \sqrt{t}\, v_i, v_i \text{ real}, \Sigma v_i^2 = 1\} \subset U_t = \{u = (u_1, \ldots, u_n)/f(u) = t\}$.

For $t \neq 0$, $u \in U_t$, if one writes $\frac{1}{\sqrt{t}} u_i = x_i + i y_i$, one has :

1.1.2
$$u \in U_t \Longleftrightarrow \left| \begin{array}{l} \Sigma\, x_i^2 - y_i^2 = 1,\ \Sigma\, x_i y_i = 0 \\ \Sigma\, x_i^2 - y_i^2 \leq \frac{1}{|t|} \end{array} \right.$$

and this represents U_t as a disk bundle in the tangent bundle of $S^{n-1}_{\sqrt{t}}$.

1.2. The boundaries of the U_t's over the disk $\Delta_{\frac{1}{2}}$ of radius $\frac{1}{2}$ form a compact C^∞ fibration, and the U_t's themselves form a fibration into varieties with boundary over the circle $S_{\frac{1}{2}} = \partial\Delta_{\frac{1}{2}}$. One can construct a trivialization of the pull-back $U'_{[02\pi]}$ of this last fibration to $[02\pi] \xrightarrow{\frac{1}{2}\exp i\theta} S_{\frac{1}{2}}$, such that the induced trivialization of the boundaries descend to $\partial U_{S_{\frac{1}{2}}}$ and extends over the disk $\Delta_{\frac{1}{2}}$. The resulting homeomorphism

$$\begin{array}{cccc} \Phi : & U'_0 & \simeq & U'_{2\pi} \\ & \| & & \| \\ & U_{\frac{1}{2}} & \rightrightarrows & U_{\frac{1}{2}} \end{array}$$

satisfies then : $\Phi_{|\partial U_{\frac{1}{2}}} = \mathrm{Id}$.

The monodromy along the circle $S_{\frac{1}{2}}$ is the action of the map Φ on the pair $(U_{\frac{1}{2}}, \partial U_{\frac{1}{2}})$. It is in fact described by the maps :$\alpha_k : H_k(U_{\frac{1}{2}}, \partial U_{\frac{1}{2}}) \to H_k(U^0_{\frac{1}{2}})$ where $U^0_{\frac{1}{2}}$ is the interior set of $U_{\frac{1}{2}}$, and α_k is defined by : $\alpha_k(\gamma) = \Phi(\gamma) - \gamma$, for γ a k-chain with boundary in $\partial U_{\frac{1}{2}}$.

Now the action of Φ being trivial on H_0, the only non trivial α_k is α_{n-1} which necessarily sends the generator of $H_{n-1}(U_{\frac{1}{2}}, \partial U_{\frac{1}{2}})$ (called the transverse cycle and represented by a fiber of the disk bundle $U_{\frac{1}{2}} \to S^{n-1}_{\sqrt{\frac{1}{2}}}$) to a multiple of the generator δ of $H_{n-1}(U^0_{\frac{1}{2}})$ (called the vanishing cycle and represented by the sphere $S^{n-1}_{\sqrt{\frac{1}{2}}}$). One can check by explicit computation that the missing coefficient is 1, if one gives compatible orientations of these generators, using the natural real orientation of the tangent disk bundle $U_{\frac{1}{2}}$. In other words, we have the local Picard-Lefschetz formula :

1.2.1. $\alpha_{n-1}(\gamma) = (\gamma, \delta)\delta$ where the product $(,)$ is the intersection between $H_{n-1}(U_{\frac{1}{2}}, \partial U_{\frac{1}{2}})$ and $H_{n-1}(U^0_{\frac{1}{2}})$ given by the orientation above.

1.3. Now if we have a family of compact complex varieties $\mathcal{X} \xrightarrow[\varphi]{} \Delta$ such that φ is smooth over Δ^*, and $X_0 = \varphi^{-1}(0)$ has only one node at x_0, that is φ behaves as in 1.1 at x_0, one chooses a neighbourhood U of x_0 as in 1.1, one assumes that Δ is small (say $\Delta = \Delta_{\frac{1}{2}}$ as in 1.2) so that $\varphi_{|\bigcup_{t \in \Delta_{\frac{1}{2}}} \partial U_t}$ is a fibration, and one extends the trivialization (1.2) of $\varphi_{|\bigcup_{t \in \Delta_{\frac{1}{2}}} \partial U_t}$

to $\mathcal{X}\setminus \bigcup_t U_t^0$. Restricting this to $S_{\frac{1}{2}}$ and taking the pull-back to $[02\pi]$, we glue this last trivialization with the trivialization of $U'_{[02\pi]}$, and obtain a map :

$$\begin{array}{cccc} \Phi : & X'_0 & \simeq & X'_{2\pi} \\ & \| & & \| \\ & X_{\frac{1}{2}} & \rightrightarrows & X_{\frac{1}{2}} \end{array}$$

satisfying : $\Phi = \mathrm{Id}$ outside $U_{\frac{1}{2}}$.

The action ρ of Φ on $H_{\cdot}(X_{\frac{1}{2}})$ is called the monodromy action (of the positive generator of $\pi_1(\Delta^*)$) and because $\Phi = \mathrm{Id}$ outside $U_{\frac{1}{2}}$, $\Phi_* - \mathrm{Id}$ clearly factors as :

$$H_k(X_{\frac{1}{2}}) \underset{\text{excision}}{\rightarrow} H_k(U_{\frac{1}{2}}, \partial U_{\frac{1}{2}}) \underset{\alpha_k}{\rightarrow} H_k(U_{\frac{1}{2}}^0) \underset{\text{inclusion}}{\rightarrow} H_k(X_{\frac{1}{2}}).$$

Finally we have to note that the orientation of $U_{\frac{1}{2}}$ used in 1.2 differs from the complex orientation by a factor $(-1)^{n(\frac{n+1}{2})}$ coming from the rearrangement of variables in 1.1.2 $(x_i, y_i)_{i=1\ldots n} \rightarrow (y_1, \ldots, y_n, x_1, \ldots, x_n)$. So if we use intersection on $H_{n-1}(X_{\frac{1}{2}})$ instead of the product (,) of 1.2, the Picard-Lefschetz formula now reads :

1.3.1. $\rho(\gamma) = \gamma + (-1)^{n(\frac{n+1}{2})} < \gamma, \delta > \delta$, for $\gamma \in H_{n-1}(X_{\frac{1}{2}})$, the monodromy being trivial on the other homology groups.

2.

2.1. Returning to the local situation 1.1, 1.2, there is a retraction of $\bigcup_{t\in\Delta_{\frac{1}{2}}} U_t$ on the union of $U_{\frac{1}{2}}$ and the "cone over the vanishing cycle" $\Gamma_\delta := \bigcup_{t\in[0\frac{1}{2}]} S^{n-1}_{\sqrt{t}}$, which is equal on $\bigcup_{t\in\Delta_{\frac{1}{2}}} \partial U_t$ to the retraction given by the trivialization of the boundaries in 1.2. The boundary of Γ_δ is $S^{n-1}_{\sqrt{\frac{1}{2}}}$ because S^{n-1}_0 is shrinked to a point. In the global situation $\mathcal{X} \rightarrow \Delta$, we can glue this retraction with the one given by the trivialization of $\mathcal{X}\setminus U$ and we obtain : $\mathcal{X}$ retracts on $X_{\frac{1}{2}} \cup \Gamma_\delta$, where Γ_δ is an n-disk glued on the sphere $S^{n-1}_{\sqrt{\frac{1}{2}}} \subset X_{\frac{1}{2}}$.

2.2. Suppose now that X is a smooth projective variety of dimension n and $(X_t)_{t\in\mathrm{P}^1}$ is a Lefschetz pencil of hypersurfaces, that is X_t has at most one node, which is not on the base locus. Assume X_∞ is smooth, and let $\widetilde{X}$ be the blow-up of the base locus $X_0 \cap X_\infty$. Then $\mathcal{X} = \widetilde{X}\setminus X_\infty$ admits a map φ to $\mathrm{C} = \mathrm{P}^1\setminus\infty$ which satisfies locally the assumptions in 1.3. One fixes a regular value 0 of φ and for each critical value t_i of φ, one chooses a path γ_i from 0 to t_i, in such a way that the γ_i's meet only at 0. The plane C retracts on the union of the γ_i's so by smoothness of φ outside $U\Delta_i$ (Δ_i a small disk around t_i), $\mathcal{X}$

retracts first on $\varphi^{-1}(U\Delta_i \cup \gamma_i)$. Each $\mathcal{X}_{\Delta_i}$ retracts then on $X_{t'_i} \underset{S^{n-1}_{t'_i}}{\cup} \Gamma_{\delta_i}$, where t'_i is the intersection of γ_i with the $\partial\Delta_i$, and finally $\varphi^{-1}[t'_i 0]$ is naturally isomorphic to $X_0 \times [t'_i, 0]$. So $\mathcal{X}$ has the homotopy type of X_{t_0} with disks $\Gamma_{t'_i}$ glued over spheres $S'^{n-1}_{t'_i} \subset X_0$, the images of the spheres $S^{n-1}_{t'_i}$ in X_0, via the induced isomorphism $X_{t'_i} \simeq X_0$.

2.3. It follows that the relative homology $H_n(\mathcal{X}, X_0)$ is generated by the disks $\Gamma_{t'_i}$, and that the kernel of the map $H_{n-1}(X_0) \to H_{n-1}(\mathcal{X})$ is generated by the classes δ_i of the spheres $S'^{n-1}_{t'_i}$. In fact one can deduce from the hard Lefschetz theorem that this result remains true for X instead of $\mathcal{X}$, [1] (here one needs the assumption that X_∞ is smooth).

2.4. As a corollary of this, and the Picard-Lefschetz formula 1.3, one finds a particular case of the global invariant cycle theorem proved in general by Deligne [6] using the degeneration of Leray spectral sequence (lecture 2) and the theory of mixed Hodge structures :

Theorem. In the situation of 2.2, if $\gamma \in H_{n-1}(X_0, \mathbb{Q}) \simeq H^{n-1}(X_0, \mathbb{Q})$ is invariant under the monodromy action $\rho : \pi_1(\mathbb{P}^1 \backslash \{t_1, \ldots, t_N\}, 0) \to \operatorname{Aut} H^{n-1}(X_0, \mathbb{Q})$, then γ is in the image of the restriction map $j^* : H^{n-1}(X, \mathbb{Q}) \to H^{n-1}(X_0, \mathbb{Q})$.

This comes from $\operatorname{Im} j^* = (\operatorname{Ker} j_*)^\perp = <\delta_i>^\perp$, and the following consequence of 1.3: $\gamma \perp \delta_i \Leftrightarrow \gamma$ is invariant under the monodromy action, generated by the Picard-Lefschetz reflections $\gamma \mapsto \gamma \pm <\gamma, \delta_i> \delta_i$.

We explain now two applications of the Picard-Lefschetz theory : The first one is :

2.5. The Noether-Lefschetz theorem : Let S be a general surface in $\mathbb{P}^3$ of degree $d \geq 4$. Then $\operatorname{Pic} S = \mathbb{Z}$, generated by $\mathcal{O}_S(1)$.

One notes first the following : If S is general, the Néron-Severi group of S is globally invariant under the monodromy action. This holds because a line bundle L on a general S is defined on the universal surface $\mathcal{S} \to U$, where $U \to V$ is a finite Galois cover of the moduli space of S. So the monodromy action on $c_1(L)$ just exchanges $c_1(L)$ with $c_1(L_\gamma)$ where L_γ is obtained from L by action of the Galois group of $U \to V$. Now if $c_1(L)$ is primitive and non zero, for a Lefschetz pencil of surfaces $(S_t)_{t \in \mathbb{P}^1}$ with $S_0 = S$, there is a vanishing cycle δ_i such that $c_1(L)$. $\delta_i \neq 0$ because the intersection form $\langle , \rangle$ on S is non degenerate on $H^2(S)^0$, generated by the δ_i's. So if $c_1(L) \neq 0$, $NS(S) \otimes \mathbb{Q}$ contains one δ_i by the Picard-Lefschetz formula. Finally we have :

Sublemma. The monodromy acts transitively on the set of vanishing cycles.

Admitting this, we see that the hypotheses would imply that $NS(S) \otimes \mathbf{Q}$ contains all the δ_i's, that is contains $H^2(S,\mathbf{Q})^0$. This is absurd if $H^2(\mathcal{O}_S) \neq 0$ (see lecture one), that is when $d \geq 4$.

For the proof of the sublemma, one uses the inclusion of $\mathbf{P}^1$ into the space $\mathbf{P}(H^0(\mathcal{O}_{\mathbf{P}^3}(d))) = \mathbf{P}^k$. One knows that $\pi_1(\mathbf{P}^1 \backslash \{t_1, \dots, t_N\}, 0\} \twoheadrightarrow \pi_1(\mathbf{P}^k \backslash \mathcal{D})$ where $\mathcal{D}$ is the discriminant hypersurface. $\mathcal{D}$ being irreducible it is easy to see that all the $\gamma_i^{-1}.\delta\Delta_i.\gamma_i$'s are conjugate in $\pi_1(\mathbf{P}^k \backslash \mathcal{D})$ hence in $\pi_1(\mathbf{P}^1 \backslash \{t_1, \dots, t_N\}, 0\})$ which proves the lemma, using the Picard-Lefschetz formula.

2.6. The second application can be found in [9] : Let us consider the family $\mathcal{X} \underset{\pi}{\to} U$ of smooth automorphism free hypersurfaces of degree d in $\mathbf{P}^{2m}$, modulo $\mathbf{P}G\ell$ action. There is the associated family of intermediate Jacobians $J \to U$, with sheaf of holomorphic sections given by $\mathcal{J} : \mathcal{H}^{2m-1}/F^m\mathcal{H}^{2m-1} \oplus H_{\mathbf{Z}}^{2m-1}$, where $H_{\mathbf{Z}}^{2m-1} = R^{2m-1}\pi_*\mathbf{Z}$, $\mathcal{H}^{2m-1} = H_{\mathbf{Z}}^{2m-1} \otimes \mathcal{O}_U$ and F^kH^{2m-1} is the Hodge filtration. A normal function ν is a section of $\mathcal{J}$. (In principle, one should impose to it infinitesimal conditions and growth conditions at the boundary but we don't need it). We will say that ν is flat if it is locally the projection in $\mathcal{J}$ of a flat section $\tilde{\nu}$ of $\mathcal{H}^{2m-1}$. By infinitesimal considerations one can show that such $\tilde{\varphi}$ is then unic up to a section of $H_{\mathbf{Z}}^{2m-1}$, and this remain true on an tale cover of U.

Now we have :

2.6. Proposition : Let $V \underset{r}{\to} U$ be an tale cover of U and let ν be a flat normal function on V. Then ν is a torsion section of $\mathcal{J}$.

Proof : Fix $0 \in V$. Then $r_* : \pi_1(V,0) \to \pi_1(U,0)$ has image of finite index, say N. Let $\tilde{\varphi}$ be a flat lifting of ν near 0 and let $\tilde{\varphi}(0) \in H^{2m-1}(X_0, \mathbf{C})$ be its value at 0. We have to check that $\tilde{\varphi}(0) \in H^{2m-1}(X_0, \mathbf{Q})$. If $\gamma : [01] \to V$ is a loop based at 0, ν being locally flat we can follow $\tilde{\varphi}$ along γ and get near $\gamma(1)$ a new lifting $\tilde{\varphi}'$ of ν which is flat. By the unicity statement, we have $\tilde{\varphi}'(0) - \tilde{\varphi}(0) \in H^{2m-1}(X_0, \mathbf{Z})$. But by definition of the monodromy on the local system $H_{\mathbf{C}}^{2m-1}, \tilde{\varphi}'(0) - \tilde{\varphi}(0) = \rho(\gamma)(\tilde{\varphi}'(0)) - \tilde{\varphi}'(0)$. It remains to prove :

2.6.1. If $\eta \in H^{2m-1}(X_0, \mathbf{C})$ satisfies : $\forall \gamma \in \pi_1(V,0), \rho(\gamma)(\eta) - \eta \in H^{2m-1}(X_0, \mathbf{Z})$, then $\eta \in H^{2m-1}(X_0, \mathbf{Q})$. To see this, one chooses a Lefschetz pencil in U, with loops γ_i acting by Picard-Lefschetz reflections associated to the δ_i's, the vanishing cycles, which generate $H^{2m-1}(X_0, \mathbf{Z})$ by 2.3. Then $\gamma_i^N \in r_*(\pi_1(V,0))$ and acts by the transformation $\eta \to \eta \pm N < \eta, \delta_i > \delta_i$. So the assumption implies : $\forall i, < \eta, \delta_i > \in \mathbf{Q}$, hence $\eta \in H^{2m-1}(X_0, \mathbf{Q})$, because $\langle , \rangle$ is non degenerate and defined over $\mathbf{Q}$.

3. We want now to explain the Hodge theory on the central fiber of a Lefschetz degeneration. The results depend on the parity of n. This follows from the Picard-Lefschetz formula which shows that the monodromy is of order two for $n-1$ even, and of infinite order for $n-1$ odd (or trivial).

However, we will use in both cases Griffiths' arguments which are not completely general but give quickly a concrete description of the limit Hodge structure, and are well adapted to the Lefschetz degenerations. For the general case one should work with normal crossing model of X_0 and introduce the logarithmic log complex [10] [11].

We will work with families $X_t \subset X$ where X_t is a hypersurface of X ample enough for its $(n-1)^{th}$ primitive cohomology to be realized by residues of meromorphic n-forms on X, the Hodge filtration corresponding to the pole order filtration (see the M. Green lecture on hypersurfaces). So $F^k H^{n-1}(X_t)$ will be generated by residues of n-forms with poles of order at most $n-k$ along X_t.

A) $n-1$ even : To kill the monodromy, which is of order two, one makes a base change $t = u^2$. The spheres $S^{n-1}_{\sqrt{t}} = S^{n-1}_u$ give then a univalued locally constant section δ_u of $H^{n-1}_{\mathbf{Z}} = H_{n-1,\mathbf{Z}}$ on the punctured disk with coordinate u. One verifies using the description of the U_t's in 1.1, that $\delta_u^2 = \pm 2$ (in particular $\delta_u \neq 0$) so over $\mathbf{Q}$ one has a splitting $H^{n-1}_{\mathbf{Q}} = <\delta_u> \oplus <\delta_u>^{\perp}$ and $<\delta_u>^{\perp}$ is by the Picard-Lefschetz formula the invariant part of $H^{n-1}_{\mathbf{Z}}$ on Δ_t^*. Cycles γ_t in $<\delta_u>^{\perp} \subset H_{n-1}(X_t, \mathbf{Z})$ can be represented by chains in $X_t \backslash U_t$ (see 2.1), hence have a limit γ_0 in $X_0 \backslash U_0$.

Clearly if γ_t is such a cycle, and ω_t / f_t^k is a holomorphically varying family of n-forms on X with pole of order $\leq k$ along X_t one has $\lim_{t \to 0} \int_{\gamma_t} \mathrm{Res}_{X_t}(\omega_t/f_t^k) = \int_{\gamma_0} \mathrm{Res}_{X_0}(\omega_0/f_0^k)$ where the last term makes sense because γ_0 is supported away from $\mathrm{Sing}\, X_0$.

It remains to study the behaviour of $\int_{\delta_u} \mathrm{Res}_{X_t}(\omega_u/f_t^k)$, which is a local problem, since the δ_u are supported near x_0. The result is then :

3.1. Proposition : Assume $\frac{df_t}{dt}\big|_{t=0}$ does not vanish at x_0 (Lefschetz assumption). Then for $2(n-k) > n-1$, $\lim_{t \to 0} \int_{\delta_u} \mathrm{Res}_{X_t}(\omega_u/f_t^k) = 0$ and for $2(n-k) = n-1$, $\lim_{u \to 0} \int_{\delta_u} \mathrm{Res}_{X_t}(\omega_u/f_t^k)$ exists and vanishes if and only if ω_0 vanishes at x_0.

This shows concretely how to extend the Hodge filtration $F^m H^{n-1}(X_0)$ over 0, for $2m \leq n-1$: Using $H^{n-1}(X_{u^2}) \simeq H_{n-1}(X_{u^2})^*$ one defines $F^m H^{n-1}(X_{u^2})$ as the space generated by the limits $\lim_{u \to 0} \int_{\gamma_u} \mathrm{Res}_{X_{u^2}}(\omega_u/f_t^k)$, for $k = n-m$, where γ_u is any locally constant section of $H_{n-1,\mathbf{Z}}$ over Δ_u. One needs supplementary assumptions to check that this filtration has correct rank (satisfied for X_t ample enough), and one extends it to a Hodge filtration on $H_0^{n-1} = \lim_{t \to 0} H^{n-1}(X_n, \mathbf{Z}) \otimes \mathbf{C}$ using complex conjugaison. The fact that this limit filtration really puts a Hodge structure on H_0^{n-1} remains to be proved, and one has more generally the following theorem, which is proved by the decreasing distance property of the period map :

3.2. Theorem : [8] For a polarized variation of Hodge structures $(H_{\mathbf{Z}}^k, F^\ell \mathcal{H}^k)$ on Δ^* without monodromy, the period map extends at 0 and defines a pure Hodge structure on $H_{\mathbf{Z}(0)}^k$.

We note finally that by proposition 3.1, the limit δ_0 of the vanishing cycle annihilates $F^{m+1}H_0^{n-1}$, where $2m = n-1$, hence is a Hodge class. A geometric way of interpreting it as an algebraic cycle is to construct the normal crossing model of X_0 (after base change $t = u^2$) as the union of the minimal desingularization $\tilde{X}_0$ of X_0 and a $(n-1)$ quadric Q which intersect $\tilde{X}_0$ along the exceptional divisor, which is identified to a hyperplane section Q' of Q. Then one can check that the spheres S_u converge to the generator of $H_{n-1}(Q \backslash Q')$, that is the difference of the two rullings of Q.

B) $n-1$ odd, $n = 2m$. In this case the vanishing cycle δ_u may have trivial homology class in X_0. Then there is no monodromy and by theorem 3.2 there is a pure Hodge structure on the fiber H_0^{n-1}. More generally, one can consider a central fiber with several nodes, and define the defect of X_0 as the number of relations between the homology classes of the associated vanishing cycles. Under some vanishing assumptions on X, one can identify this defect to the corank of the restriction map $H^0(K_X(mX_0)) \to H^0(K_X(mX_0)_{|Z})$, where Z is the singular locus of X_0 [5].

This is strongly related to the following analog of Prop. 3.1 :

3.3. Proposition : 1) for $k \leq m$, $\lim_{t \to 0} \int_{\delta_t} \mathrm{Res}_{X_t}(\omega_t / f_t^k)$ exists and is equal to zero for $k < m$, and is a non zero multiple of $\omega_0(x_0)$ for $k = m$.

(Notice that now the δ_t's give an invariant section of $H_{n-1,\mathbf{Z}}$ over Δ^*).

2) For γ a multivalued section of $H_{n-1,\mathbf{Z}}$ over Δ^*, and $k \leq m$, $\int_{\delta_t} \mathrm{Res}_{X_t}(\omega_t / f_t^k)$ has a logarithmic growth near zero and its monodromy is described by 1) and the Picard-Lefschetz formula.

So for $< \gamma_t, \delta_t >= 0$, that is γ_t has a limit γ_0 which is supported in $X_0 \backslash \{x_0\}$, one can define the limit periods $\int_{\gamma_0} \mathrm{Res}_{X_0}(\omega_0 / f_0^k)$, $k \leq m$, where under some vanishing assumptions on (X, X_t), the $\mathrm{Res}_{X_0}(\omega_0/f_0^k) = \lim_{t\to 0} \mathrm{Res}_{X_t}(\omega_t/f_t^k)$ generate the fiber at 0 of the extended Hodge bundle $F^m\mathcal{H}^{2m-1}$, which is characterized by the growth condition in 3.3.2).

The intermediate jacobian $J(X_t)$ of X_t is the compact complex torus $F^mH^{2m-1}(X_t)^*/H_{2m-1}(X_t,\mathbf{Z})$ and the generalized intermediate jacobian $J(X_0)$ of X_0 is defined as the partial torus $(F^mH_0^{2m-1})^*/$ periods.

This torus has for quotient the intermediate jacobian of $\tilde{X}_0$, which under the same assumptions is realized by projecting $(F^mH_0^{2m-1})^*$ to $((F^mH_0^{2m-1})^0)^*$, where $(F^mH_0^{2m-1})^0$ is the hyperplane generated by the residues $\mathrm{Res}(\omega_0/f_0^m)$, with $\omega_0(x_0) = 0$. (We now assume that δ is non zero). By 3.3 i) $\int_\delta$ project to zero in $((F^mH_0^{2m-1})^0)^*$ so $J(X_0) \to J(\tilde{X}_0)$ represents $J(X_0)$ as an extension of $J(\tilde{X}_0)$ by $\mathbf{C}^*$. Such an extension is classified by $J(\tilde{X}_0)^\nu/\pm 1 = J(\tilde{X}_0)/\pm 1$ and one has the following :

3.4. Theorem : [2], [4], [5] When the vanishing cycle has non zero homology class in X_t, the two rullings of the exceptional divisor of $\tilde{X}_0$ are homologous and the image of their difference in $J(\tilde{X}_0)$ by the Abel-Jacobi map describes the extension :

$$0 \to \mathbf{C}^* \to J(X_0) \to J(\tilde{X}_0) \to 0.$$

References

[1] A. Andreotti, T. Frenkel, The second Lefschetz theorem on hyperplane sections, Global analysis, A symposium in honor of K. Kodaira, Princeton University Press, Princeton (1969) 1-20.

[2] F. Bardelli, Polarized mixed Hodge structures, On irrationality of threefolds via degenerations, Ann. Mat. Pura Appl. (4) 137 (1984) 287-369.

[3] J. Carlson, The Geometry of the extension class of a mixed Hodge structure, Proceedings of Symposia in Pure Mathematics, Vol. 46 (1987).

[4] H. Clemens, Degeneration technics in the study of threefolds in *Algebraic threefolds*, Lecture Notes in Math. n° 947, Springer Verlag (1981).

[5] H. Clemens, double solids, Advances in Mathematics 47, 107-230 (1983).

[6] P. Deligne, Théorie de Hodge II, Publ. Math. IHES (40) (1971) 5-57.

[7] P. Griffiths, On the periods of certain rational integrals I and II, Annals of Math. 90 (1969) 460-541.

[8] P. Griffiths, Periods of integrals on algebraic manifolds III, Publ. Math. IHES 38 (1970) 125-180.

[9] P. Griffiths, A theorem concerning the differential equations satisfied by normal functions associated to algebraic cycles, Amer. J. Math. 101 (1979) 94-131.

[10] D. Morrison, The Clemens-Schmid exact sequence and applications, in *Topics in transcendental Algebraic Geometry*, ed. by P. Griffiths, Ann. of Math. Studies, Study 106, Princeton 1984.

[11] J. Steenbrink, Limits of Hodge structures, Inv. Math. 31 (1976) 229-257.

[12] P. Griffiths - J. Harris, On the Noether-Lefschetz theorem and some remarks on codimension two cycles, Math. Ann. 271, 31-51 (1985).

Lecture 5. 0-cycles I.

Lectures 5 and 6 are devoted to 0-cycles modulo rational equivalence, especially for surfaces. This subject does not seem a priori much related to transcendental aspects of Hodge theory, and in fact all that we will explain belongs to algebraic geometry, even if for simplicity we use at some places the usual topology and the Betti cohomology of complex varieties. However, the relation with our main topic is the discovery by Mumford that the non-representability of Chow groups (here we will be concerned with CH_0), is related to the transcendental character of the corresponding Hodge theory. In the first lecture, we have shown that the Hodge theoretic objects related to divisors are the Picard torus and the set of Hodge classes in H^2. The first one is an abelian variety, and the second one generalizes to the "Tate-Hodge structures", which are made only of Hodge classes. These two kinds of objects are the only algebro-geometric objects that can be extracted from a Hodge structure, and this gives a more transcendental character to the remaining part of the Hodge theory of a variety. In this lecture, we explain the following now classical results : Mumford theorem on infinite dimensionality of the CH_0 group of a surface having non zero holomorphic two-form, subsequent generalizations of it by Roitman, and we present Bloch's conjecture on correspondences between surfaces. We sketch also the argument of Bloch-Kas-Lieberman, which gives for surfaces not of general type the following consequence of Bloch's conjecture : $H^0(K_S) = 0 \Rightarrow CH_0^0(S) = \mathrm{Alb}\, S$. We also sketch Roitman's proof of his two fundamental theorems : CH_0 finite dimensional $\Rightarrow CH_0^0 = \mathrm{Alb}$, and $\mathrm{tors}(CH_0) \simeq \mathrm{tors}(\mathrm{Alb})$. (However we follow largely [2] for the proof of the later statement).

1.

1.1. Let S be a smooth projective surface, and let G be a smooth projective variety. Let $\begin{matrix} Z & \xrightarrow{q} & S \\ p\downarrow & & \\ G & & \end{matrix}$ be a zero correspondence between G and S, that is $Z \subset G \times S$ is a reduced algebraic subset and p is finite; then for $\omega \in H^0(K_S)$, one can construct a holomorphic two form $Z[\omega]$ on G, either by defining (carefully) the trace $p_*(q^*\omega)$, either by the construction of the associated map $\varphi_Z : G \to S^{(N)}$, such that $\varphi_Z(g) = q(p^{-1}(g))$, where $N = d^0 p$, and $S^{(N)}$ is the symmetric product (unfortunately singular) of S, and by showing that the symmetric two-form $\sum_{i=1}^N pr_i^*\omega$ on $S^{(N)} \setminus \mathrm{sing}\, S^{(N)}$ has a non singular pull-back to G.

For a non reduced Z, $Z = nZ'$, one defines $Z[\omega] = nZ'[\omega]$. The main theorem is then:

1.2. Theorem : ([4]). If Z_1 and Z_2 are two correspondences between G and S, satisfying : $\forall g \in G$, $\varphi_{Z_1}(g)$ and $\varphi_{Z_2}(g)$ are rationally equivalent zero-cycles on S, then $\forall \omega \in H^0(K_S)$, $Z_1[\omega] = Z_2[\omega]$ in $H^0(\Omega^2_G)$.

The general idea is the following : by countability of the Hilbert scheme of rational curves in $S^{(k)}$, $k \in \mathbb{N}$, and by definition of rational equivalence, one may assume that there is an etale map $G' \underset{r}{\rightarrow} G$, a map $\Psi : G' \rightarrow S^{(k)}$ and a map $\Phi : \mathbb{P}^1 \times G' \rightarrow S^{(N+k)}$ such that: for $g' \in G'$, $\Phi(0, g') = \Phi_{Z_1}(g') + \Psi(g')$, and $\Phi(\infty, g') = \varphi_{Z_2}(g') + \Psi(g')$, where we use "+" for the obvious map $S^{(N)} \times S^{(k)} \rightarrow S^{(N+k)}$. Now, for $\omega \in H^0(K_S)$, there is as before a pull-back $\Phi^*(\omega) \in H^0(\Omega^2_{\mathbb{P}^1 \times G'})$ which necessarily is of the form $\pi^*\alpha$, where $\pi : \mathbb{P}^1 \times G' \rightarrow G'$ is the projection to G', and $\alpha \in H^0(\Omega^2_{G'})$. By the obvious additivity with respect to Z of the traces $Z[\omega]$, and by restriction of the above equalities to $\{0\} \times G'$ and $\{\infty\} \times G'$ we find $r^*Z_1[\omega] + \Psi^*[\omega] = \pi^*\alpha_{|\{0\}\times G'} = \alpha = r^*Z_2[\omega] + \Psi^*[\omega]$, that is $Z_1[\omega] = Z_2[\omega]$.

Now $CH_0(S)$ contains for each N the quotient of $S^{(N)}$ by the relation of rational equivalence, which is described in $S^{(N)} \times S^{(N)}$ by a countable union of algebraic subsets. For a cycle $Z \in S^{(N)}$, one defines the dimension d_Z of its orbit 0_Z under rational equivalence as the maximal dimension of an algebraic component of $\{Z' \in S^{(N)} / Z' \underset{rat}{\equiv} Z\}$, and clearly d_Z is a constant for a general cycle Z. 0_Z being roughly the fiber through Z of the map $R^N : S^{(N)} \rightarrow CH_0(S)$, one defines $\operatorname{Im} R^N := 2N - d_Z$, for a general cycle Z.

Mumford applies then 1.2 to show :

1.3. Theorem : [4]. If $H^0(K_S) \neq \{0\}$, $\lim_{N\rightarrow\infty} \dim \operatorname{Im} R^N = \infty$, that is $CH_0(S)$ is not finite dimensional.

The point is the following : Let Z be general in $S^{(N)}$ and choose a component G_Z of 0_Z of maximal dimension d_Z; then one may assume that $Z \in G_Z$, G_Z is smooth at Z, and that Z is made of distinct points $\{z_1, \ldots, z_N\}$. If $\omega \in H^0(K_S)$ is non zero, one may also assume that $\omega(z_i) \neq 0$; by theorem 1.2, the form $\Omega = \Sigma\, pr_i^*\omega$ on the smooth variety $S^{(N)} \setminus \operatorname{sing} S^{(N)} \ni Z$ has to vanish on G_Z near Z, and $\omega(z_i) \neq 0$ implies that Ω is non degenerate (as a two form) at Z; it follows that $\dim_Z G_Z \leq \frac{1}{2} \dim S^{(N)}$. Hence $\dim \operatorname{Im} R^N \geq N$, which implies 1.3.

1.4. It follows a posteriori [5] that in fact $\dim \operatorname{Im} R^N = 2N$, when $H^0(K_S) \neq \{0\}$, that is, a general cycle Z has a zero dimensional orbit 0_Z. A proof of this can be checked as follows : Assume a general cycle moves in its orbit O_Z and fix an ample curve $C \subset S$. Then 0_Z will meet $C + S^{(N_1)} \subset S^{(N)}$. It follows by induction on k that for a general cycle Z in $S^{(N+k)}$ its orbit will meet $C^{(k+1)} + S^{(N-1)}$, and because the image of $C^{(k)}$ in $CH_0(S)$ has dimension bounded by g = genus of C, this would show that $\forall k$, $\dim R^{(N+k)} \leq g+2(N-1)$ in contradiction with 1.3.

We refer to [5] for generalizations and refinements of this kind of results for higher dimensional varieties.

1.5. Let us consider now smooth surfaces Σ, S, and let $Z \subset \Sigma \times S$ be a codimension two algebraic cycle, say $Z = \Sigma n_i Z_i$ with Z_i irreducible and generically finite over Σ. Then as in 1.1, Z gives a map $Z : \omega \to Z[\omega] = \Sigma n_i Z_i[\omega], H^0(K_S) \to H^0(K_\Sigma)$. (We don't need that $Z_i \underset{p_1}{\to} \Sigma$ be finite because it has positive dimensional fibers only over a codimension two subset of Σ, so we can use Hartog's theorem).

Now it is easily seen from the definition that the map Z_i is the $(2,0)$ Hodge component of the map $Z_i : H^2(S) \to H^2(\Sigma)$ which is a morphism of Hodge structures and can be constructed alternatively as :

i) consider the Hodge class $[Z_i] \in H^4(\Sigma \times S, \mathbf{Z})$; then its Künneth $(2,2)$ component lies in $H^2(\Sigma) \otimes H^2(S) \simeq \mathrm{Hom}(H^2(S), H^2(\Sigma))$, which gives our Z_i.

ii) Choose a desingularisation $\widetilde{Z}_i \to Z_i$ of Z_i : then one has $\widetilde{p}_1 : \widetilde{Z}_i \to \Sigma, \widetilde{p}_2 : \widetilde{Z}_i \to S$ and one can define Z_i as the composite :

$$H^2(S) \underset{\widetilde{p}_2^*}{\to} H^2(\widetilde{Z}_i) \simeq H_2(\widetilde{Z}_i) \underset{\widetilde{p}_{1*}}{\to} H_2(\Sigma) \simeq H^2(\Sigma)$$

, where $\simeq$ means Poincaré duality isomorphism. Consider the splitting (over $\mathbf{Q}$) of $H^2(S, \mathbf{Q})$ into $NS(S) \otimes \mathbf{Q}$ and $TH^2(S) = NS(S)^\perp$. Let $\varphi : H^2(S) \to H^2(\Sigma)$ be a morphism of Hodge structures which vanishes on $H^{2,0}(S)$; then $\mathrm{Ker}\,\varphi \cap TH^2(S)$ is a sub-Hodge structure of $TH^2(S)$, and contains $H^{2,0}(S)$. Its orthogonal for $<,>_{|TH^2(S)}$ is defined over $\mathbf{Q}$ and perpendicular to $H^{2,0}(S)$ hence is contained in $TH^2(S) \cap NS(S) \otimes \mathbf{Q} = \{0\}$. So φ vanishes in fact on $TH^2(S)$. From this and theorem 1.2 follows :

1.6. Proposition : Let $Z \subset \Sigma \times S$ be a codimension two cycle; then the induced map of Hodge structures $TH^2(S) \to H^2(\Sigma)$ vanishes if the map $p_{1*}^*(p_2^*().Z) : CH_0(\Sigma) \to CH_0(S)$ induced by Z is zero.

One can refine 1.6 as follows : coming back to the Mumford argument, one sees easily that if the map $p_{1*}(p_2^*().Z) : CH_0(\Sigma) \to CH_0(S)$ is zero on the set of cycles of degree zero and in the kernel of the Albanese map of Σ, then $Z : H^{2,0}(S) \to H^{2,0}(\Sigma)$ vanishes, hence $Z : TH^2(S) \to H^2(\Sigma)$ also vanishes. Bloch [1] has conjectured the converse of the last statement :

1.7. Conjecture : Let $Z \subset \Sigma \times S$ be a codimension two cycle whose (2,2) Künneth component lies in $NS(\Sigma) \otimes NS(S)$ (equivalently Z vanishes on $TH^2(S)$). Then $p_{2*}(p_1^*().Z) : CH_0^0(\Sigma) \to CH_0(S)$ factors through the Albanese variety of Σ.

Here $CH_0^0(\Sigma)$ is the set of degree 0 cycles on Σ. We refer to J.P. Murre's lectures for the construction of the degree and Albanese map on CH_0).

A particular case of Bloch's conjecture is given by the diagonal $\Delta_S \subset S \times S$ of a surface with $h^0(K_S) = 0$. Bloch's conjecture implies in this case :

1.8. Subconjecture : If $h^0(K_S) = 0$, $CH_0^0(S) = \operatorname{Alb} S$.

We will explain in the next lecture a proof of this for Godeaux type surfaces.

Using classification of surfaces, Bloch-Kas-Lieberman have proved 1.8 for surfaces with p_g $(= h^0(K_S)) = 0$ and not of general type : In fact classification will imply with some work that for minimal surfaces there are essentially three cases to consider :

i) $q = \dim \operatorname{Alb} S = 1$, and alb: $S \to E$ is a smooth fibration, so $S = E' \times F/G$, G acting on E' by a finite group of translations. Let $\pi : E' \times F \to E' \times F/G$ be the quotient map.

ii) $q = \dim \operatorname{Alb} S = 1$, and alb: $S \to E$ has elliptic fibres. Then one shows that the associated jacobian fibration, which has isomorphic CH_0 group, fails into i).

iii) $q = 0$ and S has an elliptic pencil $S \to \mathbf{P}^1$. Then they show that the associated jacobian fibration $S' \to \mathbf{P}^1$ is a rational surface, hence has $CH_0 = \mathbf{Z}$, using the Castelnuovo criterion : $q(S') = h^0(K_{S'}^{\otimes 2}) = 0 \Rightarrow S'$ is rational. Now it is easy to see that $CH_0(S) = CH_0(S')$.

For case i), one uses the Roitman theorem (2.2) which implies that $\operatorname{Ker}(\text{alb}) \subset CH_0^0(S)$ has no torsion. So it suffices to check that $\pi^*(\operatorname{Ker} \text{alb}) = 0$ in $CH_0^0(E' \times F)$ mod torsion. But if $Z \in \operatorname{Ker}(\text{alb})$, $\pi^* Z = \sum_{i,g} n_i(\tilde{e}_i + g, g.x_i)$, with $\Sigma n_i e_i = 0$ in $\operatorname{Alb} S = E$. Now $h^0(K_S) = 0 \Leftrightarrow F$ has no non zero 1-form invariant under $G \Leftrightarrow F/G = \mathbf{P}^1$. So $\forall_i, \sum_g gx_i = h = \text{const}$. On the other hand, up to torsion, that we don't consider, $\sum_{i,g} n_i(\tilde{e}_i + g, g.x_i) = \sum_{i,g} n_i(\tilde{e}_i, g.x_i) = \sum_i n_i(\tilde{e}_i * (\sum_g g.x_i))$ in $CH_0(E' \times F)$. So $\pi^* Z = (\sum_i n_i \tilde{e}_i) * h$ up to torsion in $CH_0(E' \times F)$. Since $\sum_i n_i \tilde{e}_i =$ torsion point in $CH_0^0(E')$, $\pi^* Z = 0$ up to torsion, and we are done. (The product $*$ that we used between $CH_0(E')$ and $CH_0(F)$ is such that $Z * Z' = \Sigma n_i m_j (z_i, z'_j)$ for $Z = \Sigma n_i z_i$, $Z' = \Sigma m_j z'_j$).

2. We turn now to the proof of the following fundamental theorems of Roitman :

2.1. Theorem : [5]. Let X be a projective variety, such that $CH_0(X)$ is finite dimensional: then $CH_0^0(X) \stackrel{\text{alb}}{\to} \text{Alb}\, X$ is an isomorphism.

2.2. Theorem : [2], [6]. Let X be a projective variety; then the Albanese map induces an isomorphism alb : $\text{Tors}\, CH_0(X) \simeq \text{Tors}(\text{Alb}\, X)$.

For the definition of finite dimensionality in 2.1, we have (over the uncountable field $\mathbf{C}$) different equivalent characterizations. (We use the notations of 1.1 - 1.3 with S replaced by X).

i) There is an integer $d(X)$ such that $\forall N \in \mathbf{N}$, $\dim \text{Im}\, R^N \leq d(X)$.

ii) There exists an integer N such that the difference map : $X^{(N)} \times X^{(N)} \to CH_0^0(X)$ is surjective.

The equivalence follows from the argument sketched in 1.4. If i) holds, one chooses N such that $N \dim X - d(X) \geq \dim X$. Then if $C \subset X$ is an ample curve and $Z \in X^{(N)}$ is a general cycle, 0_Z meets $C \times X^{(N-1)}$, so any cycle in $X^{(N)}$ is rationally equivalent to a cycle in $C + X^{(N-1)}$. As in 1.4 it follows that any cycle in $X^{(N+k)}$ is rationally equivalent to a cycle in $C^{(k+1)} + X^{(N-1)}$, and since $C^{(g)} \times C^{(g)} \underset{\text{difference map}}{\longrightarrow} CH_0^0(C)$ is surjective, ii) follows. ii) $\Rightarrow$ i) comes also from the fact that rational equivalence between cycles in $S^{(K)}$ is described by a countable union of algebraic sets in $S^{(K)} \times S^{(K)}$.

2.3. Now the argument for theorem 2.1 goes as follows :

Step 1 : If $CH_0(X)$ is finite dimensional, there exists an abelian variety A, and a family of cycles of degree zero $Z \subset A \times X$ inducing a surjective map of groups $A \underset{f}{\to} CH_0^0(X)$, $f(a) = Z(a)$.

Proof : For any abelian variety A, and for any family of zero-cycles of degree 0 $Z \subset A \times X$ inducing a map of groups $f : A \to CH_0^0(X)$, $f(a) = Z(a)$, the kernel is a countable union of algebraic subsets of A, and it is a group. So A', its connected component through 0, is an abelian variety and we define $\dim f(A) = \dim A/A'$. Let $d(X)$ be as in i). Then one checks $\dim f(A) \leq d(X)$. So there exists (A, Z) as above such that $\dim f(A)$ is maximal. Adding to A the jacobian of any curve $C \subset X$, it is then easy to show that f is surjective.

Step 2 : One may assume that $A \underset{f}{\to} CH_0^0(X)$ has a countable kernel.

Proof : The kernel of f is a countable union of algebraic subsets and it is a subgroup. So an algebraic component of it passing through 0 is an abelian subvariety $B \subset A$, and we can work with an abelian subvariety $A' \subset A$, isogenous to A/B, and with the restricted cycle $Z_{|A'}$.

Step 3 : In $X \times A$, the set $\{(x,a)/x - x_0 \equiv Z(a) = f(a) \text{ in } CH_0^0(X)\}$ is a countable union of algebraic subsets, which projects onto X. So one of its components, say R, projects onto X. Such R is finite over X, because $f : A \to CH_0^0(X)$ has countable kernel. So R gives rise to a zero correspondence between X and A, and we have the following commutative diagram :

2.3.1.
$$\begin{array}{ccc} X & \xrightarrow{g} & A \\ & k \searrow \quad \swarrow f & \\ & CH_0^0(X) & \end{array}$$

where $k(x) = n(x - x_0)$, for $n = \deg(R/X)$, and $g(x) = \text{alb}(A) \circ R(x)$.

$\text{Im}\, g$ generates A as a group, otherwise $\dim(\text{Im}\, g) < \text{Im}\, A$ and $\dim A = \dim CH_0(X)$ by the fact that f has a countable kernel, would contredict the fact that $CH_0^0(X)$ is generated by $k(X)$. (Here we define $\dim CH_0(X) = \dim \text{Im}\, R^N$ (cf. 1.2, 1.3)). By the universal property of the Albanese map, we find now a surjective map $g'' : \text{Alb}\, X \to A$, such that the following diagram commutes :

2.3.2.
$$\begin{array}{ccccc} X & \xrightarrow{g} & A & \xrightarrow{f} & CH_0^0(X) \\ & \text{alb} \searrow & \nearrow g'' & & \downarrow \text{alb} \\ & & \text{Alb}\, X & \xrightarrow[n \times \text{Id}]{} & \text{Alb}\, X \end{array}$$

It follows that the kernel of f is in fact finite, so $CH_0^0(X)$ is an abelian variety A', as a finite quotient of A. Also the map $\text{alb} : A' \to \text{Alb}\, X$ is an algebraic map of abelian varieties, and induces by Theorem 2.2 an isomorphism on torsion points, so it is an isomorphism.

Proof of Theorem 2.2. : We follow partially [2], because there is a point which is not clear in Roitman's paper [6]. Notice that the surjectivity is clear because there is a curve C in X such that $JC \to \text{Alb}\, X$ is surjective with connected fibers, by the weak Lefschetz theorem. So we find that k-torsion $(CH_0^0(C)) = k - \text{torsion}(JC) \twoheadrightarrow k - \text{torsion}(\text{Alb}\, X)$. For the injectivity, Bloch does the following :

Step 1 : Let $Z = \Sigma n_i Z_i$ be a k-torsion cycle in X. By definition of rational equivalence, one can find curves C_i in X and rational functions φ_i on C_i such that $\Sigma \,\text{div}\, \varphi_i = kZ$. By birational invariance of CH_0, one can blow up X to $\widetilde{X}$, and replace UC_i by a stable curve C [(1)]. One can choose a smooth surface S containing C such that $\text{Alb}\, S \to \text{Alb}\, \widetilde{X} = \text{Alb}\, X$

[(1)] An important point here is the fact that one can connect the local components of the proper transform of UC_i by rational curves.

is an isomorphism. (By the weak Lefschetz theorem, it suffices to choose for S a complete intersection of ample divisors, and C being stable, smoothness of S is possible). Now, assuming $\mathrm{alb}_X(Z) = 0$, one has $\mathrm{alb}_S(Z) = 0$, and it suffices to show that Z is rationally equivalent to zero in S.

Finally we may add components to C, so that the new curve C' is a very ample divisor in S.

Now the technical point in Bloch's proof (this is curve theory) is the following

Step 2 : We may move Z on C' up to rational equivalence in such a way that it is supported on the smooth part of C' and determines a k-torsion line bundle L_k on C', or a k-torsion point in the generalized jacobian of C'.

Admitting this, the end of the proof goes as follows :

Step 3 : We can deform C' to a smooth curve C'' in the same linear system on S, and the line bundle L_k has accordingly a deformation L''_k as a k-torsion line bundle on C'.

Now one can use here the Roitman argument : Consider the pencil $\mathbf{P}^1$ on S determined by C' and C'' and let $D \to \mathbf{P}^1$ be the covering parametrizing k-torsion line bundles on fibers $C_t)_{t\in\mathbf{P}^1}$. Then the map $D \to CH_0^0(S)$ given by $d \in D \to L_d$ on $C_t \to j_{t*}(L_d) \in CH_0^0(S)$ is constant on connected components of D.

This is because for a component D' of D, $CH_0^0(D')$ is divisible and the image of the induced map $CH_0^0(D') \to CH_0^0(S)$ is contained in the k-torsion of $CH_0^0(S)$, so is 0.

Step 4 : ([2], [6]). Now we have a smooth curve C very ample on S, and may assume that it belongs to a Lefschetz pencil on S. We have a k-torsion line bundle on it, which sends to 0 in $\mathrm{Alb}\, S$, via $j_* : JC \to \mathrm{Alb}\, S$, and we want to show that it goes to zero in $CH_0^0(S)$, via $j_* : JC \to CH_0^0(S)$. The kernel of $j_* : JC \to \mathrm{Alb}\, S$ is a finite quotient of the connected abelian variety $(JC)^0 = (H^0(\Omega_C)^*)^0/H_1(C,\mathbf{Z})^0$ where $(H^0(\Omega_C)^*)^0 = \mathrm{Ker}(j_* : H^0(\Omega_C)^* \to (H^0(\Omega_S))^*)$, and $H_1(C,\mathbf{Z})^0 = \mathrm{Ker}(j_* : H_1(C,\mathbf{Z}) \to H_1(S,\mathbf{Z}))$. The connectedness follows from the surjectivity of this last j_* (Lefschetz). k-torsion points in $\mathrm{Ker}\, j_*$ lift to k'-torsion points in $(JC)^0$ for some k'.

One knows by Lefschetz theory that $H_1(C,\mathbf{Z})^0$ is generated by the vanishing cycles of the pencil (see Lecture 4) and that if the discriminant hypersurface for the linear system associated to C is irreducible, which is true for very ample C, the vanishing cycles are all conjugate under the monodromy action.

The k'-torsion of $(JC)^0$ is generated by the $\frac{1}{k'}\,\delta_i$ modulo $H_1(C,\mathbf{Z})^0$ and by the Roitman argument in step 3 they all have the same image in $CH_0^0(S)$. More precisely, it follows from this argument that the map $j_* : k' - \text{torsion}((JC)^0) \to CH_0^0(S)$ is invariant under monodromy.

So for $v \in \frac{1}{k'}H_1(C,\mathbf{Z})^0$ and δ_i a vanishing cycle, the Picard-Lefschetz formula gives $j_*(\bar{v}) = j_*(\overline{v + (\delta_i|v).\delta_i})$ where $\bar{\ }$ means reduction modulo $H_1(C,\mathbf{Z})^0$. Hence :

2.4.1. $(k'v|\delta_i) \,.\, j_*(\overline{\tfrac{1}{k'}\delta_i}) = 0.$

Finally, the last trick in Bloch's proof is :

2.4.2. Given δ a vanishing cycle, the integers $(\delta|\delta_i)$ have no common multiple, where δ_i runs through the set of vanishing cycles.

Otherwise by the Picard-Lefschetz formula the cycles δ' obtained from δ by monodromy action would satisfy $\delta' = \delta$ modulo $dH_1(C,\mathbf{Z})^0$, which is absurd because $H_1(C,\mathbf{Z})^0$ is generated by those δ''s and, if it is non zero (which one may assume !), it has rank ≥ 2.

Using 2.4.1, with $v = \frac{1}{k'}\,\delta$, one deduces from 2.4.2 that there exists integers m_i, with $\Sigma m_i = 1$, such that ${}_i m_i j_*(\overline{\tfrac{1}{k'}\delta_i}) = 0$, and because all $j_*(\overline{\tfrac{1}{k'}\delta_i}) = 0$ are equal, we find that j_* vanishes on the k'-torsion of $(JC)^0$.

References;

[1] S. Bloch : Lectures on algebraic cycles, Lecture one, Duke University Mathematics series IV, Durham 1980.

[2] S. Bloch : Torsion algebraic cycles and a theorem of Roitman, Compo. Math. 39 (1979) 107-127.

[3] S. Bloch, A. Kas, D. Lieberman : Zero-cycles on surfaces with $p_g = 0$, Compo. Math. 33 (1976) 135-145.

[4] D. Mumford : Rational equivalence of zero-cycles on surfaces, J. Math. Kyoto Univ. 9 (1969) 195-204.

[5] A.A. Roitman : Rational equivalence of zero-cycles, Math. USSR, Sbornik 18 (1972) 571-588.

[6] A.A. Roitman : The torsion of the group of 0-cycles modulo rational equivalence, Annals of Mathematics 111 (1980) 553-569.

Lecture 6 : Zero-cycles II

0. We continue on zero-cycles and turn now to more recent contributions. We first explain the ideas of Bloch-Srinivas [2], which are based on an elementary lemma, but which shed a new light on Mumford's theorem. In Mumford's approach the accent was put on the (local) relation zero-cycles $\leftrightarrow$ holomorphic forms. In the Bloch-Srinivas approach correspondences are considered as global objects which are shown to be essentially controlled by their action on the CH_0 group. This approach leads more immediately to global results on the effect of a correspondence on the Hodge theory of a variety (more precisely on its "motive"), and we shall describe a few applications of it. The rest of the lecture is devoted to a proof [6] of Bloch's "subconjecture" (see lecture 5) for Godeaux surfaces, which are surfaces of the general type with $q = p_g = 0$ obtained as quotients of complete intersection surfaces by a finite group, and to a generalization of the Mumford criterion, in the case of families of surfaces [7]. There we consider a family of 0-cycles $(Z_b)_{b\in B}$ in a family of surfaces $(S_b)_{b\in B}$, and we give a Hodge theoretic criterion for Z_b to be rationally equivalent to zero in S_b, $\forall\, b \in B$. We explain the two following applications: (Here the restriction to surfaces in $\mathbf{P}^3$ is not essential and simply motivated by the fact that the algebra related to infinitesimal variations of Hodge structures is well understood in this case (see M. Green lectures)):

1) If $C \subset S$ is a general plane section of a general surface of degree $d \geq 5$ in $\mathbf{P}^3$, $\mathrm{Ker}\, j_* : JC \to CH_0^0(S)$ is equal to the torsion of JC.

2) If $S \subset \mathbf{P}^3$ is a general surface of degree $d \geq 7$, two distinct points of S are not rationally equivalent.

It should be noticed that for the second statement, we come back to the local approach of Mumford, in the sense that we study restrictions of holomorphic forms to families of rationally equivalent cycles.

1) 1.1. Let X, Y be smooth algebraic varieties over $\mathbf{C}$, and $V \subset Y$ an algebraic subset. Let $Z \subset X \times Y$ be a zero correspondence between X and Y, so $Z = \Sigma n_i Z_i$ with $Z_i \xrightarrow[pr_1]{} X$ generically finite. There is a common algebraically closed field k of definition of $X, Y, V Z_i$, and we may assume that k has finite transcendance degree over $\mathbf{Q}$. Then the function field $k(X)$ and any algebraic extension of it admits embeddings to $\mathbf{C}$, extending a given embedding $k \subset \mathbf{C}$. The Z_i can be considered as points of Y defined over $k(Z_i)$, and we can find a Galois extension L of $k(X)$ containing all $k(Z_i)$. For each inclusion γ_ℓ of $k(Z_i)$ in L over $k(X)$, $\ell = 1, \ldots, d^0 k(Z_i)/k(X)$, one has the L point Z_i^ℓ of Y obtained from

Z_i by extension of scalars $k(Z_i) \xrightarrow[\gamma_i]{} L$. One has then a zero-cycle of Y, defined over $L: Z_L = \sum_{i,\ell} n_i Z_i^\ell$.

Let us choose compatible embeddings

$$\begin{array}{ccc} k(X) & \hookrightarrow & \mathbf{C} \\ & \searrow \quad \nearrow. & \\ & L & \end{array}$$

Then Z_L gives a zero-cycle $Z_{\mathbf{C}} \in Y_{\mathbf{C}}$ by extension of scalars $L \hookrightarrow \mathbf{C}$, and the diagonal $\Delta \subset X \times X$, seen as a point of X defined over $k(X)$, gives a point $\Delta_{\mathbf{C}} \in X_{\mathbf{C}}$. Clearly $Z_{\mathbf{C}} = Z(\Delta_{\mathbf{C}}) = pr_{2*}(pr_1^*(\Delta_{\mathbf{C}}).Z)$.

Now assume that X, Y, V satisfy the following property:

1.1.1. $\forall\, x \in \Delta_{\mathbf{C}}$, $Z(x)$ is rationally equivalent to zero in $Y_{\mathbf{C}} \backslash V_{\mathbf{C}}$.

It follows that $Z_{\mathbf{C}}$ is rationally equivalent to zero in $Y_{\mathbf{C}} \backslash V_{\mathbf{C}}$. Then Z_L is rationally equivalent to zero in $Y \backslash V$, over a finite extension of L, which means that there exists a Zariski open set $V \subset X$, and a finite, flat and proper morphism $\varphi : U' \to U$ such that $(\varphi \times \mathrm{Id})^*(Z_{|U \times Y \backslash V})$ is rationally equivalent to zero. It follows that $(\varphi \times \mathrm{Id}_{Y \backslash V})_*(\varphi \times \mathrm{Id})^*(Z_{|U \times Y \backslash V}) = d^0\varphi(Z_{|U \times Y \backslash V})$ is rationally equivalent to zero.

Let $D = X \backslash U$. Then the exact sequence:

$$CH(D \times Y \cup X \times V) \longrightarrow CH(X \times Y) \longrightarrow CH(U \times Y \backslash V) \longrightarrow 0$$

shows that $d^0\varphi.Z$ is rationally equivalent to a cycle supported on $D \times Y \cup X \times V$. So we have

1.2. Proposition: Let X, Y, V, Z be as before, and assume that 1.1.1 holds: $\forall\, x \in X_{\mathbf{C}}$, $Z(x) \equiv 0$ in $Y_{\mathbf{C}} \backslash V_{\mathbf{C}}$. Then there exists a divisor D of X such that a multiple of Z is supported on $D \times Y \cup X \times V$, modulo rational equivalence.

(Note that in 1.2, one can obviously allow Z to have components of codimension $\dim Y$ which are not finite over X).

Let us give an easy but useful corollary:

1.2.1. Corollary: Let X, Y be two varieties of the same dimension n and $Z \subset X \times Y$ be a codimension n cycle. Then if Z satisfies property 1.1.1, for some proper algebraic subset V of Y, ${}^tZ \subset Y \times X$ also satisfies this property. In particular if X and Y are surfaces $Z : CH_0^0(X) \to CH_0^0(Y)$ factors through $\mathrm{Alb}\, X$, if and only if ${}^tZ : CH_0^0(Y) \to CH_0^0(X)$ factors through $\mathrm{Alb}\, Y$.

For the last statement, one notes the existence of a curve $C \subset X$ such that $JC \twoheadrightarrow \mathrm{Alb}\, X$. Obviously there is a curve $C' \subset Y$ such that $Z(CH_0^0(C))$ is supported on C'; then one takes for V the curve C', and there exists a curve $D \subset X$ and an integer N such that NZ is supported up to rational equivalence in $D \times Y \cup X \times C'$. It follows that N^tZ factors through a map (given by an algebraic correspondence $(\Gamma : CH_0^0(Y) \to CH_0^0(D)$, hence by the universal property of the Albanese map, $N^tZ_{|CH_0^0(Y)}$ factors through $\mathrm{Alb}\, Y$. The same is true for ${}^tZ$ by divisibility of $\mathrm{Ker}(CH_0^0(Y) \xrightarrow{\mathrm{alb}} \mathrm{Alb}\, 4)$. Notice that divisibility of CH_0^0 is also used in the proof of the first statement.

For the applications in Bloch-Srinivas one considers the diagonal cycle $\Delta \subset X \times X$. Proposition 1.2 then gives:

1.3. Proposition: Assume that $\exists\, V \subset X$ such that $CH_0(X_{\mathbf{C}} \backslash V_{\mathbf{C}}) = 0$ then for some integer $N, N\Delta$ is up to rational equivalence supported on $D \times X \cup X \times V$, for some divisor D of X. We will write $N\Delta \equiv \Gamma_1 + \Gamma_2$, with $\Gamma_1 \subset D \times Y$ and $\Gamma_2 \subset X \times V$.

Remark: It is a very interesting problem to decide whether the integer N can be set equal to 1. This is the case if X is a rational variety, and the minimal such N is a birational invariant.

1.3.1. The correspondence Δ acts on all Chow groups as the identity. The action is given by $\gamma \mapsto pr_{2*}(pr_1^*\gamma.\Delta)$ or by $\gamma \mapsto pr_{1*}(pr_2^*\gamma.\Delta)$. Replacing $N\Delta$ by $\Gamma_1 + \Gamma_2$, one finds that this decomposition, which is obtained only by the consideration of the action of Δ on CH_0, has many implications on the other Chow groups. Let us first recover the Mumford-Roitman theorem:

1.4. Proposition: With the notations of proposition 1.3, one has

$$H^{k,0}(X) = 0, \text{ for } k > \dim V.$$

Proof: Let $\widetilde{D} \xrightarrow[\tilde{j}]{} X$ be a desingularization of $D \underset{j}{\hookrightarrow} X$. There exists a cycle $\widetilde{\Gamma}_1 \subset \widetilde{D} \times Y$ such that $N\Delta = \tilde{j}_*\widetilde{\Gamma}_1 + \Gamma_2$ with $\Gamma_2 \subset X \times V$. The action of ${}^t\Gamma_2$ on $H^k(X)$, ${}^t\Gamma_2 = pr_{1*}(pr_2^*(\ ).\Gamma_2)$, factors through the restriction to V, hence annihilates $H^{k,0}(X)$, for $k >$

$\dim V$, so we find: $\forall\, \omega \in H^{k,0}(X)$, $N\omega = j_*({}^t\widetilde{\Gamma}_1(\omega))$. But j_* is a morphism of Hodge structures of bidegree $(1,1)$, so its image does not contain a non zero element of type $(k,0)$, and we find: $N\omega = 0$.

Now Bloch and Srinivas give the analogous consequences on Chow groups. As an example they assume $\dim V \leq 3$ and give the following consequences of 1.3 on $CH^2(X)$:

1.5. Theorem: i) If $\dim V \leq 3$, the Hodge conjecture for rational $(2,2)$ classes on X holds.

ii) If $\dim V \leq 2$, homological equivalence and algebraic equivalence coincide on $CH^2(X)$.

iii) If $\dim V \leq 1$, $CH^2(X)_{\text{hom}}$ is isomorphic to $J^2(X)$ via the Abel-Jacobi map Φ_X.

We will show ii) and iii) only up to torsion, the argument being then very easy. For the analysis of the torsion in the Chow groups we refer to J. Murre's lectures.

Proof of 1.5. i) Let α be a $(2,2)$ integral class in $H^4(X)$. We want to show that a multiple of α is algebraic, and by 1.3.1 we need only to show that ${}^t\Gamma_1(\alpha)$, ${}^t\Gamma_2(\alpha)$ are algebraic. Using desingularizations of D and V, we have ${}^t\Gamma_1(\alpha) = \tilde{j}_*(\beta)$, where β is a $(1,1)$ integral class on $\widetilde{D}$, and ${}^t\Gamma_2(\alpha) = {}^t\widetilde{\Gamma}_2(\alpha_{|\widetilde{V}})$, where $\widetilde{\Gamma}_2$ is the desingularized correspondence Γ_2, between X and $\widetilde{V}$. By the Lefschetz theorem on $(1,1)$ classes, β is algebraic. From $\dim V \leq 3$, we conclude that a multiple of $\alpha_{|\widetilde{V}}$ is also algebraic, because the Hodge conjecture is true in degree 4, for varieties of dimension less than 3. (By the hard Lefschetz theorem, the Lefschetz theorem on $(1,1)$ classes implies the Hodge conjecture in degree $2\dim_{\mathbb{C}}(\ ) - 2$). So ${}^t\Gamma_1(\alpha)$ and ${}^t\Gamma_2(\alpha)$ are algebraic, and i) is proved.

ii) Let Z be a codimension two cycle homologous to zero. Then $\dim V \leq 2 \Rightarrow Z.\widetilde{V}$ is algebraically equivalent to zero on $\widetilde{V}$. Also ${}^t\widetilde{\Gamma}_1(Z) \subset \widetilde{D}$ is a divisor in $\widetilde{D}$ homologous, hence algebraically equivalent, to zero. So $NZ = j_*({}^t\widetilde{\Gamma}_1(Z)) + {}^t\widetilde{\Gamma}_2(Z.\widetilde{V})$ is algebraically equivalent to zero.

iii) $\dim V \leq 1 \Rightarrow {}^t\widetilde{\Gamma}_2$ vanishes on $CH^2(X)$. So we have $N\,\mathrm{Id} = j_* \circ {}^t\widetilde{\Gamma}_1$ on $CH^2(X)$, and on $CH^2(X)_{\text{hom}}$, we have the following diagram:

1.5.1.
$$\begin{array}{ccccc} CH^2(X)_{\text{hom}} & \xrightarrow{{}^t\widetilde{\Gamma}_1} & CH^1(\widetilde{D})_{\text{hom}} & \xrightarrow[j_*]{} & CH^2(X)_{\text{hom}} \\ \Phi_X \downarrow & & \Phi_{\widetilde{D}} \downarrow & & \downarrow \Phi_X \\ J^3(X) & \xrightarrow[{}^t\widetilde{\Gamma}_1]{} & J^1(\widetilde{D}) & \xrightarrow[j_*]{} & J^3(X). \end{array}$$

We refer to [6] for the commutativity of 1.5.1. On the last line ${}^t\widetilde{\Gamma}_1$ and j_* are the morphisms between abelian varieties corresponding to the morphisms of Hodge structures ${}^t\widetilde{\Gamma}_2$ and j_* ($J^3(X)$ is an abelian variety by 1.4 and (Lecture 2)). Now ${}^t\widetilde{\Gamma}_2$ annihilates $H^3(X)$, as $\dim V \leq 1$, so on the last line, $j_* \circ {}^t\widetilde{\Gamma}_1 = N \times \mathrm{Id}$. $\Phi_{\widetilde{D}}$ being an isomorphism it follows immediately from 1.5.1 that Φ_X is surjective with kernel contained in the N-torsion of $CH^2(X)$.

Remark: 1.5. i) generalizes [3], and 1.5. ii) generalizes [1] and other works on Fano threefolds.

2) 2.1. We explain now the method of [7] to prove the conjecture 1.8 of lecture 5 for the following type of surfaces:

1) Consider $G = \mathbf{Z}/5\mathbf{Z}$ acting on $\mathbf{P}^3$ by $g_\zeta^*(X_0, \ldots, X_3) = (\zeta X_0, \ldots, \zeta^4 X_3)$, for ζ a primitive 5th root of unity. For a generic $F \in H^0(\mathcal{O}_{\mathbf{P}^3}(5))$, satisfying $g_\zeta^*(F) = F$, $S = V(F)$ is smooth, G acts freely on it, $\Sigma := S/G$ is of general type, and $H^0(K_\Sigma) = H^0(K_S)^{\mathrm{inv}} = 0$, where "inv" means "invariant part under G".

2) Consider $G = \mathbf{Z}/8\mathbf{Z}$ acting on $\mathbf{P}^6$ by $g_\zeta^*(X_0, \ldots, X_6) = (\zeta X_0, \ldots, \zeta^7 X_6)$. Let $Q_i \in H^0(\mathcal{O}_{\mathbf{P}^6}(2))$ be general quadrics satisfying $g_\zeta^* Q_i = \zeta^{2i} Q_i$, $i = 1, \ldots, 4$. Then $S = \cap Q_i$ satisfies the same conclusion as in 1. We shall prove:

2.2. Theorem: [7] for $\Sigma = S/G$ as in 1) or 2), $CH_0^0(\Sigma) = 0$.

Let us first give the argument for case 1):

Step 1: The linear system H made of G-invariant quintic polynomials on $\mathbf{P}^3$ has no base point. So S is covered by smooth curves $C = S \cap S'$, with $S' = V(F')$; $F' \in H$. If $x, y \in S$ are generic there is such a C containing x and y. Let $\varphi : S \to \Sigma$ be the quotient map. By the Roitman theorem, and $\mathrm{Alb}(\Sigma) = 0$, $CH_0^0(\Sigma)$ has no torsion and it suffices to prove $\varphi^* CH_0^0(\Sigma) = 0$. $\varphi^* CH_0^0(\Sigma)$ is generated by cycles $Z = \sum_{g \in G} g.x - g.y$, for generic $x, y \in S$. Let $C \overset{j}{\hookrightarrow} S$ be a curve as above containing x and y. Then $\sum_{g \in G} g.x - g.y$ is a G-invariant 0-cycle Z' of degree zero on C and $Z = j_* Z'$. So it suffices to prove:

2.2.1. For C as above, the map $j_* : (JC)^{\mathrm{inv}} \xrightarrow[j_*]{} CH_0(S)$ is 0, where $(JC)^{\mathrm{inv}}$ is the invariant part of JC under G.

Step 2: Let us consider the pencil $(S_t)_{t \in \mathbf{P}^1}$ determined by S and S'. Each S_t is defined by a G-invariant quintic polynomial so has no invariant holomorphic two form.

By the Lefschetz theorem on $(1,1)$ classes, it follows that $H^2(S_t,\mathbf{Z})^{\rm inv}$ is generated by classes of G-invariant line bundles on S_t, and because $H^2(S_t,\mathbf{Z})^{\rm inv}$ is finitely generated, we conclude:

2.2.2. There exists a smooth ramified cover $D \xrightarrow[r]{} \mathbf{P}^1$, such that D parametrizes G-invariant line bundles on fibers S_t (i.e. to $d \in D$ corresponds a line bundle L_d on S_t, $t = r(d)$), and such that, over the open set U of $\mathbf{P}^1$ parametrizing smooth S_t's, the map $\alpha : r_*\mathbf{Z}_{|U} \to (R^2\pi_*\mathbf{Z})^{\rm inv}{}_{|U}$, which sends 1_d to $c_1(L_d)$, is surjective. Here $\pi : \mathcal{S} \to \mathbf{P}^1$ is the family of surfaces $(S_t)_{t\in\mathbf{P}^1}$.

We have a natural map $\beta : D \to (\mathrm{Pic}\, C)^{\rm inv}$, which associates $j_t^*(L_d)$ to $d \in D$, where j_t is the inclusion of C in S_t. We also write β for the induced map $JD \to (\mathrm{Pic}^0\, C)^{\rm inv} = (JC)^{\rm inv}$. Now we show that:

2.2.3. $(JC)^{\rm inv}$ is generated by $\mathrm{Im}\,\beta$ and by the various $j_{t_i}^*(\mathrm{Pic}^0(\widetilde{S}_{t_i})^{\rm inv})$ where the S_{t_i}'s are the singular surfaces of the pencil and $\widetilde{S}_{t_i} \to S_{t_i}$ is a G-invariant desingularization.

To prove 2.2.3, we blow up C in $\mathbf{P}^3$, so that $\mathcal{S} = \widetilde{\mathbf{P}}^3$ and G acts on $\widetilde{\mathbf{P}}^3$. It is well known that there is a natural isomorphism

2.2.4. $$JC \simeq J^3(\widetilde{\mathbf{P}}^3), \quad (JC)^{\rm inv} \simeq J^3(\widetilde{\mathbf{P}}^3)^{\rm inv}.$$

The surjective map $\alpha : r_*\mathbf{Z}_{|V} \to (R^2\pi_*\mathbf{Z})^{\rm inv}{}_{|U}$ induces a surjective map $\alpha : H^1(r_*\mathbf{Z}_{|V}) \to H^1\big((R^2\pi_*\mathbf{Z})^{\rm inv}{}_{|U}\big)$. By the Leray spectral sequence for $\pi : V := \pi^{-1}(U) \to U$ one has: $H^3(V,\mathbf{Q})^{\rm inv} = H^1\big((R^2\pi_*\mathbf{Q})^{\rm inv}{}_{|V}\big)$. So we have:

$$\alpha : H^1(r^{-1}(U),\mathbf{Q}) \twoheadrightarrow H^3(V,\mathbf{Q})^{\rm inv}.$$

The map $\beta : JD \to (JC)^{\rm inv} = J^3(\widetilde{\mathbf{P}}^3)^{\rm inv}$ induces a map $\beta_* : H^1(D,\mathbf{Z}) \to H^3(\widetilde{\mathbf{P}}^3,\mathbf{Z})^{\rm inv}$ and it is not difficult to check that the following diagram commutes:

2.2.5. $$\begin{array}{ccccc} \alpha & : & H^1(r^{-1}(U),\mathbf{Q}) & \longrightarrow & H^3(V,\mathbf{Q})^{\rm inv} \\ & & \uparrow \text{restriction} & & \uparrow \text{restriction} \\ \beta_* & : & H^1(D,\mathbf{Q}) & \longrightarrow & H^3(\widetilde{\mathbf{P}}^3,\mathbf{Q})^{\rm inv}. \end{array}$$

α and β_* are morphisms of mixed Hodge structures [4], and it follows from the strictness of such maps for the W-filtration that the surjectivity of α implies that of:

2.2.6. $$\beta_* : H^1(D,\mathbf{Q}) \longrightarrow H^3(\widetilde{\mathbf{P}}^3,\mathbf{Q})^{\rm inv} / \mathrm{Ker}(\text{restriction}).$$

On the other hand, the kernel of the restriction map $H^3(\widetilde{\mathbf{P}}^3,\mathbf{Q})^{\text{inv}} \to H^3(V,\mathbf{Q})^{\text{inv}}$ is equal to:

2.2.7.
$$\sum_i \widetilde{k}_{t_i *} H^1(\widetilde{S}_{t_i},\mathbf{Q})^{\text{inv}},$$

where $\widetilde{S}_{t_i} \xrightarrow[\widetilde{k}_{t_i}]{} \widetilde{\mathbf{P}}^3$ is composed of the desingularization map and of the inclusion $k_{t_i} : S_{t_i} \to \widetilde{\mathbf{P}}^3$. $\widetilde{k}_{t_i}$ is induced on rational homology by the corresponding maps:

2.2.8.
$$\widetilde{k}_{t_i *} : \mathrm{Pic}^0(\widetilde{S}_{t_i})^{\text{inv}} \longrightarrow CH_1(\widetilde{\mathbf{P}}^3)^{\text{inv}}_{\text{hom}} \xrightarrow[\text{Abel-Jacobi}]{} J^3(\widetilde{\mathbf{P}}^3)^{\text{inv}},$$

which can also be identified to $j^*_{t_i} : \mathrm{Pic}^0(\widetilde{S}_{t_i})^{\text{inv}} \to (JC)^{\text{inv}}$. From 2.2.5, 2.2.7 and 2.2.8 we conclude that the map of abelian varieties $\beta \oplus \sum_i j^*_{t_i} : JD \oplus \bigoplus_i \mathrm{Pic}^0(\widetilde{S}_{t_i})^{\text{inv}} \to JC$ induces a surjective map on rational homology, hence is surjective.

Step 3. We have shown that $(JC)^{\text{inv}}$ is generated by line bundles of degree 0 in $\bigoplus_{t\in\mathbf{P}^1} j^*_t : \bigoplus_{t\in\mathbf{P}^1} \mathrm{Pic}(\widetilde{S}_t)^{\text{inv}} \to \mathrm{Pic}\, C$, (where $\widetilde{S}_t = S_t$ if S_t is non singular).

To conclude that $j_* : (JC)^{\text{inv}} \to CH^0_0(S)^{\text{inv}}$ is 0 it suffices to note the commutativity of the following diagram: (for S_t singular or not)

2.2.9.
$$\begin{array}{ccc} \mathrm{Pic}\,\widetilde{S}_t & \xrightarrow{k_{t*}} & CH_1(\mathbf{P}^3) \\ \downarrow j^*_t & & \downarrow k^* \\ \mathrm{Pic}\, C & \xrightarrow[j_*]{} & CH_0(S)\,, \end{array}$$

where $k_t : \widetilde{S}_t \to \mathbf{P}^3$ is the inclusion, eventually composed with desingularization. If $\Sigma n_t\, j^*_t(L_t)$, $L_t \in \mathrm{Pic}\,\widetilde{S}_t$, has degree 0 on C, $\Sigma n_t k_{t*}(L_t)$ is homologous to zero in $\mathbf{P}^3$, so is rationally equivalent to 0, hence $j_*(\Sigma n_t j^*_t(L_t)) = k^*(\Sigma n_t k_{t*}(L_t)) = 0$ in $CH^0_0(S)$, and 2.2.1 is proved.

2.3. The second case is treated similarly, replacing $\mathbf{P}^3$ by $X = Q_1 \cap Q_2 \cap Q_3$ which is a Fano threefold with a representable CH_1^{hom} group , and H by the linear system $\{Q_4\}$. Going thru the proof one concludes by the analog of the diagram 2.2.9 that the map $k^* : CH_1(X)_{\text{hom}} \to CH^0_0(S)^{\text{inv}}$ is surjective. So $CH^0_0(\Sigma)$ is finite dimensional, hence is zero by $\mathrm{Alb}\,\Sigma = 0$, and by Roitman's theorem (lecture 5).

3) 3.1. We explain now a generalization of Mumford's criterion for 0-cycles in a family of surfaces: [8]

We consider a family of smooth regular projective surfaces over a smooth quasi-projective basis $B : \mathcal{S} \underset{\pi}{\longrightarrow} B$. Let $Z \subset \mathcal{S}$ be a codimension two cycle flat over B such that $\forall\, b \in B$, Z_b has degree zero on S_b. The cycle Z has a class in $H^2(\Omega^2_{\mathcal{S}})$ (see [5]). Hence there is an induced section δ_Z of $\operatorname{Ker} H^0(R^2\pi_*\Omega^2_{\mathcal{S}}) \to H^0(R^2\pi_*\Omega^2_{\mathcal{S}/B})$. We write now the exact sequence: $0 \to \pi^*\Omega_B \to \Omega_{\mathcal{S}} \to \Omega_{\mathcal{S}/B} \to 0$, which gives:

3.1.1. a) $0 \to K \to \Omega^2_{\mathcal{S}} \to \Omega^2_{\mathcal{S}/B} \to 0$ defining K,

b) $0 \to \pi^*\Omega^2_B \to K \to \pi^*\Omega_B \otimes \Omega_{\mathcal{S}/B} \to 0$.

It follows that δ_Z identifies to a section of $R^2\pi_*K$, and then to a section of $\Omega^2_B \otimes R^2\pi_*\mathcal{O}_{\mathcal{S}}/\Psi(\Omega_B \otimes R^1\pi_*\Omega_{\mathcal{S}/B})$ where Ψ is obtained by the long exact sequence associated to 3.1.1. b). In terms of variations of Hodge structure, we have on B the "VHS" $H^2_{\mathbb{Z}} = R^2\pi_*\mathbb{Z}$, $\mathcal{H}^2 = H^2_{\mathbb{Z}} \otimes \mathcal{O}_B$, with Hodge filtration $F^i\mathcal{H}^2$, such that $\mathcal{H}^{0,2} = \mathcal{H}^2/F^1\mathcal{H}^2 = R^2\pi_*\mathcal{O}_{\mathcal{S}}$, $\mathcal{H}^{1,1} = F^1\mathcal{H}^2/F^2\mathcal{H}^2 = R^1\pi_*(\Omega_{\mathcal{S}/B})$. The infinitesimal variation of Hodge structure gives $\overline{\nabla} : \mathcal{H}^{1,1} \to \Omega_B \otimes \mathcal{H}^{0,2}$, and then $\overline{\nabla}_2 : \mathcal{H}^{1,1} \otimes \Omega_B \to \Omega^2_B \otimes \mathcal{H}^{0,2}(\overline{\nabla}_2(\alpha \otimes \omega) = \omega \wedge \overline{\nabla}\alpha)$. Griffith's description of $\overline{\nabla}$ gives that $\overline{\nabla}_2 = \Psi$.

So we have associated to Z the infinitesimal invariant $\delta_Z \in \mathcal{H}^{0,2} \otimes \Omega^2_B/\operatorname{Im}\overline{\nabla}_2$.

3.2. Suppose now that Z satisfies:

3.2.1. $\forall\, b \in B$, Z_b is rationally equivalent to zero in S_b: arguing as in Bloch-Srinivas, we see that up to torsion, modulo rational equivalence, Z is supported over a proper algebraic subset of B. By [5] it follows then that the class of Z vanishes over the complementary of this subset and we conclude:

3.3. Proposition. Assume $\mathcal{H}^{0,2} \otimes \Lambda^2\Omega_B/\operatorname{Im}\overline{\nabla}_2$ has constant rank. Then, if Z satisfies 3.2.1, δ_Z vanishes on B.

Notice that 3.3 is exactly Mumford's theorem in the case where $\mathcal{S}$ does not vary, i.e. $\mathcal{S} = S \times B$, because δ_Z identifies to the trace map $Z : H^0(K_S) \to H^0(\Omega^2_B)$ in this case.

For the applications, we give two descriptions of δ_Z:

3.4. Assume now that Z is a divisor of relative degree 0 in $\mathcal{C}$, where $\mathcal{C} \to B$ is a smooth family of curves over B, and that $\mathcal{C} \underset{j}{\hookrightarrow} \mathcal{S}$ is an inclusion over B.

Then Z has an infinitesimal invariant $\delta\nu_Z \in \mathcal{H}_C^{0,1} \otimes \Omega_B / \operatorname{Im} \overline{\nabla}_C$ (see M. Green's lectures), and if the maps $\overline{\nabla}_C : \mathcal{H}_C^{1,0} \to \mathcal{H}^{0,1} \otimes \Omega_B$, $\overline{\nabla}_{2,C} : \mathcal{H}_C^{1,0} \otimes \Omega_B \to \mathcal{H}_C^{0,1} \otimes \Lambda^2 \Omega_B$ (infinitesimal variation of Hodge structure of the family $\mathcal{C}$) are injective one has:

3.4.1. $\delta\nu_Z = 0 \Rightarrow \exists$ locally a flat section of $\mathcal{H}_C^1$, projecting onto ν_Z (the normal function associated to Z), unique up to a section of $H^1_{\mathbf{Z}}$. Now we have:

3.5. Proposition: There exists a natural map $j_* : \mathcal{H}_C^{0,1} \otimes \Omega_B / \operatorname{Im} \overline{\nabla}_C \longrightarrow \mathcal{H}^{0,2} \otimes \Omega_B^2 / \operatorname{Im} \overline{\nabla}_2$ such that $j_*(\delta\nu_Z) = \delta_Z$, where δ_Z is the invariant of 3.1 for the cycle $j(Z)$.

3.6. Assume finally that $Z \subset \mathcal{S}$ is given by $\sum_i n_i \sigma_i(B)$, with $\sigma_i : B \to \mathcal{S}$, sections of π, and $\Sigma n_i = 0$.

Then if $N = \dim B$ one has an isomorphism at $0 \in B$:

3.6.1. $\mathcal{H}_{(0)}^{0,2} \otimes \Omega_{B(0)}^2 / \operatorname{Im} \overline{\nabla}_{2(0)} = \left[H^0(\Omega_{\mathcal{S}}^N \otimes \pi^* K_{B|S_0}^{-1}) / H^0(\pi^*(\Omega_B^N \otimes K_B^{-1})_{|S_0}) \right]^*$.

Then one proves:

3.7. Proposition: $\delta Z_{(0)}$, as an element of the dual of $H^0(\Omega_{\mathcal{S}}^N \otimes \pi^* K_{B|S_0}^{-1}) / H^0(\pi^* \Omega_B^N \otimes K_B^{-1}{}_{|S_0})$, is equal to

$$\Sigma n_i \sigma_i^* : H^0(\Omega_{\mathcal{S}}^N \otimes \pi^* K_B^{-1}{}_{|S_0}) \longrightarrow H^0(\Omega_{B(0)}^N \otimes K_{B(0)}^{-1}) \simeq \mathbf{C}.$$

(Notice that $\Sigma n_i = 0 \Rightarrow \Sigma n_i \sigma_i^*$ vanishes on $H^0(\pi^* \Omega_B^N \otimes K_B^{-1}{}_{|S_0})$). Now we explain two applications of 3.3, 3.5, 3.7:

3.8. Theorem: [8] Let $S \subset \mathbf{P}^3$ be a general surface of degree $d \geq 5$, and $C \subset S$ a general plane section. Then $\operatorname{Ker} j_* : JC \to CH_0^0(S)$ is equal to the torsion of JC.

Sketch of proof: Let B be the moduli space of the pair (C, S) (or its smooth part). We have to show: If $U \to B$ is étale and

$$\begin{array}{ccccc} Z_U & \subset & \mathcal{C} & \underset{j}{\subsetneq} & \mathcal{S} \\ & \searrow & \downarrow & \swarrow & \\ & & U & & \end{array}$$

is as in 3.4 and satisfies $\forall\, u \in U,\ j_*(Z_u)$ is rationally equivalent to zero in S_u, then Z_u is of torsion in JC_u. Now we use the following (this uses the nice properties of the jacobian rings describing the variation of Hodge structure of the pair (C, S)):

3.8.1. For $d \geq 5$, the map j_* of 3.5 is injective, for

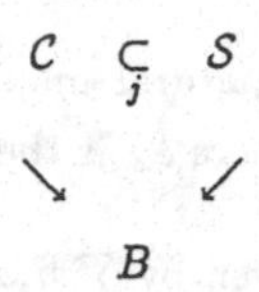

as above, when B is the universal deformation of (C, S). So if we have such Z_U, we find that $\delta\nu_{Z_u} = 0$, by 3.3 and 3.5, and then from $\overline{\nabla}^C_U$, $\overline{\nabla}^C_{2,U}$ injective, which is also a consequence of the general properties of jacobian rings, we conclude that ν_{Z_U} is a locally flat normal function. We use then the monodromy argument of lecture 4 to conclude that ν_{Z_U} is a torsion normal function, that is: $Z_u \in \mathrm{Tors}(JC_u)$ for $u \in U$.

3.9. Theorem: [8] Let S be a surface in $\mathbf{P}^3$ general of degree $d \geq 7$. Then if $p \neq q$ are points of S, they are not rationally equivalent in S.

Sketch of proof: Again by standard arguments it suffices to show: Let $U \to B$ be an étale cover of the moduli space of S. Let $\sigma_1,\ \sigma_2 : U \to \mathcal{S}_U$ be two distinct sections of the universal surface $\mathcal{S}_U$. Then for u general in U, $\sigma_1(u)$ and $\sigma_2(u)$ are not rationally equivalent in S_u.

If this is not the case, we apply 3.3 and 3.7 and this gives: For any $0 \in U$, the maps $\sigma_1^* : H^0(\Omega^N_{\mathcal{S}} \otimes \pi^* K_U^{-1}{}_{|S_0} \to \mathbf{C}$ and $\sigma_2^* : H^0(\Omega^N_{\mathcal{S}} \otimes \pi^* K_U^{-1}{}_{|S_0}) \to \mathbf{C}$ are equal ($N = \dim U$). For $\sigma_1(0) \neq \sigma_2(0)$, this contradicts the following:

3.10. Proposition: [8] For $d \geq 7$, $\mathcal{S} \xrightarrow[\pi]{} B$ the local universal deformation of S_0, $N = \dim B$, the vector bundle $\Omega^N_{\mathcal{S}|S_0}$ is very ample on S_0.

From 3.8, one deduces another proof of a theorem of Xu: for $d \geq 5$, a general surface S in $\mathbf{P}^3$ of degree d contains no rational curve.

From 3.10, one has also the following geometric corollary: let $d \geq 7$, and C be a fixed curve ; then for a general surface S of degree d, there is no non constant map from C to S.

References:

[1] S. Bloch - J.P. Murre: On the Chow groups of certain types of Fano threefolds. Comp. Math. Vol. 39 (1979), 47–105.

[2] S. Bloch - V. Srinivas: Remarks on correspondences and algebraic cycles, Amer. J. Math. 105 (1983), 1235–1253.

[3] A. Conte - J.P. Murre: The Hodge conjecture for fourfolds admitting a covering by rational curves, Math. Ann. 238 (1978), 79–88.

[4] P. Deligne: Théorie de Hodge II, Publ. Math. IHES (40) (1971), 5–57.

[5] F. Elzein: Complexe dualisant et applications à la classe fondamentale d'un cycle, Bull. Soc. Math. France, Mémoire 58 (1978), 4–66.

[6] F. Elzein, S. Zucker: Extendability of normal functions associated to algebraic cycles, in *Topics in transcendental algebraic geometry*, ed. by P. Griffith, Annals of mathematics studies, Study 106, Princeton (1984), 269–288.

[7] C. Voisin: Sur les zéro-cycles de certaines hypersurfaces munies d'un automorphisme, Annali della Scuola Norm. Sup. di Pisa, Vol. 29 (1993), p. 473-492.

[8] C. Voisin: Variations de structure de Hodge et zéro-cycles des surfaces générales, preprint IHES (1993).

Lecture 7 : Griffiths group.

This lecture is devoted to another phenomenon which holds only for cycles of codimension ≥ 2 : The non triviality of the Griffiths group {cycles homologous to zero modulo cycles algebraically equivalent to zero}, [6], and more spectacularly in contrast with the divisor case, its non finite generation [3].

This last fact was discovered only recently, and this is related to the difficulty of constructing interesting cycles of codimension at least two, (excepted for the zero-cycles, for which the above-mentioned phenomena do not hold). We will devote the next lecture to this problem.

For codimension two cycles, it is at the moment conjectured that the Griffiths group can be detected using the Abel-Jacobi map, more precisely its projection in the transcendental part of the intermediate jacobian J^3 ([6], [8]). However, in the paper [8], it is shown that for higher codimension cycles, the Griffiths group can be non trivial modulo torsion, even if the corresponding intermediate jacobian is trivial.

The lecture is organized as follows : We first describe Griffiths' argument [6] for the non triviality of the Griffiths group, and continue with a sketch of Clemens' method to get infinite generation of the Griffiths group. We follow then [8], and explain the application of "Hodge theoretic connectivity" to non triviality in the Griffiths group of cycles restricted to general complete intersection subvarieties. We finally describe the ideas of Bloch and Ogus, and present, in a non rigorous way, the Bloch-Ogus resolution and the Bloch-Ogus formula for the group of cycles modulo algebraic equivalence [2].

1.

1.1. Griffiths [6] worked with quintic hypersurfaces in $\mathbf{P}^4$. This is the smallest degree for which these hypersurfaces have $h^{3,0} \neq 0$. These varieties also satisfy the following property, which suggests that they have an interesting CH^2 group, by analogy of what was known previously for cubics and quartics :

1.1.1. Fact : A generic quintic in $\mathbf{P}^4$ has a finite number $N > 1$ of (rigid) lines.

Now Griffiths proved, using 1.1.. that the group $\mathrm{Hom}^2 / \mathrm{Alg}^2$ of cycles of codimension two homologous to zero modulo algebraic equivalence is generally a non torsion group :

1.2. Theorem : Let $X \subset \mathbf{P}^4$ be a general quintic 3-fold. $\ell_1 \neq \ell_2$ two distinct lines of X. Then $\ell_1 - \ell_2$ is a non-torsion element of $\mathrm{Hom}^2 / \mathrm{Alg}^2(X)$.

Step I : One has $H_2(X,\mathbf{Z}) = H^2(X,\mathbf{Z})^* = H^2(\mathbf{P}^4,\mathbf{Z})^*$ by weak Lefschetz theorem, and ℓ_1 and ℓ_2 have the same degree, so they are homologous. So one can use Φ_X, the Abel-Jacobi map of X, which gives

$$\Phi_X(\ell_1 - \ell_2) \in J^3(X) = H^3(X,\mathbf{C})/F^2H^3(X) \oplus H^3(X,\mathbf{Z}).$$

As explained by M. Green, the image of Φ_X on cycles algebraically equivalent to zero is contained in $J(X)^{\mathrm{alg}}$:= the maximal abelian subvariety of $J^3(X)$, with tangent space contained in $H^{1,2}(X) \subset H^3(X,\mathbf{C})/F^2H^3(X,\mathbf{C})$.

From the injectivity of $\overline{\nabla}$: $H^{1,2}(X) \to \mathrm{Hom}(H^1(T_X), H^3(\mathcal{O}_X))$ (see M. Green, Lecture 4), one deduces now :

1.2.1. If X is general [(1)], no non zero integral class $\alpha \in H^3(X,\mathbf{Z})$ is contained in $F^1H^3(X)$.

It follows that $J(X)^{\mathrm{alg}} = 0$, for general X. So we conclude that Theorem 1.2 follows from :

1.2.2. Proposition : If X is general, $\Phi_X(\ell_1 - \ell_2)$ is a non torsion point of $J^3(X)$, for $\ell_1 \neq \ell_2$ two lines in X.

Step II : Let $X \subset Y$, with Y a smooth quintic in $\mathbf{P}^5$ containing two planes P_1 and P_2 such that $P_1 \cap X = \ell_1$, $P_2 \cap X = \ell_2$. Let $(X_t)_{t\in\mathbf{P}^1}$ be a Lefschetz pencil of hyperplane sections of Y, such that $X_0 = X$.

On the open set $U \subset \mathbf{P}^2$ parametrizing smooth X_t's, there is a holomorphic section ν of the fibration $J \to U$ of intermediate jacobians with fiber $J_t = J^3(X_t)$, given by :

1.2.3. $\quad \nu(t) = \Phi_{X_t}(\ell_1^t - \ell_2^t)$, where $\ell_i^t = P_i \cap X_t$ is a line in X_t, $i = 1,2$.

1.2.2 is then equivalent to :

1.2.4. Proposition : ν is not a torsion section of J.

Step III : The sheaf of holomorphic sections of J over U is given by : $\mathcal{J} = \mathcal{H}^3/F^2\mathcal{H}^3 \oplus H^3_{\mathbf{Z}}$, where if $\mathcal{X}_U \xrightarrow{\pi} U$ is our family of threefolds, as usual $H^3_{\mathbf{Z}} = R^3\pi_*\mathbf{Z}, \mathcal{H}^3 = H^3_{\mathbf{Z}} \otimes \mathcal{O}_U$, and $F^2\mathcal{H}^3_{(t)} = F^2H^3(X_t) \subset H^3(X_t,\mathbf{C})$.

---

(1) Here "general" has not the usual sense : it means : locally outside a countable union of analytic subsets, instead of : outside a countable union of algebraic subsets.

Using the exact sequence :

1.2.5. $0 \to H^3_{\mathbf{Z}} \to \mathcal{H}^3/F^2\mathcal{H}^3 \to \mathcal{J} \to 0$. $\nu \in H^0(\mathcal{J})$ gives a natural element $[\nu] \in H^1(U, H^3_{\mathbf{Z}}) = \mathrm{Ker}(H^4(\mathcal{X}_U, \mathbf{Z}) \to H^4(X_t, \mathbf{Z}))$.

Then an important fact is :

1.2.6. Proposition : Let $\widetilde{Y}$ be the blow-up of Y along the base-locus of the pencil; let $\pi : \widetilde{Y} \to \mathbf{P}^1$ and $\tau : \widetilde{Y} \to Y$ be the natural maps. Then $\mathcal{X}_U = \pi^{-1}(U) \hookrightarrow \widetilde{Y}$, and we have: $[\nu] =$ restriction to $\mathcal{X}_U$ of $\tau^*([P_1 - P_2]) \in \mathrm{Ker}\, H^4(\widetilde{Y}) \to H^4(X_t)$.

Now 1.2.4 follows from 1.2.6, because $[P_1 - P_2] \in H^4(Y, \mathbf{Z})$ is not of torsion (this follows from the computation of the intersection $(P_1 - P_2)^2$), and more precisely no multiple of it is supported on some fiber of the pencil, by the Lefschetz assumption. This implies easily that the restriction of $\tau^*[P_1 - P_2]$ to $\mathcal{X}_U$ is not of torsion, so ν is not of torsion.

Finally, 1.2.6 is proved as follows : Consider the restrictions π_1, π_2 of π to the proper transforms $\widetilde{P}_1, \widetilde{P}_2$ of P_1, P_2 in $\widetilde{Y}$. There is a finite number of points $s_1, \ldots, s_k$ of U, over which $\pi_1 \cup \pi_2$ fails to be a fibration. Let $V = U \backslash \{s_1, \ldots, s_k\}$. The restriction $H^1(U, R^3\pi_*\mathbf{Z}) \to H^1(V, R^3\pi_*\mathbf{Z})$ being injective it suffices to prove 1.2.6 on V. Let $0 \in V$, and $G = \pi_1(V, 0)$. Let $L_0 = H^3(X_0, \mathbf{Z})$. Then G acts on L_0 by the monodromy representation, and $H^1(V, R^3\pi_*\mathbf{Z}) = H^1(G, L_0)$ is represented by cocycles $g \to \alpha_g \in L_0$, well defined up to coboundary $\alpha_g = {}^g\beta - \beta$, $\beta \in L_0$. To find a representative of $\tau^*[P_1 - P_2]_{|\mathcal{X}_V}$ in $H^1(G, L_0)$, one notes that on the universal cover $\widetilde{V}$ of V, one can trivialize the pulled-back family of pairs $((\widetilde{P}_1 \cup \widetilde{P}_2)_{\widetilde{V}}, X_{\widetilde{V}})$. Because $(P_1 - P_2).X_0$ is homologous to zero in X_0, one can then find on $\widetilde{V}$ a continuously varying family of real 3-chains $(v \to \gamma_v \subset X_v)_{v \in \widetilde{V}}$, such that $\forall v \in \widetilde{V}, \partial\gamma_v = P_1 \cap X_v - P_2 \cap X_v = \ell_1^v - \ell_2^v$. If $\widetilde{0} \in \widetilde{V}$ is a point of $\widetilde{V}$ over 0, for any $g \in G$, there is a canonical isomorphism $g : X_{g^{-1}(\widetilde{0})} \simeq X_{\widetilde{0}}$, such that $g(\ell_i^{g^{-1}\widetilde{0}}) = \ell_i^{\widetilde{0}}$, for $i = 1, 2$, and we can construct the cocycle $g \mapsto g.\gamma_{g^{-1}(\widetilde{0})} - \gamma_{\widetilde{0}} \in H_3(X_{\widetilde{0}}, \mathbf{Z}) \equiv H^3(X_{\widetilde{0}}, \mathbf{Z}) = L_0$. This gives our representative. On the other hand, by definition of the Abel-Jacobi map, the γ_v's give a global lifting $\widetilde{\nu}$ of ν on $\widetilde{V}$ to $F^2\mathcal{H}^3_{\widetilde{V}}$, if one defines $\widetilde{\nu}(v) = \int_{\gamma_v} \in F^2H^3(X_v)^*$ for $v \in \widetilde{V}$. Then clearly, by the definition of $[\nu]$ in 1.2.5, a representative of $[\nu]$ in $H^1(G, L_0)$ is given by the cocycle : $g \mapsto g.\widetilde{\nu}(g^{-1}\widetilde{0}) - \widetilde{\nu}(\widetilde{0}) \in H_3(X_{\widetilde{0}}, \mathbf{Z}) \subset F^2H^3(X_{\widetilde{0}})^*$, that is by : $g \mapsto \int_{g.\gamma_{g^{-1}(\widetilde{0})} - \gamma_{\widetilde{0}}}$, so 1.2.6 is proved.

2. Griffiths' discovery left open the possibility that the group $\mathrm{Hom}^2 / \mathrm{Alg}^2$ (which is in any case a countable group, because there are only countably many components of Chow varieties parametrizing effective cycles of codimension two in a given variety) is a finitely generated group. However, Clemens, working also with the quintic hypersurfaces, has shown that this is not the case, even modulo torsion ([3]) : (See also [7] for generalization to other K-trivial complete intersections.

2.1. Theorem : Let X be a general quintic threefold in $\mathbf{P}^4$; then $\mathrm{Hom}^2 / \mathrm{Alg}^2(X)$ has infinite rank over $\mathbf{Q}$. More precisely its image in JX (see 1.2, Step I), tensorized by $\mathbf{Q}$ has infinite rank over $\mathbf{Q}$.

The proof is somewhat delicate, and we will only sketch the main ideas. First of all, there is an interesting statement, of independent interest from the point of view of the geometry of Calabi-Yau threefolds :

2.2. Theorem : (Clemens) If X is general as above, X contains infinitely many rigid rational curves.

This step is done as follows : One construct to begin with a surface $S \subset \mathbf{P}^3$ smooth of degree 4, having infinitely many smooth rational curves L_n. Then if X is a generic quintic containing S, one shows that singularities of X are nodes which are not on the L_n's, and that the normal bundle of L_n in X is $\mathcal{O}_{\mathbf{P}^1}(-1) \oplus \mathcal{O}_{\mathbf{P}^1}(-1)$ ($L_n \simeq \mathbf{P}^1$). Such curves then deform with X, by the Kodaira stability theorem.

The second geometric point is to show, again working with the surface S, that one can specialize X to X_0 having a node on any given L_{n_0}, and no node on the other L_n's.

Finally a careful study of the generic normal bundle of L_{n_0} in the desingularization $\widetilde{X}_0$ of X_0 shows that under deformations of X_0 smoothing the node on L_{n_0}, L_{n_0} deforms with X_0, only when one takes a double cover B_2 of the basis B ramified along the discriminant locus (the locus where the node is preserved).

Consider now

$$\begin{array}{ccc} \mathcal{X} & \to & B \\ \uparrow & & \uparrow \\ \mathcal{X}_2 & \to & B_2 \end{array},$$

and let i be the involution of B_2 over B. Then on B_2 one has the cycle $L_{n_0}(b) - L_{n_0}(ib) \subset X_b$. Now, the most difficult part of Clemens' argument is :

2.3. The family of intermediate jacobians $J^3(X_b)$ for smooth X_b, $b \in B_2$, extends over the discriminant locus; the central fiber $J^3(X_0)$ has two components, and the normal function $\nu_{L_{n_0}}(b) = \Phi_{X_b}(L_{n_0}(b) - L_{n_0}(ib))$ extends over 0 in the component which does not pass through 0. (This uses essentially Theorem 3.4 of Lecture 5).

Admitting this, the rest of the argument goes as follows : Let C be a plane section of X. For each L_n, of degree d_n, $5L_n - d_nC$ is homologous to zero in X, so we have $\Phi_X(5L_n - d_nC) \in J^3(X)$. Assume there is for general X a relation :

(*) $\Sigma_n c_n \Phi_X(5L_n - d_nC) = 0$, $c_n \in \mathbf{Z}$. Then going to X_0 having a node on L_{n_0}, and letting the monodromy act on (*) by $L_{n_0}(b) \to L_{n_0}(ib)$, $L_n(b) \to L_n(b)$, for $n \neq n_0$, one concludes that c_{n_0} must be even. So all c_n must be even, and it remains only to prove :

2.4. Let $G \subset J^3(X)$ be a countable subgroup such that $G \otimes \mathbf{Z}/2\mathbf{Z}$ has infinite rank. Then $G \otimes \mathbf{Q}$ has infinite rank.

We will give in the next lecture a somewhat different approach to the infinite generation of $\Phi_X(CH^2(X)\,\mathrm{hom})$.

3. Now we turn to the results of Nori, which are essentially new for codimension ≥ 3 cycles. One of the most striking consequences of [8] is :

3.1 Theorem : [8]. For $d \geq 3$, $n > d$. there exist varieties X of dimension n with a cycle $Z \subset X$ of codimension d. homologous to zero and Abel-Jacobi equivalent to zero, such that no multiple of Z is algebraically equivalent to zero.

This results in fact of the following more precise statement :

3.2 Theorem : [8] Let X be a variety and Z be a cycle of codimension d, satisfying $[Z] \neq 0$ in $H^{2d}(X, \mathbf{Q})$. Then for $n > d$ and $Y \subset X$ a general complete intersection of sufficiently ample divisors in X, such that $\dim Y = n > d$, $Z \cap Y$ is not algebraically equivalent to zero in Y.

3.2 implies 2.1 because we can take in 3.2 a variety X of dimension $2d$, having no odd dimensional cohomology, with an algebraic cycle Z of codimension d, primitive with respect to an ample divisor L, and consider a complete intersection $Y \subset X$ of dimension n satisfying $d < n \leq 2d - 2$ (here one needs $d \geq 3$), of divisors in (L^{n_i}), satisfying 3.9. Then $J^{2d-1}(Y) = 0$, and $Z_{|_Y}$ satisfies 3.1. This way we get all (n, d) satisfying $d \geq 3$, $2d - 2 \geq n > d$. We can take products with a projective space to obtain the other (n, d)'s.

The theorem 3.2 is a consequence of the following Theorem 3.3, the proof of which will be given by M. Green.

Let X be a projective variety of dimension $n + k$, L_i $i = 1, \ldots, k$ be ample divisors on X. Fix positive integers $n_1, \ldots, n_k$ and let $S = \bigoplus_1^k H^0(X, L^{n_i})$.

Let $Y_S \subset S \times X$ be the universal complete intersection. Then :

3.3 Theorem : For $n_i \gg 0$, $m \leq 2n$, and any smooth base change $T \to S$, one has $H^m(T \times X, Y_T, \mathbb{Q}) = 0$.

3.4. Now we show how Theorem 3.3 implies Theorem 3.2. So, under the assumptions of Theorem 3.1, and Y being a general complete intersection of divisors in $|L^{n_i}|$, n_i as in 3.3, assume that $Z \cap Y$ is algebraically equivalent to zero.

Then there exists an etale morphism $S' \to S$, a smooth family of curves $\mathcal{C} \xrightarrow[\varphi]{} S'$, a divisor D on $\mathcal{C}$, homologous to zero on the fibers of φ, a cycle Γ of codimension d in $\mathcal{C} \times_{S'} Y_{S'}$, such that, writing j for the natural map $Y_{S'} \to X$, one has :

3.4.1. $j^*Z = p_{2*}(\Gamma . p_1^* D)$. in $CH^d(Y_{S'})$ (this is the definition of algebraic equivalence, put in family). Here we work with the diagram :

$$(*) \qquad \begin{array}{ccccc} & & \mathcal{C} \times_{S'} Y_{S'} & & \\ & p_1 \swarrow & & \searrow p_2 & \\ \mathcal{C} & & & & Y_{S'} \xrightarrow[j]{} X \end{array}$$

Now let $T = \mathcal{C}$. So $\mathcal{C} \times_{S'} Y'_S = Y_T$. The cycle Γ has a class γ in $H^{2d}(Y_T, \mathbb{Q})$, and by Theorem 3.3 and $2d < 2n$, one finds that this class extends to $\tilde{\gamma} \in H^{2d}(T \times X, \mathbb{Q})$. Consider then :

3.4.2. $\beta = \varphi_*(p_1^*[D].\tilde{\gamma}) \in H^{2d}(S' \times X, \mathbb{Q})$, where we consider now the following diagram :

$$(**) \qquad \begin{array}{ccccc} & & T \times X & & \\ & p_1 \swarrow & & \searrow \varphi & \\ T & & & & S' \times X \end{array},$$

which contains $(*)$. Then by 3.4.1, $\beta_{|Y_{S'}} = j^*[Z] \in H^{2d}(Y_{S'}, \mathbb{Q})$, where j is as before the natural map $Y_{S'} \to X$. But by Theorem 3.3 applied to S', the restriction map $H^{2d}(S' \times X, \mathbb{Q}) \to H^{2d}(Y_{S'}, \mathbb{Q})$ is injective, so we find :

3.4.3. $\beta = p_2^*[Z]$, where p_2 is the second projection $S' \times X \to X$.

But if one chooses $s \in S'$, one finds that in $H^{2d}(X, \mathbb{Q})$, $\beta_{|s \times X} = [Z] = \varphi_{s*}(p_1^*[D_s].\tilde{\gamma}_{|C_s \times X})$, where one works with the diagram :

$$(***) \qquad \begin{array}{ccccc} & & C_s \times X & & \\ & p_1 \swarrow & & \searrow \varphi_s = \varphi_{|C_s} & \\ C_s & & & & s \times X \end{array}.$$

So $\beta_{|_{\bullet\times X}} = 0$, because $[D_s] = 0$ and $[Z] = 0$ in $H^{2d}(X, \mathbb{Q})$, hence Theorem 3.2 is proved.

4. To conclude this lecture we want to describe the Bloch-Ogus resolution and Bloch-Ogus formula for the group of cycles modulo algebraic equivalence [2].

Recall the following classical result :

4.1. Lemma : Let X be a smooth projective variety and $D = \Sigma n_i D_i$ be a divisor on X. Then a multiple of D is homologous to zero if and only if there exists a (necessarily closed) holomorphic one form ω with logarithmic poles along the support $|D|$ of D, and such that $\mathrm{Res}_{|D|}\,\omega = \Sigma n_i 1_{D_i}$.

The notion of logarithmic pole is well defined when $|D|$ is a divisor with normal crossings, which we can achieve by blowing up X. Then locally $|D|$ admits for equation $z_1, \ldots, z_k$, in some coordinate systems $(z_1, \ldots, z_n)$ on X, and ω has logarithmic pole if it is of the form : $\sum_1^k f_i \dfrac{dz_i}{dz_i} + \sum_{i>k} g_i dz_i$, for holomorphic functions f_i, g_i.

The residues of ω are then the fonctions $\frac{1}{2i\pi} f_i$ on D_i. They are constant if ω is closed.

The lemma follows from the exact sequence :

4.1.1. $\quad 0 \to \Omega_X \to \Omega_X(\log|D|) \xrightarrow{\mathrm{Res}} \bigoplus_i \mathcal{O}_{D_i} \to 0$. See [9].

It is easy to show that the induced map : $\bigoplus_i H^0(\mathcal{O}_{D_i}) \to H^1(\Omega_X)$ sends $\Sigma n_i 1_{D_i}$ to the De Rham class of D in $H^1(\Omega_X)$. So if $C_1^{\mathbb{R}}(\Sigma n_i D_i) = 0$, $\Sigma_i n_i 1_{D_i} = \mathrm{Res}\,\omega$, as we wanted. Such a form ω is closed by the degeneration at E_1 of the Hodge-DeRham spectral sequence for the logarithmic complex $\Omega_X^0(\log|D|)$. See [9]).

4.2. More generally, if X is smooth not necessarily compact and $Y \subset X$ is a smooth hypersurface, one has a residue $\mathrm{Res}_Y : H^k(X\backslash Y) \to H^{k-1}(Y)$ see [6], and if one has $Y_1, Y_2 \subset X$ intersecting transversally, one has :

4.2.1

$$H^k(X\backslash Y_1 \cup Y_2) \xrightarrow{\mathrm{Res}_{Y_2}} H^{k-1}(Y_2\backslash Y_1 \cap Y_2) \xrightarrow{\mathrm{Res}_{Y_1\cap Y_2}} H^{k-2}(Y_1 \cap Y_2)$$

$$H^k(X\backslash Y_1 \cup Y_2) \xrightarrow[\mathrm{Res}_{Y_1}]{} H^{k-1}(Y_1\backslash Y_2 \cap Y_1) \xrightarrow[\mathrm{Res}_{Y_2\cap Y_1}]{} H^{k-2}(Y_1 \cap Y_2)$$

Now the very important point is the following fact :

4.2.2. i) $\mathrm{Res}_{Y_1\cap Y_2} \circ \mathrm{Res}_{Y_2} = -\mathrm{Res}_{Y_2\cap Y_1} \circ \mathrm{Res}_{Y_1}$

ii) $\mathrm{Res}_Y\, H^k(X) = 0$.

Furthermore Lemma 4.1 has now the following integral version :

4.2.3. $\Sigma n_i D_i$ is homologous to zero $\Longleftrightarrow \Sigma n_i 1_{Y_i} = \Sigma\, \mathrm{Res}_{Y_i}(\eta)$, for some $\eta \in H^1(X \setminus \cup Y_i, \mathbf{Z})$.

4.2.2 is the essential point for the construction of the Bloch-Ogus resolution, although there are in fact big difficulties coming from the singularities of the subvarieties we consider.

4.3. Define the Zariski sheaf $\mathcal{H}^k$ on X as associated to the presheaf $U \to H^k(U, A)$ (where A may be integral, rational or complex coefficients), and define $H^k(\mathsf{C}(X)) = \varinjlim_{V \subset X} H^k(X)$. Then $\mathcal{H}^k$ is also $R^k j_*(A)$ where $j : X^{an} \to X^{zar}$ is the continuous map from the usual topology to the Zariski topology.

4.2.2 (with a lot of work) gives a complex :

$$0 \to \mathcal{H}^k \to H^k(\mathsf{C}(X)) \to \bigoplus_{\substack{\mathrm{cod}\, D=1 \\ D \mathrm{irred}}} H^{k-1}(\mathsf{C}(D)) \to \bigoplus_{\substack{\mathrm{cod}\, Z=2 \\ Z \mathrm{irred}}} H^{k-2}(\mathsf{C}(Z)) \to$$

4.3.1
$$\ldots \to \bigoplus_{\substack{\mathrm{cod}\, Z=k \\ Z \mathrm{irred}}} A_Z \to 0.$$

Here one sees $H^\ell(\mathsf{C}(Z))$ as a constant sheaf supported on the irreducible subvariety Z.

The main result of Bloch-Ogus is then :

4.4. Theorem : [2]. 4.3.1 is exact.

So we have a resolution of $\mathcal{H}^k$ by acyclic sheaves, for the Zariski topology, which has length $\leq k$.

If one considers the end of this resolution, for $A = \mathbf{Z}$, one finds :

4.5.
$$H^k_{\mathrm{zar}}(\mathcal{H}^k_{\mathbf{Z}}) = \bigoplus_{\substack{\mathrm{cod}\, Z=k \\ Z \mathrm{irred}}} (\mathbf{Z}_Z) / \mathrm{Res}\Big(\bigoplus_{\substack{\mathrm{cod}\, D=k-1 \\ D \mathrm{irred}}} H^1(\mathsf{C}(D), \mathbf{Z})\Big).$$

But now by 4.2.3, one sees that $\Sigma n_i 1_{Z_i}$. codim $Z_i = k$ defines a zero element in $H^k_{\mathrm{zar}}(\mathcal{H}^k_{\mathbf{Z}})$ if and only if it is a combinaison of cycles homologous to zero in varieties of codimension $k-1$, that is iff $\Sigma n_i Z_i$ is algebraically equivalent to zero. In other words, one has the Bloch-Ogus formula :

4.6. $\quad CH^k(X)/\mathrm{alg.eq} \simeq H^k_{\mathrm{zar}}(\mathcal{H}^k)$.

4.7. This fascinating formula did not prove very useful for the understanding of the groups CH^k/alg.eq., but the Bloch-Ogus theory has very interesting consequences on the spectral sequence associated to the map j of 4.3. An important consequence of Theorem 4.4 is :

4.7.1. $H^p_{\mathrm{zar}}(\mathcal{H}^q) = 0$ for $p > q$.

For $p+q = 3, 4$ it follows that the spectral sequence $H^p_{\mathrm{zar}}(\mathcal{H}^q) \Rightarrow H^{p+q}(X)$ degenerates quickly, and gives rise to the following exact sequence.

4.7.2. $H^3(X) \to H^0_{\mathrm{zar}}(\mathcal{H}^3) \xrightarrow{d_2} H^2_{\mathrm{zar}}(\mathcal{H}^2) \to H^4$. where the first map is given by the restrictions $H^3(X) \to H^3(U, \mathbf{Z}), U \subset X$, and is in many cases injective, and the last one identifies to the cycle class defined on $CH^2(X)$/alg.eq.

For complex coefficients a section of $\mathcal{H}^3$ can be understood as a meromorphic form which is locally the sum of a holomorphic form and of an exact form. The obstruction for such a form to define a global cohomology class lies in the group $(\mathrm{Hom}^2 / \mathrm{Alg}^2(X)) \otimes \mathbf{C}$, which is generally non trivial by the Griffiths-theorem 1.2 (see [1], [2]).

References :

[1] M.F. Atiyah, W.V.D. Hodge : Integrals of the second kind on an algebraic variety, Annals of Maths. Vol. 62, n°1 (1955) 56-91.

[2] S. Bloch, A. Ogus. Ann. Scient. Ec. Norm. Sup. 4ème Série, t.7 (1974) 181-202.

[3] H. Clemens : Homological equivalence, modulo algebraic equivalence, is not finitely generated, Publ. Math. IHES 58 (1983) 19-38.

[4] H. Clemens : Neron models for families of intermediate jacobians, Publ. Math. IHES 58 (1983) 5-18.

[5] N. Coleff, M. Herrera, D. Lieberman : Algebraic cycles as residues of meromorphic forms, Math. Ann. (1980) 73-87.

[6] P.A. Griffiths : On the periods of certain rational integrals I, II, Ann. of Math. 90 (1969) 460-541.

[7] K.H. Paranjape : Curves on threefolds with trivial canonical bundle, Proc. Indian Acad. Sci. Math. 101 (1991) n°3. 199-213.

[8] M.V. Nori : Algebraic cycles and Hodge theoretic connectivity, Inv. Math. 111(2) 349-373 (1993).

[9] S. Zucker : Degeneration of Hodge bundles (after Steenbrink) in "Topics in transcendental algebraic geometry" ed. by P. Griffiths, Annals of Mathematics Studies, Study 106, Princeton (1984) 121-141.

Lecture 8 : Application of the Noether-Lefschetz locus to threefolds.

This last lecture describes the applications of the criterion of M. Green (see Lecture 3) for the existence and density of the Noether-Lefschetz locus, to the study of one-cycles on threefolds, which were worked out in [15], [16], [17].

The simplest application is a purely infinitesimal (hence algebraic) method to solve the Hodge conjecture for the algebraic part of the intermediate jacobian of certain threefolds [15]. Of course to apply the method, one needs information which depend on the particular case we consider, but the method applies well in some cases where the geometry is so intricated that the existence of interesting codimension two cycles is not obvious.

We also sketch an infinitesimal proof of Clemens' theorem [17]. Here we don't use the full power of M. Green's criterion, because we specialize to quintics containing the Fermat surface S of degree five : A crucial point to get the "infinite generation" is the fact that $\mathrm{rank}(\mathrm{Pic}\, S)^{\mathrm{prim}} \geq 2$, and the existence of such surface is not predicted by this criterion. (Note that one can always construct surfaces S with $\mathrm{rank}(\mathrm{Pic}\, S)$ large in a given threefold, but they will not correspond to components of the Noether-Lefschetz locus of the right codimension, a condition which is important for all computations).

The most convincing application is the generalization of the Griffiths theorem ([6], and Th. 1.2 of Lecture 7) to any non-rigid Calabi-Yau threefold, that is a smooth projective threefold X satisfying $H^1(\mathcal{O}_X) = H^2(\mathcal{O}_X) = \{0\}$ and K_X trivial) : we prove :

Theorem [16] : Let X be a Calabi-Yau threefold satisfying $H^1(T_X) \neq 0$. Then for a general deformation X' of X, one has : $\mathrm{Hom}^2 / \mathrm{Alg}^2(X')$ is not a torsion group. More precisely, the Abel-Jacobi map $\Phi_{X'}$ of X' is not of torsion.

1.

1.1. For complete intersection threefolds X which have a negative canonical bundle, it is known how to parametrize the intermediate jacobian $J^3(X)$ using the Abel-Jacobi map, as predicted by the Hodge conjecture (see J.P. Murre's lecture on the Hodge conjecture): These varieties contain a positive dimensional family D of lines, and the Abel-Jacobi map $\Phi_X : \mathrm{Alb}\, D \rightarrow J^3(X)$ is surjective. Bardelli [11] has constructed an example of a complete intersection X with effective canonical bundle and with an involution i without fixed point such that $(JX)^-$ is contained in the algebraic part of JX, that is $H^{3,0}(X)^- = 0$, and has constructed geometrically interesting codimension two cycles in X, which give a parametrization of $(JX)^-$ by algebraic cycles.

1.2. We sketch now an argument which works for other varieties X of dimension 3 with a group of automorphisms G, such that the invariant part of $J^3(X)$ under G, or its orthogonal is contained in $J^3(X)^{\mathrm{alg}}$ [15]. Assuming that one of these two possibilities occurs, we will refer to them respectively as case i) and case ii).

Case i) is then equivalent to $H^{3,0}(X)^{\mathrm{inv}} = 0$, case ii) is equivalent to G acts trivially on $H^{3,0}(X)$.

1.3. Now assume that X carries a G-invariant line bundle with enough G-invariant sections, and such that :

Case i) for a smooth $S = V(\sigma)$, $\sigma \in H^0(L)^{\mathrm{inv}}$, the number $h^{2,0}(S)^{\mathrm{inv}} := \dim H^{2,0}(S)^{\mathrm{inv}}$ satisfies : $h^{2,0}(S)^{\mathrm{inv}} < h^0(N_S X)^{\mathrm{inv}}$.

Case ii) for a smooth $S \in V(\sigma)$, $\sigma \in H^0(L)^{\mathrm{inv}}$, the number $h^{2,0}(S)^\sharp := \dim(h^{2,0}(S)^\sharp := \dim(H^{0,2}(S)^{\mathrm{inv}\,\perp})$ satisfies : $h^{2,0}(S)^\sharp < h^0(N_S X)^{\mathrm{inv}}$.

These assumptions are made plausible by the adjunction formula, which gives :

$$H^0(K_S)^{\mathrm{inv}} = H^0(K_X(L)_{|S})^{\mathrm{inv}} \hookleftarrow H^0(K_X)^{\mathrm{inv}} \otimes H^0(N_S X)^{\mathrm{inv}} \ (= 0 \text{ in case i)})$$

$$H^0(K_S)^\sharp = H^0(K_X(S)_{|S})^\sharp \hookleftarrow H^0(K_X)^\sharp \otimes H^0(N_S X)^{\mathrm{inv}} \ (= 0 \text{ in case ii)})$$

Now as explained in Lecture 3, $(h^{2,0})^{\mathrm{inv}}$ (resp. $h^{2,0}(S)^\sharp$ is the expected codimension of the components of the Noether-Lefschetz loci corresponding to invariant (resp. skew) classes λ, on the space of G-invariant surfaces. So assumption 1.3 means that in both cases, the Noether-Lefschetz locus, if non empty has positive dimensional components. For (S,λ) in a component S_λ of the Noether-Lefschetz locus, with $\lambda \in H^2(S,\mathbf{Z})^{\mathrm{inv}} \cap H^{1,1}(S)$ in case i), $\lambda \in H^2(S,\mathbf{Z})^\sharp \cap H^{1,1}(S)$ in case ii), one considers λ as an element of $CH_1(S)^{\mathrm{inv}}$ (resp. $CH_1(S)^\sharp$ in case ii), where a one-cycle Z is skew, if $\sum_{g\in G} gZ = 0$). Here one assumes that $H^1(\mathcal{O}_S) = 0$, for simplicity.

If $j : S \hookrightarrow X$ is the inclusion, $j_*\lambda$ gives an element of $CH_1(X)^{\mathrm{inv}}$ (resp. $CH_1(X)^\sharp$ in case ii).

Using theAbel-Jacobi map Φ_X, one deduces a map :

1.3.1. $S_\lambda \underset{\alpha_\lambda}{\rightarrow} (JX)^{\mathrm{inv}}$ (resp. $(JX)^\sharp$) in case ii), depending on the choice of a base point, that is well defined up to constant, and S_λ being positive dimensional, one can hope that the image of α_λ generates $(JX)^{\mathrm{inv}}$ (resp. $(JX)^\sharp$).

1.4. Now we explain how one can make this work with only algebraic assumptions on the infinitesimal variation of Hodge structure of $S \subset X$.

Let $U \subset \mathbf{P}(H^0(L)^{\mathrm{inv}})$ parametrize the smooth invariant surfaces $S \subset X$, so $TU_{(S)} = H^0(N_S X)^{\mathrm{inv}} = H^0(L_{|s})^{\mathrm{inv}}$. On U, we consider in case i) the invariant variation of Hodge structure $H^{2\,\mathrm{inv}}_{\mathbf{Z}}, \mathcal{H}^{2\,\mathrm{inv}}, F^k\mathcal{H}^{2\,\mathrm{inv}}$, with infinitesimal variation described at $(S) \in U$ by $=$

$$\begin{array}{cccc} \overline{\nabla}: & H^1(\Omega_S)^{\mathrm{inv}} & \to & \mathrm{Hom}(TU_{(S)}, H^2(\mathcal{O}_S)^{\mathrm{inv}}) \\ & \| & & \\ & \mathcal{H}^{1,1\,\mathrm{inv}}_{(S)} & \to & \mathrm{Hom}(TU_{(S)}, \mathcal{H}^{0,2\,\mathrm{inv}}_{(S)}) \end{array}$$

and we write $\lambda \in H^1(\Omega_S)^{\mathrm{inv}} \to \overline{\nabla}(\lambda) \in \mathrm{Hom}(TU_{(S)}, H^2(\mathcal{O}_S)^{\mathrm{inv}})$. In case ii) we work with $H^{2\sharp}_{\mathbf{Z}}$, and the associated VHS. Now to apply M. Green's Lemma, we refine assumption 1.3 as follows :

1.4.1. For S generic in $U, \exists\lambda \in H^1(\Omega_S)^{\mathrm{inv}}$ (resp. $H^1(\Omega_S)^{\sharp}$ in case ii)), such that $\overline{\nabla}(\lambda) : TU_{(S)} \to H^2(\mathcal{O}_S)^{\mathrm{inv}}$ is surjective (resp. $\overline{\nabla}(\lambda) : TU_{(S)} \twoheadrightarrow H^2(\mathcal{O}_S)^{\sharp}$) in case ii)).

Then M. Green's Lemma says that for generic S and generic λ in $H^1(\Omega_S)^{\mathrm{inv}}_{\mathbf{R}}$ (resp. $H^1(\Omega_S)^{\sharp}_{\mathbf{R}}$) we can approximate it by (S_n, λ_n) such that : $\lambda_n \in H^2(S_n, \mathbf{Q})^{\mathrm{inv}} \cap H^{1,1}(S_n)$, (resp. $\lambda_n \in H^2(S_n, \mathbf{Q})^{\sharp} \cap H^{1,1}(S_n)$).

Also for large enough n, (λ_n, S_n) also satisfies assumption 1.4.1 so that S_{λ_n} is smooth of maximal codimension at S_n. (See Lecture 3). This is the existence step for our construction.

Next we explain which infinitesimal condition is needed for the Abel-Jacobi map α_{λ_n} of 1.3.1 to be non trivial.

For $\lambda \in H^1(\Omega_S)^{\mathrm{inv}}$ (resp. $H^1(\Omega_S)^{\sharp}$ in case ii)) one defines : $TU_\lambda := \mathrm{Ker}\,\overline{\nabla}(\lambda)$: $TU_{(S)} \to H^2(\mathcal{O}_S)^{\mathrm{inv}}$ (resp. $H^2(\mathcal{O}_S)^{\sharp}$). Then let $\alpha_{\lambda_*} : TU_\lambda \to H^2(\Omega_X)$ be defined as follows : One has $TU_{(S)} \subset H^0(L_{|s})^{\mathrm{inv}}$, hence a map :

$$\mu_\lambda : TU_{(S)} \to H^1(\Omega_S(L))^{\mathrm{inv}} \text{ (resp. } H^1(\Omega_S(L))^{\sharp}).$$

Consider the exact sequence : $0 \to \mathcal{O}_S(-L) \to \Omega_{X|S} \to \Omega_S \to 0$, which gives $\delta : H^1(\Omega_S(L))^{\mathrm{inv}} \to H^2(\mathcal{O}_S)^{\mathrm{inv}}$ (resp. $\delta : H^1(\Omega_S(L))^{\sharp} \to H^2(\mathcal{O}_S)^{\sharp}$). Then one has

(See M. Green's lecture on hypersurfaces) :

$$\delta \circ \mu_\lambda = \overline{\nabla}(\lambda)$$

One concludes that $\mu_\lambda(TU_\lambda)) \subset H^1(\Omega_X(L)_{|s})^{\text{inv}}$ (resp. $H^1(\Omega_X(L))_{|s})^\sharp$), using $H^1(\mathcal{O}_S) = 0$, and this goes naturally to $H^2(\Omega_X)^{\text{inv}}$ (resp. $H^2(\Omega_X)^\sharp$), using the exact sequence $0 \to \Omega_X \to \Omega_X(L) \to \Omega_X(L)_{|s} \to 0$.

This defines our α_{λ_*}, and one can check :

1.4.2. Lemma : If $\lambda \in H^1(\Omega_S) \cap H^2(S, \mathbf{Z})$, α_{λ_*} is the differential of the map α_λ of 1.3.1, where one identifies $TU_{(\lambda)}$ to $TS_{\lambda(S)}$. (See Lecture 3).

If λ is only rational, $N\lambda \in H^2(S, \mathbf{Z}) \cap H^1(\Omega_S)$, for some $N \in \mathbf{N}$, and one has obviously: $\alpha_{\lambda_*} = \frac{1}{N}$ (differential of $\alpha_{N\lambda}$).

Consider now the following assumption (a formal assumption on $IVHS$).

1.4.3. For generic $S \in U$, $\lambda \in H^1(\Omega_S)^{\text{inv}}$ (resp. $H^1(\Omega_S)^\sharp$), the map $\alpha_{\lambda *} : TU_{(\lambda)} \to H^2(\Omega_X)$ is non zero.

If 1.4.3 is true for (S, λ), satisfying assumption 1.4.1, it is also true for some (S, λ), with λ real satisfying 1.4.1. Then we approximate (S, λ) by (S_n, λ_n) as in 1.4.1. For n large enough, $TU_{(\lambda_n)}$ has the minimal dimension, so 1.4.3 will also hold at (S_n, λ_n). According to 1.4.2, this means that $(\alpha_{\lambda_n})*$ is non zero at (S_n, λ_n) and S_{λ_n} being smooth at S_n, it follows that α_{λ_n} is non zero.

One can generally conclude that α_{λ_n} is in fact surjective, by checking that $(JX)^{\text{inv}}$ or $(JX)^\sharp$ is a simple abelian variety. Hence we have shown under the formal assumptions 1.4.1 and 1.4.3 that the Hodge conjecture holds for $(JX)^{\text{inv}}$ or $(JX)^\sharp$.

1.5. Examples and variant i) ([15]). Consider a generic quintic polynomial in $\mathbf{P}^4$ invariant under the involution i acting by $i^*(X_0, \dots, X_4) = (-X_0, -X_1, X_2, X_3, X_4)$ then $h^{3,0}(X)^- = 0$. One considers the quintic fourfolds $Y \subset \mathbf{P}^5$ invariant under the extended involution i acting on $\mathbf{P}^5$: $i^*(X_0, \dots, X_5) = (-X_0, -X_1, X_2, \dots, X_5)$, and containing X. These quintics have $h^{4,0} = 0$, and by Bloch-Srinivas the Hodge conjecture is true for them in degree 4. So the method of the Noether-Lefschetz locus can be applied and it is verified in [15] that the analogs of 1.4.1, 1.4.3 hold true.

ii) More generally, the method works for all K-trivial complete intersections with an involution, excepted the one constructed by Bardelli.

2. We explain now how one can prove Clemens' theorem by infinitesimal methods [17].

2.1 Theorem : Let X be a general quintic threefold in $\mathbf{P}^4$. Then the one-cycles on X homologous to zero, and supported on a hyperplane section of X generate by the Abel-Jacobi map a countable subgroup of JX of infinite rank over $\mathbf{Q}$.

Let $S \subset X$ be a hyperplane section, and let $\lambda \in H^1(\Omega_S)^{\text{prim}} \cap H^2(S, \mathbf{Z})$ be an algebraic class.

Note that by K_X trivial, and the adjunction formula, one has $\dim H^0(N_S X) = \dim H^0(K_S) = \dim H^2(\mathcal{O}_S)$.

Working with the same notations as in 1, with $U \subset \mathbf{P}(H^0(\mathcal{O}_X(1))$ the Zariski open set parametrizing smooth surfaces, we suppose now :

2.2. Assumption : The map $\overline{\nabla}(\lambda) : H^0(N_S X) \to H^2(\mathcal{O}_S)$ is an isomorphism.

Then consider $V \subset \mathbf{P}(H^0(\mathcal{O}_{\mathbf{P}^4}(5)))$ the open set parametrizing smooth quintics, and let $\mathcal{U} \xrightarrow{\pi} V$ be the fibration with fiber $\mathcal{U}_t$ = open set in $\mathbf{P}(H^0(\mathcal{O}_{X_t}(1)))$ parametrizing smooth surfaces in X_t. (S, X) is a point of $\mathcal{U}$, and one can consider the component $\mathcal{S}_\lambda \subset \mathcal{U}$ of the Noether-Lefschetz locus passing through (S, X). By the surjectivity of the map $\overline{\nabla}(\lambda)$ in 2.2 $\mathcal{S}_\lambda$ is smooth of codimension $h^{2,0}$ at (S, X), and the kernel of $\pi_* : T\mathcal{S}_{\lambda(S,X)} \to TV_{(X)}$ is clearly equal to $\text{Ker}(\overline{\nabla}(\lambda) : H^0(N_S X) \to H^2(\mathcal{O}_S))$. Also π_* is surjective by the surjectivity of $\overline{\nabla}(\lambda)$. So we conclude :

2.3 Lemma : Assumption 2.2 implies that $\mathcal{S}_\lambda \xrightarrow{\pi} V$ is etale in a neighbourhood of (S, X).

2.4. Consider the Fermat quintic surface $S : \sum_0^3 X_i^5 = 0$. It has a primitive Picard group of rank $\rho_{\text{prim}} \geq 2$. Let us denote by $NS^{1,1}_{\text{prim}} \subset H^1(\Omega_S)^{\text{prim}}$ the subspace of $H^1(\Omega_S)^{\text{prim}}$ generated over $\mathbf{C}$ by algebraic classes.

Using an easy algebraic characterization of $NS^{1,1}_{\text{prim}}$, and the jacobian description of the $IVHS$ of S (cf. M. Green's Lecture), one checks now :

2.5 Lemma. Let X be a generic quintic threefold containing S, and let λ be a generic element of $NS^{1,1}_{\text{prim}}$. Then the map $\overline{\nabla}(\lambda) : H^0(N_S X) \to H^2(\mathcal{O}_S)$ is an isomorphism.

Because $NS^{1,1}_{\text{prim}}$ is a $\mathbf{C}$-vector space with a $\mathbf{Q}$-structure given by $NS_{\text{prim}} \otimes \mathbf{Q}$, and because the asumption on λ is Zariski open, it follows that the conclusion of 2.5 holds for infinitely many elements of $\mathbf{P}(NS_{\text{prim}} \otimes \mathbf{Q})$.

In other words, assumption 2.2 holds for infinitely many non proportional algebraic classes in S, for some fixed X containing S.

2.6. By 2.3 and 2.5, we have now in a neighbourhood of a generic $X \in V$ containing S, countably many components $\mathcal{S}_{\lambda_n}$, tales over V and to each of them we associate a holomorphic family of one-cycles homologous to zero in the fibers $(X_v)_{v \in V}$: to $v \in V$ we associate $j_{v*}(\lambda_{n,v})$, where $v \to S_v \underset{j_v}{\hookrightarrow} X_v$ is a local section of $\mathcal{S}_{\lambda_n} \underset{\pi}{\to} V$, and $\lambda_{n,v} \in H^2(S_v, \mathbf{Z})_{\text{prim}} \cap H^{1,1}(S_v) \simeq CH_1(S_v)_{\text{prim}}$, because $S_v \in \mathcal{S}_{\lambda_n}$.

We consider the associated germs of normal functions $\nu_n \in \mathcal{J}(V)$, defined as $\nu_n(v) = \Phi_{X_v}(j_{v*}(\lambda_{n,v}))$.

To each such germ, we associate its infinitesimal invariant at X (See M. Green's Lecture on normal functions), that is :

2.6.1 : $\quad \delta\nu_{n(X)} \in H^{1,2}(X) \otimes \Omega_{V(X)}/\overline{\nabla}(H^{2,1}(X)).$

Now the technical point is the following :

2.6.2 Lemma : Let S be the Fermat quintic. Then there exists two non proportional algebraic classes $\lambda_1, \lambda_2 \in NS(S)_{\text{prim}}$, satisfying : For X general containing S, there exists an infinite number of non proportional combinations $\alpha\lambda_1 + \beta\lambda_2$, $\alpha, \beta \in \mathbf{Z}$, such that :

i) $\alpha\lambda_1 + \beta\lambda_2$ satisfies 2.2, for the inclusion $S \subset X$.

ii) the associated germs of normal functions $\nu_{\alpha\lambda_1+\beta\lambda_2}$ on V near X have infinitesimal invariants $\delta\nu_{\alpha\lambda_1+\beta\lambda_2}(X)$ which are independent over $\mathbf{Q}$, in $H^{1,2}(X) \otimes \Omega_{V(X)}/\overline{\nabla}(H^{2,1}(X))$.

Now the infinitesimal proof of Theorem 2.1 is almost finished. Assume there is for a general point near X a relation with a finite number of integral coefficients between the normal functions $\nu_{\alpha\lambda_1+\beta\lambda_2}$, defined in a common open set of V containing X. By a countability argument one may assume that a fixed relation holds in this open set. But this relation will then also hold between the associated $\delta\nu_{\alpha\lambda_1+\beta\lambda_2}$ at X which contredicts their independence over $\mathbf{Q}$.

2.7 Remark : This method has been successfully applied in [12], where the authors use to begin with the Noether-Lefschetz locus and conclude with Clemens argument, and in [10] where theorem 2.1 is proved for cycles of dimension 3 in cubic hypersurfaces of dimension 7.

3. We conclude our lecture with the following theorem, which makes a full use of M. Green's criterion :

3.1 Theorem : [16] Let X be a non-rigid Calabi-Yau threefold. Then for a general deformation X' of X, the Abel-Jacobi map $\Phi_{X'}$ of X' is not of torsion.

As in Griffiths' theorem, this implies that $\mathrm{Hom}^2 / \mathrm{Alg}^2(X')$ is not of torsion, because for a Calabi-Yau threefold, the map $\overline{\nabla} : H^{1,2} \to \mathrm{Hom}(H^1(T_X), H^3(\mathcal{O}_X))$ is an isomorphism (which identifies, via K_X trivial $\Rightarrow T_X \simeq \Omega^2_X$, to the Serre duality) and its injectivity implies as in Lecture 7 that JX' has no algebraic part, for general X'.

The first step is done in [18], where one checks M. Green's criterion for sufficiently ample surfaces in X :

3.2 Theorem : Let X be a Calabi-Yau threefold, and $L \to X$ an ample line bundle. Then for n large enough, there is a smooth $S \in |L^n|$, and a $\lambda \in H^1(\Omega_S)^0 := \mathrm{Ker}(H^1(\Omega_S) \to H^2(\Omega^2_X))$ satisfying :

3.2.1. $\overline{\nabla}(\lambda) : H^0(N_S X) \to H^2(\mathcal{O}_S)$ is an isomorphism.

Notice, and this point was also crucial for the proof of 2.1 that "K_X trivial" $\Rightarrow$ $\dim H^0(K_S) = \dim H^0(N_S X)$, and this fact suggested the use of the Noether-Lefschetz locus instead of the rigid rational curves of Clemens, because one expected from it that the Noether-Lefschetz locus for surfaces in X is essentially 0-dimensional.

In fact, applying M. Green's criterion and using 3.2, we find now :

3.3 Corollary : For $n \gg 0$, the Noether-Lefschetz locus in $U \subset \mathbb{P}(H^0(X, L^n))$ (the open set parametrizing smooth surfaces) has countably many reduced 0-dimensional components, which are dense in U.

Note also the following very important refinement of 3.3, which is part of M. Green's Lemma :

The couples (S_n, λ_n) where λ_n is rational algebraic on S_n and $\{S_n\}$ is a zero-dimensional reduced component of S_{λ_n} are dense in the total space of $\mathcal{H}^{1,1}_{\mathbf{R},\mathrm{prim}}$ on U. (See Lecture 3 for notations).

3.4. Now we do the same construction as in §2.

3.4.1. One knows that the local universal deformation V of X is smooth. For n large enough, let $\mathcal{U} \underset{\pi}{\to} V$ parametrizing smooth surfaces $S_v \in |L^n_v|$ on X_v, $v \in V$. Let $S \subset X$, $\lambda \in H^2(S, \mathbb{Z})^0 \cap H^{1,1}(S)$ be an algebraic class satisfying 3.2.1. Then as in 2.3 it follows that $\mathcal{S}_\lambda \subset \mathcal{U}$ is smooth of codimension $h^{2,0}(S) = \dim$ (fibers of $\pi : \mathcal{U} \to V$), and that $\pi_{|_{\mathcal{S}_\lambda}} : \mathcal{S}_\lambda \to V$ is etale over a neighbourhood of (X) in V.

3.4.2. As in 2.6, we conclude that we have now countably many germs of families $(Z_\lambda(v))$ on V near (X), of 1-cycles homologous to zero in the fibers X_v, and countably many corresponding germs of normal functions $\nu_\lambda(v)$ on V near (X), which are in bijection with the countably many 0-dimensional reduced components of the Noether-Lefschetz locus of $U = \mathcal{U}_{(X)}$.

The Theorem 3.1 then follows from :

3.5 Theorem : There is at least one ν_λ which is not of torsion.

We want to sketch the proof of this. The invariant used here is a refinement of the infinitesimal invariant of a normal function, and we want to describe it.

3.5.1. Let $S \underset{j}{\hookrightarrow} X$ be smooth, and $\lambda \in H^2(S, \mathbb{Z})^0 \cap H^{1,1}(S)$. Then $\lambda \in CH_1(S)^0$ $(H^1(\mathcal{O}_S) = 0)$, so we have $j_*(\lambda) \in CH_1(X)_{\text{hom}}$, and we have also $\Phi_X(j_*\lambda) \in J^3(X)$.

Now using the Deligne description of the Abel-Jacobi map [14], $\Phi_X(j_*\lambda)$ can be described as follows :

Let $Y = X \backslash S$. Then there is an exact sequence :

3.5.2. $0 \to H^3(X, \mathbb{Z}) \to H^3(Y, \mathbb{Z}) \underset{\text{Res}}{\longrightarrow} H^2(S, \mathbb{Z})^0$, and $H^3(Y, \mathbb{C})$ has a Hodge filtration $F^i H^3(Y)$, [13], such that :

3.5.3. $\text{Res}\, F^i H^3(Y) = F^{i-1} H^2(S)^0$, $F^i H^3(Y) \cap H^3(X) = F^i H^3(X)$. By 3.5.2, λ admits a lifting $\lambda_{\mathbb{Z}} \in H^3(Y, \mathbb{Z})$. By 3.5.3, λ admits a lifting λ_F in $F^2 H^3(Y)$. Then $\lambda_{\mathbb{Z}} - \lambda_F \in H^3(X, \mathbb{C})$ and is well defined modulo $H^3(X, \mathbb{Z})$ and $F^2 H^3(X)$. This gives our point $\Phi_X(j_*\lambda) \in J^3(X) = H^3(X, \mathbb{C}) / F^2 H^3(X) \oplus H^3(X, \mathbb{Z})$.

As a corollary, we have :

3.5.2. $\Phi_X(j_*\lambda)$ is a torsion point of $J^3(X) \Leftrightarrow \lambda$ admits a lifting in $H^3(Y, \mathbb{Q}) \cap F^2 H^3(Y)$.

Consider $W \subset H^3(Y, \mathbb{R})$ defined as $W = F^2 H^3(Y) \cap H^3(Y, \mathbb{R})$. Then by the residue, W is isomorphic to $H^{1,1}_{\mathbb{R}}(S)^0 \ni X$ and Corollary 3.5.4 rewrites as :

3.5.5. $\Phi_X(j_*\lambda)$ is of torsion $\Leftrightarrow$ the lifting $\tilde{\lambda}$ of λ in W lies in $H^3(Y, \mathbb{Q}) \subset H^3(Y, \mathbb{R})$.

We put this in family : consider the normal function $\nu_\lambda(v) = \Phi_{X_v}(j_{v*}(\lambda_v)) \in J^3(X_v)$. Let $Y_v = X_v \backslash S_v$, where $S_v \underset{j_v}{\hookrightarrow} X_v$ is the point of $\mathcal{S}_\lambda \underset{\text{local isom.}}{\longrightarrow} V$ over v near S.

Then we have a $\mathcal{C}^\infty$ real bundle $\mathcal{W}$ on $\mathcal{S}_\lambda$, with fiber $W_{(v)} = H^3(Y_v), \mathbf{R}) \cap F^2H^3(Y_v)$, and $\mathcal{W}$ is isomorphic to $(\mathcal{H}_{\mathbf{R}}^{1,1})^0$. On $\mathcal{S}_\lambda$, λ gives a section of $(\mathcal{H}_{\mathbf{R}}^{1,1})^0$ and has a unic lifting $\tilde{\lambda}$ in $\mathcal{W} \subset \mathcal{H}^3_{\mathbf{R},Y}$ (the flat real $\mathcal{C}^\infty$-bundle with fiber $H^3(Y_v, \mathbf{R})$). 3.5.5 gives now :

3.5.6. If ν_λ is of torsion, $\tilde{\lambda}_v$ is in fact in $H^3(Y_v, \mathbf{Q})$ for any $v \in V$, so in particular, via the inclusion $\mathcal{W} \subset \mathcal{H}^3_{\mathbf{R},Y}$, $\tilde{\lambda}$ is a flat section of $\mathcal{H}^3_{\mathbf{R},Y}$.

3.5.7. The infinitesimal variation of the Hodge filtration of the open sets Y_v over $\mathcal{S}_\lambda$, and more generally of the sets $Y_t = X_v \backslash S_t$, for $v \in V$, $t \in \mathcal{U}$, $\pi(t) = v$, gives the following diagrams :

3.5.8.

$$\begin{array}{cccc} \nabla^Y & : & F^2\mathcal{H}^3_Y & \to & F^1\mathcal{H}^3_Y \otimes \Omega_{\mathcal{U}} \\ & & \downarrow & & \downarrow \\ \overline{\nabla}^Y & : & F^2/F^3\mathcal{H}^3_Y & \to & F^1/F^2\mathcal{H}^3_Y \otimes \Omega_{\mathcal{U}} \end{array},$$

which restricts on $\mathcal{S}_\lambda$ to :

$$\begin{array}{cccc} \nabla^Y & : & F^2\mathcal{H}^3_Y & \to & F^1\mathcal{H}^3_Y \otimes \Omega_{\mathcal{S}_\lambda} \\ & & \downarrow & & \downarrow \\ \overline{\nabla}^Y & : & F^2/F^3\mathcal{H}^3_Y & \to & F^1/F^2\mathcal{H}^3_Y \otimes \Omega_{\mathcal{S}_\lambda} \end{array},$$

$\tilde{\lambda}$ is a $\mathcal{C}^\infty$ section of $F^2\mathcal{H}^3_Y$ on $\mathcal{S}_\lambda$, which is flat. One deduces then from 3.5.8 :

3.5.9. $\forall v \in \mathcal{S}_\lambda$, $\overline{\nabla}^Y(\overline{\lambda}_v)$ vanishes in $F^1/F^2\mathcal{H}^3_Y \otimes \Omega_{\mathcal{S}_\lambda}(v)$, where $\overline{\lambda}_v$ is the value at v of the projection in $F^2/F^3\mathcal{H}^3_Y$ of $\tilde{\lambda}$. Note that $\overline{\lambda}_v$ belongs to the image $\overline{W}_v$ of $W \otimes \mathbf{C}$ in $F^2/F^3H^3(Y_v)$, and that via Res, $\overline{W}_v$ is isomorphic to $H^1(\Omega_{S_v})^0$.

We conclude now the proof of 3.5.

3.5.10. If all the ν_λ's of 3.4.2 were of torsion, one would have, using 3.5.9 and the density statement in 3.3. : For generic $S \subset X$, $S \in |L^n|$, for generic $\lambda \in H^1(\Omega_S)^0$, the natural lifting $\overline{\lambda}$ of λ in $\overline{W}(Y_S)$ ($Y_S = X \backslash S$) satisfies :

3.5.11. $\overline{\nabla}^Y(\overline{\lambda}) \in \Omega_{\mathcal{U}} \otimes F^1/F^2H^3(Y_S)$ vanishes in $(\operatorname{Ker} \overline{\nabla}(\lambda))^* \otimes F^1/F^2H^3(Y_S)$.

It is shown in [16], using the results of [18], that 3.5.11 is not true. Note that 3.5.11 is an algebraic statement except for the data of the space $\overline{W} \subset F^2/F^3H^3(Y_S)$ which does not vary holomorphically with S. We prove that 3.5.11 is false for any subspace $\overline{W}$ of $F^2/F^3H^3(Y_S)$ isomorphic to $H^1(\Omega_S)^0$ via Res, (for S generic). So 3.5 is proved.

References.

We use the same references as for Lecture 7, and add the following :

[10] A. Albano-A. Collino : On the Griffith group of the cubic sevenfold, Math. Ann., Vol. 299, No. 4 (1994) 715-726

[11] F. Bardelli : On Grothendieck generalized conjecture for a family of threefolds with geometric genus one. Proc. of the Conf. Alg. Geometry, Berlin 1985, 14-23, Teubner texte Math 92, Leipzich (1986).

[12] F. Bardelli - S. Müller-Stach : Algebraic cycles on certain Calabi-Yau threefolds, Math. Z., 215, No. 4 (1994) 569-582

[13] P. Deligne : Théorie de Hodge II, Publ. Math. IHES 40 (1972) pp 5-57.

[14] F. El Zein - S. Zucker : Extendability of normal functions associated to algebraic cycles. In *Topics in transcendental Algebraic Geometry*, ed. by P. Griffiths, Annals of Mathematics Studies, Study 106, Princeton (1984) 269-288.

[15] C. Voisin : Sur les zero cycles de certaines hypersurfaces munies d'un automorphisme, Annali della Scuola Norm. Sup di Pisa, Vol. 29 (1993) 473-492.

[16] C. Voisin : Sur l'application d'Abel-Jacobi des variétés de Calabi-Yau de dimension 3, to appear in Ann. de l'ENS Paris.

[17] C. Voisin : Une approche infinitésimale du théorème de H. Clemens sur les cycles d'une quintique générale de $\mathbf{P}^4$, J. Algebraic Geometry 1 (1992) 157-174.

[18] C. Voisin : Densité du lieu de Noether-Lefschetz pour les sections hyperplanes des variétés de Calabi-Yau, International J. of Math., Vol. 3, n° 5 (1992) 699-715.

The infinitesimal invariant of C^+-C^-

Gian Pietro Pirola(*)

Introduction

Our aim is to compute the infinitesimal invariant of a normal function associated to a very natural cycle: the curve in its Jacobian. We have to solve an exercise along some lines explained on these Cime lectures. This "exercise" is a joint work with Alberto Collino [CP] with a big hint given by F. Bardelli and S. Muller Stach that we will explain below.

The normal function constructed from the basic cycle C^+-C^- is a natural object and appears in different areas. An example of this is given for instance by the work about the harmonic integral (cf. [H1], [H2], [Hain1] and [Pu]). An interesting connection between the cycle and the work of Johnson (cf. [J1] and [J2]) on the mapping class group has been recently found by R. Hain (cf. [Hain2] and see (1.11) below) and therefore some possible relations with the syzygies of the canonical curve was pointed out to us by C. Voisin.

The Griffiths infinitesimal invariant was introduced as a tool (cf. [Gr1]) to decide when a normal function is locally constant. The work of M. Green (cf. [G]) and C. Voisin (cf. [V1]) made possible, in many cases, to compute it, so the infinitesimal invariant has become a main tool for the study of the normal functions.

We have found that our infinitesimal invariant carries important information and, in genus 3, it determines the curve (see (2.6)). In [Gr1] a similar Torelli theorem is proved. Griffiths considered the normal function defined by the difference of the two $g_3^{\ 1}$ on the Jacobian of a curve of genus 4. This work and in particular the treatment of the Schiffer variations was our model.

Our computations allows us to improve Ceresa's theorem by making possible to study the algebraic dependence of C^+ and C^- when C varies in subvarieties of the moduli space of curves. In this way we obtain some answers to a question asked by H.Clemens (see section 1 below).

In section 1 we recall some definitions, results and problems on the basic cycle. We introduce next, in section 2, the infinitesimal invariant and state our Torelli theorem. Sketches of the proof of it and of the main formula (3.4) are given in section 3.

(*) Partially supported by 40% project Algebraic geometry and fundings from M.U.R.S.T. and G.N.S.A.G.A. (C.N.R.), Italy .

M. Green et al.: LNM 1594, A. Albano and F. Bardelli (Eds.), pp. 223–232, 1994.

1. Curves in their Jacobians and the normal function.

Let C be a smooth complete curve of genus g and $J(C)=Pic^0(C)$ be its Jacobian. Abel-Jacobi map gives $AJ^{\pm}:C\times Pic^{-1}(C)\to J(C)$ defined by:

$$AJ^{\pm}(p,L)=L^{\pm 1}\otimes\mathcal{O}_C(p). \tag{1.1}$$

Fixing $M\in Pic^{-1}(C)$ we have the curves $C^{\pm}(L)=AJ^{\pm}(C,L)$. Let $C^{+}=AJ^{+}(C)$ and $C^{-}=AJ^{-}(C)$ denote the corresponding cycles of J(C) considered up to translations. Note that they are well defined up to algebraic equivalence. Since the multiplication by j=(-1) on J(C) acts trivially on the even homology it follows that C^{+} is homologically equivalent to C^{-}. Hence the basic cycle:

$$C^{+}-C^{-} \tag{1.2}$$

provides an element of the Griffiths group of J(C).

(1.3) Let M_g denote the moduli space of the genus g smooth curves. Clemens (see [Cl 2]) posed the difficult question to describe the sublocus M_g parametrized by the curve C where C^{+} is algebraically equivalent C^{-}. We consider the filtration of M_g:

$$H_g\subset X_g(rat)\subset X_g(alg)\subset X_g(\tau)\subset M_g.$$

where H_g is the hyperelliptic locus,

$X_g(rat)=\{[C]:\ C^{+}-C^{-}$ is rationally equivalent to zero up to translations$\}$,

$X_g(alg)=\{[C]:\ C^{+}-C^{-}$ is algebraic eq. to zero$\}$,

$X_g(\tau)=\{[C]:\ C^{+}-C^{-}$ is τ eq. to zero, i.e there is $n\in\mathbb{Z}$, $n\neq 0$ such that $n(C^{+}-C^{-})$ is alg. eq. to zero$\}$ (see the lectures of Prof. Murre).

The inclusion $H_g\subset X_g(rat)$ follows because the hyperelliptic involution induces (-1) on the Jacobian. It is not known if $H_g\neq X_g(alg)$, however Chad Schoen (cf. [Sc]) shows that $H_g\neq X_g(\tau)$. The example is provided by a curve with an automorphism group G isomorphic to $(\mathbb{Z}/3\mathbb{Z})\times(\mathbb{Z}/3\mathbb{Z})$ and such that the quotient J(C)/G, under the natural action of G on J(C), is a unirational variety.

From [Gr2], as it has been recalled in this Cime school, we learn how to construct the Abel-Jacobi mapping AJ, we expect $AJ(C^{+}-C^{-})$ to belong to

$$J^{g-1}(J(C)) = H^{2g-3}(J(C),\mathbb{C})/F^{g-1}+H^{2g-3}(J(C),\mathbb{Z})\cong(H^{2,1}+H^{3,0})^{*}/H_3(J(C),\mathbb{Z}). \tag{1.4}$$

There is however the ambiguity due to the translations. To overcome it we first fix $L\in Pic^{-1}(C)$ and define $\varphi:Pic^0(C)\to J^{g-1}(J(C))$ by:

$$\varphi(\xi)=AJ(C^{+}(L(\xi))-C^{-}(L(\xi)))-AJ(C^{+}(L)-C^{-}(L))$$

Letting $P(C)=J^{g-1}(J(C))/\varphi(Pic^0(C))$ be primitive intermediate Jacobian of J(C), we, by abuse of notations, set:

$$AJ(C^+-C^-)=AJ(C^+(L)-C^-(L) \bmod \varphi(Pic^0(C)). \tag{1.5}$$

The word "primitive" is motivated because P(C) is (cf. [CP]) isomorphic to

$$PH^{2g-1}(J(C),\mathbb{C})/PF^{g-1}+PH^{2g-1}(J(C),\mathbb{Z})\cong(P^{2,1}+P^{3,0})^*/(PH_3(J(C),\mathbb{Z})) \tag{1.6}$$

where PH stands by primitive cohomology and $P^{p,q}$ is the primitive p,q hodge space of J(C). In particular we have that $P^{3,0}=H^{3,0}$ and that $P^{2,1}$ is isomorphic to $H^{2,1}/\Theta H^{1,0}$ where Θ is the class of the theta divisor of J(C). The construction extends holomorphically to families. If

$$f:\mathcal{C}\to U \tag{1.7}$$

is a smooth family of genus g curves, we let $j(f):J(\mathcal{C})\to U$ and $p(f):P(\mathcal{C})\to U$ be respectively the associated Jacobian and primitive Jacobian (of Jacobians) fibrations. It follows that p(f) is canonically equipped with a section ν:

$$\nu(t)=AJ(f^{-1}(t)^+-f^{-1}(t)^-). \tag{1.8}$$

This is the normal function we study. Let AP(C) and TP(C) be the abelian and the transcendental part of P(C) (see Lectures of Prof. Green and Voisin), we write down a new filtration of M_g:

$$H_g\subset X_g(0)\subset X_g(ab)\subset X_g(ab+tors)\subset M_g \tag{1.9}$$

where: $X_g(0)=\{[C]\in M_g: \nu([C])=0\}$, $X_g(ab)=\{[C]\in M_g: \nu([C])\in AP(C)\}$ and $X_g(ab+tors)=\{[C]\in M_g: n\nu(C)\in AP(C), n\in\mathbb{Z}, n\neq 0)$.

By construction this filtration is larger than the one given in (1.3), i.e $X_g(rat)\subset X_g(0)$, $X_g(alg)\subset X_g(ab)$ and $X_g(\tau)\subset X_g(ab+tors)$.

(1.10) Ceresa (cf. [Ce]) proved:
A) $X(0)\neq M_3$: he showed that $\nu\neq 0$ on the boundary of M_3;
B) for a generic genus 3 curve C P(C)=TP(C), i.e AP(C)=0.
From A) + B) it follows that $X_3(alg)\neq M_3$ (and moreover $X_3(\tau)\neq M_3$). By degeneration and induction he also proved that $X_g(alg)\neq M_g$ for g>3.

We would like to recall that B. Harris by using harmonic integrals gave two proofs of the Ceresa result:
1) he studied, in [H1], the differential of ν along the directions normal to the hyperelliptic locus (this was reproved by Bardelli (cf [Ba]) in the case of genus 3 by means of the infinitesimal Abel-Jacobi mapping).

2) he computed, in [H2], the integration of the Abel-Jacobi map for the Fermat curve of degree 4 (and genus 3).
Along these lines we recall the related work on the Hodge structures of the fundamental group of a Riemann surface (cf. [Pu] and [Hain1]).

In (cf. [B]) S. Bloch showed that the quartic Fermat Jacobian is interesting for the Bloch-Beilinson conjecture, and by using a refined cycle map and reduction mod. p, that C^+-C^- is not in the torsion of the Griffiths group. Along these lines there was the thesis of Zelinsky (cf. [Z]) and Top ([Top 1]) and the joint work in progress of Buhler, Schoen and Top (see the report [Top 2]), they have found many genus 3 curves defined over number fields outside $X_3(\tau)$.

(1.11) Because all of the topics, moduli of curves, curves in their Jacobians and primitive cohomology, have already appeared in this section we would like to recall Johnson's construction (cf. [J1]). The Torelli subgroup T_g of the mapping class group F_g is the kernel of the monodromy map $F_g \to SP(2g,\mathbb{Z})$. An element of T_g can be represented by a smooth family of genus g curves over the circle, $p:\mathcal{C} \to S^1=\{z\in \mathbb{C} : |z|=1\}$, with fixed curve $C= p^{-1}(1)$ and trivial monodromy on the homology. It follows that $J(\mathcal{C}) \to S^1$ is topologically trivial: $J(\mathcal{C})=J(C)\times S^1$. The Abel-Jacobi mapping, up to translations and homotopy, defines a map $\mathcal{C} \to J(\mathcal{C})$. Next by composing with the projection $J(\mathcal{C}) \to J(C)$ we get a map $\lambda:\mathcal{C} \to J(C)$. Since $\dim(\mathcal{C})=3$ $\lambda_*[\mathcal{C}]\in H_3(J(C),\mathbb{Z})$, but due to the translations only an element of the primitive cohomology is well defined. This gives a map:

$$\eta : T_g \to PH_3(J(C),\mathbb{Z}).$$

The theorem (cf. [J2]) is that η induces, up to 2 torsion, an isomorphism between the abelianization of T_g and $PH_3(J(C),\mathbb{Z})$. The connection between Johnson's map with ν, the normal function defined in (1.8), is studied in [Hain2]. In particular R. Hain was able to deduce that up to torsion the normal function ν is the only normal function defined on M_g for g>3. The following interesting question was asked by E. Colombo: is, up to torsion ν the only normal function <u>algebraically</u> defined on M_g? For g=3 this is false (cf. [N1]).

2. Griffiths' infinitesimal invariant

We would like to define Griffiths' infinitesimal invariant of ν. To this aim let $f:\mathcal{C} \to U$ be a smooth family of curves as in (1.7), and $C=f^{-1}(p)$ where p is a point of U, but we assume now that U is simply connected. It follows that the bundle:

$$PH^{2g-3}(J(\mathcal{C}),\mathbb{C}) \to U$$

is trivial and isomorphic to $PH^{2g-3}(C,\mathbb{C})\times U$. Let

(2.1) $$\pi:PH^{2g-3}(C,\mathbb{C})\times U\rightarrow P(\mathcal{C})$$

be the quotient map and choose, by shrinking U if necessary, a (local) lifting $\mu:U\rightarrow PH^{2g-3}(C,\mathbb{C})$ of ν, therefore $\nu=\pi\cdot\mu$.

Let T_p be the tangent of U at p and consider the Koszul complex induced by cup product with Kodaira-Spencer classes:

(2.2) $$\Lambda^2T_p\otimes P^{3,0}\xrightarrow{\chi}T_p\otimes P^{2,1}\xrightarrow{\beta}P^{1,2}.$$

Let H be the cohomology of the complex and ∇ be the Gauss-Manin connection on $PH^{2g-3}(C,\mathbb{C})$. The infinitesimal Griffiths invariant, $\delta\nu$, computed at p, can be defined as the functional of H, $\delta\nu\in H^*$, given by:

(2.3) $$\delta\nu(\Sigma_i[\xi_i\otimes\omega_i])=\Sigma_i\int_{J(C)}(\nabla_{\xi_i}\mu)\cdot\omega_i\ ,\ \Sigma_i\xi_i\otimes\omega_i\in \mathrm{Ker}(\beta).$$

The infinitesimal invariant was first intrduced by Griffiths (cf [Gr.1]) as an operator on Ker(β). Next M. Green defined the infinitesimal invariant by means of the Koszul cohomology (cf. [G]) and $\delta\nu$ became the first one of a sequence of invariant. The fact that $\delta\nu$ vanishes on the image of χ is very useful in many circumstances. One instance of this is now explained.

(2.4) Fix g=3, assume that C is non-hyperelliptic and that $f:\mathcal{C}\rightarrow U$ is a Kuranishi family of C. Set $V=H^{0,1}(C)$. We have the isomorphisms: $T_p\cong H^1(C,T_C)\cong Sym^2(V)$, $H^{3,0}\cong P^{3,0}\cong\Lambda^3V^*$, $H^{3,0}\otimes H^1(C,T_C)\cong P^{2,1}$ and $P^{1,2}\cong(P^{2,1})^*$, where * stands for dual. By means of these identifications the complex (2.2) becomes:

$$\Lambda^2(Sym^2V)\otimes\Lambda^3V^*\rightarrow Sym^2V\otimes Sym^2V\otimes\Lambda^3V^*\rightarrow Sym^2V^*\otimes\Lambda^3V.$$

We finally obtain, by GL(V)-decomposition, that H is isomorphic to

$$Sym^4V\otimes\Lambda^3V\cong Sym^4(H^{0,1})\otimes H^{3,0}$$

(2.5) Remark. Playing with Koszul groups one sees, in this case, that (2.3) provides the full Green-Griffiths invariant. Similarly the analogous of (2.2) is exact in the case of a universal family of abelian varieties of dimension 4 (cf.[P1]). More generally Nori ([N2] cf. (7.5)) proved that, up to torsion, there are no interesting Abel-Jacobi maps on a generic abelian variety of dimension larger than 3.

Coming back to our case since dim(V)=3, we have that $\delta\nu$ can be seen as a homogeneus polynomial of degree 4 in 3 variables. Let $\mathbb{P}^2=\mathbb{P}(V)$ be the projectivization of V. The "hint" of Bardelli and Muller-Stach was the conjecture that $\delta\nu$ should give the "equation" of the canonical curve. Our work gives the affirmative answer.

(2.6) (Torelli) Theorem. The zero locus of $\delta\nu$ is the canonical image of C.

(2.7) It can be worthwhile to recall a long standing question raised by Griffiths (cf. [Gr 3]). He asked if it is possible to reconstruct a general quintic threefold from the infinitesimal invariant associated with the lines on it. Our Torelli theorem gives, perhaps, some indirect evidence of this. For the similarity between the quintics and Abelian threefold with respect to the Griffiths group see [Ba], [Cl1], [N2] and [V2].

As application we obtain some information about the loci defined above (cf (1.3)):

(2.8) Theorem. Let Y be a subvariety of M_3 of dimension larger than 3. If $Y \not\subset H_3$ then $Y \not\subset X_3(\tau)$. In particular C_p^+ is not algebraically equivalent to C_p^- in $J(C_p)$ for generic p of Y.

Proof. (see [CP]).

It is easy to see in the genus 3 case (cf. [CP] section 4)) that if ν is locally constant in codimension 2 then $\delta\nu$ must be the product of two polynomials. Since the canonical curve of a non hyperelliptic curve is irreducible and reduced (2.8) follows from our Torelli theorem.

3. The Main Formula.

Now we have to explain the basic result of [CP], the computation of $\delta\nu$ against decomposable tensors. We assume g>2.

Let T_C and ω_C be respectively the tangent and the canonical bundle of C. An element $\zeta \in H^1(C,T_C)$ defines an extension of lines bundle:

$$0 \to \mathcal{O}_C \to E \to \omega_C \to 0. \tag{3.1}$$

Remark that $\det(E)=\omega_C$ and that the coboundary is the cup product mapping ζ: $H^{1,0}(C)=H^0(\omega_C) \to H^1(\mathcal{O}_C)=H^{0,1}(C)$. From (3.1) we have the exact sequence: $0 \to H^0(\mathcal{O}_C) \to H^0(E) \to \ker(\zeta) \to 0$. Assume that the dimension of $\ker(\zeta)$ is larger than one and take two independent forms σ and γ in $\ker(\zeta)$. We fix $\sigma(\varepsilon)$ and $\gamma(\varepsilon)$ in $H^0(E)$ that map onto σ and γ, by means of the natural map:

$$\alpha: \Lambda^2 H^0(E) \to H^0(\det(E)) = H^0(\omega_C) \tag{3.2}$$

we define: $\alpha(\sigma(\varepsilon) \wedge \gamma(\varepsilon)) \in H^0(\omega_C)$.

Take $\omega\in H^{1,0}(C)$ and, by using the identifications $H^{1,0}(C)=H^{1,0}(J(C))$ and $H^{1,0}(C)=H^{1,0}(J(C))$, let $\Omega=\sigma\Lambda\gamma\Lambda\zeta\cdot\omega$ be the corresponding 2.1 form on J(C). We consider the tensor

(3.3) $$\zeta\otimes\Omega.$$

We have: **A)** $\Omega=\zeta\cdot(\sigma\Lambda\gamma\Lambda\omega)$ is the derivative of a (3.0) form and then is primitive, i.e. $\Omega\in P^{2,1}$, and **B)** $\zeta\cdot(\sigma\Lambda\gamma\Lambda\zeta\cdot\omega)=0$ and hence $\zeta\otimes\Omega\in\ker(\beta)$.

We can now state our result:

(3.4) Main Formula: $\delta\nu(\zeta\otimes\sigma\Lambda\gamma\Lambda\zeta\cdot\omega)=-2\int_C \alpha(\sigma(\varepsilon)\Lambda\gamma(\varepsilon))\Lambda\zeta\cdot\omega$

(3.5) If we denote by Λ the plane generated by σ and γ it follows that $\alpha(\sigma(\varepsilon)\Lambda\gamma(\varepsilon))$ is independent, modulo Λ, of the choices of $\sigma(\varepsilon)$ and $\gamma(\varepsilon)$. The same is then true for the right terms of (3.4). In fact:

$$\int_C\sigma\Lambda\zeta\cdot\omega=-\int_C\zeta\cdot\sigma\Lambda\omega=0=-\int_C\zeta\cdot\gamma\Lambda\omega=\int_C\gamma\Lambda\zeta\cdot\omega.$$

(3.6) Remark. We notice that $\delta\nu(\zeta\otimes\sigma\Lambda\gamma\Lambda\zeta\cdot\omega)\neq 0$ implies that the basic cycle is not algebraically equivalent to zero when the curve deforms in the direction of ζ. In fact the 3.0 form $\sigma\Lambda\gamma\Lambda\omega$ and its derivative $\Omega=\sigma\Lambda\gamma\Lambda\zeta\cdot\omega$ are orthogonal to the Abel-Jacobi image of algebraically trivial cycles that deform with ζ. This simple remark makes sometime possible to prove the algebraic inequivalence avoiding considerations about the transcendental part of the intermediate Jacobian.

(3.7) Before to give a proof of (3.4) we deduce the Torelli theorem (2.6) from it. Let C be a non hyperelliptic genus 3 curve. By using the isomorphisms: $H^1(C,T_C)=Sym^2(V)$, $V=H^{0,1}(C)$, we see that a rank one deformation has the form $\zeta=v\otimes v$, where v is a non zero element of V. Let (v) denote the corresponding point of $\mathbb{P}^2(V)$ and set $\ker(\zeta)=\Lambda$. Fix a bases (σ,γ) of L and take $\omega\in H^0(\omega_C)$ such that $\zeta\cdot\omega\neq 0$. We have: $\int_C\sigma\Lambda\zeta\cdot\omega=\int_C\gamma\Lambda\zeta\cdot\omega=0$ and $\int_C\omega\Lambda\zeta\cdot\omega\neq 0$. Clearly (σ,γ,ω) is a bases of $H^0(\omega_C)$. Write $\alpha(\sigma(\varepsilon)\Lambda\gamma(\varepsilon))=\lambda_1\sigma+\lambda_2\gamma+\lambda_3\omega$. From (3.4) we obtain:

$$\delta\nu(\zeta\otimes\sigma\Lambda\gamma\Lambda\zeta\cdot\omega)=-2\int_C\alpha(\sigma(\varepsilon)\Lambda\gamma(\varepsilon))\Lambda\zeta\cdot\omega=-2\lambda_3\int_C\omega\Lambda\zeta\cdot\omega$$

and hence $\quad\delta\nu(\zeta\otimes\sigma\Lambda\gamma\Lambda\zeta\cdot\omega)=0\Leftrightarrow\lambda_3=0\Leftrightarrow\alpha(\sigma(\varepsilon)\Lambda\gamma(\varepsilon))\in\Lambda.$

We have now a basic algebraic result:

(3.8) Lemma. If $|\Lambda| = |\lambda_1\sigma + \lambda_2\gamma|$ is a base points free pencil then $\alpha(\sigma(\varepsilon)\Lambda\gamma(\varepsilon)) \notin \Lambda$. (Proof cf. [CP] (1.1)).

Conclusion of the proof of (2.6): Let $k:C\to \mathbb{P}^2(V)$ be the canonical map. By identifying $\mathbb{P}^2(V)$ with the projective space of the 2 dimensional vector subspaces of $H^0(C,\omega_C)$ we have:

$$k(p)=H^0(C,\omega_C(-P)).$$

A pencil of abelian differentials of a genus 3 non-hyperelliptic curve can have just one base point. Then $\alpha(\sigma(\varepsilon)\Lambda\gamma(\varepsilon))\in \Lambda$ implies $\Lambda=H^0(C,\omega_C(-P))$, $P\in C$. This means that $(v)=k(P)$ and that $\zeta=v\otimes v$ is a Schiffer variation of P. We have seen that the degree 4 homogeneus polynomial corresponding to $\delta\nu$ can only vanish on the canonical curve k(C) and it does vanish there!

(3.9) We have to prove (3.4). Once again we follows Griffiths. We consider a 1-dimensional smooth family of curves $f:\mathcal{C}\to D$, where D is a disk, $0\in D$ and $f^{-1}(0)=C$. Let $\mathcal{C}(\varepsilon)\to\Delta=\mathrm{spec}\,\mathbb{C}[\varepsilon]/(\varepsilon^2)$ be the induced first order deformation with Kodaira-Spencer class ζ, $J(\mathcal{C})\to D$ and $J(\mathcal{C}(\varepsilon))\to\Delta$ be their jacobian fibrations. Fix an Abel-Jacobi embedding $\mathcal{C}\to J(\mathcal{C})$.

By choosing a lifting of $\partial/\partial\varepsilon$ we define a vector field ζ^+ on $J(\mathcal{C})$ normal to the central fiber. The restriction of ζ^+ to $\mathcal{C}$, that by abuse of notation we still call ζ^+, gives a trivialization of the normal bundle N of C in $\mathcal{C}$. The conormal sequence:

$$0\to N^*\to\Omega^1_{\mathcal{C}/C}\to\omega_C\to 0$$

becomes the sequence (3.1), thus it provides an isomorphism $E\cong\Omega^1_{\mathcal{C}/C}$.

Let $\Omega=\sigma\Lambda\gamma\Lambda\zeta\cdot\omega$ be the 2.1 form on J(C) where $\zeta\cdot\sigma=\zeta\cdot\gamma=0$. Take a form $\Omega(\varepsilon)$ on $J(\mathcal{C})$ such that:

$$\Omega(\varepsilon)=\sigma'(\varepsilon)\Lambda\gamma'(\varepsilon)\Lambda\phi(\varepsilon) \mod \varepsilon^2 \tag{3.10}$$

where $\phi(0)=\zeta\cdot\omega$, $\sigma'(\varepsilon)$ and $\gamma'(\varepsilon)$ are 1.0 forms on $J(\mathcal{C}(\varepsilon))$ such that $\sigma'(\varepsilon)_{/\mathcal{C}}=\sigma(\varepsilon)$ and $\gamma'(\varepsilon)_{/\mathcal{C}}=\gamma(\varepsilon)$. In particular $\Omega(0)=\Omega$. Griffiths' formula (cf. [Gr1] (6.44) p.306) says:

$$\delta\nu(\zeta\otimes\Omega)=\int_Z\xi-\int_Z i(\zeta^+)\Omega(\varepsilon) \tag{3.11}$$

where $Z=C^+-C^-$, $i(\zeta^+)\Omega(\varepsilon)$ is the contraction of the form against the normal vector field ζ^+ and ξ is a form on J(C) such that

$$\bar{\partial}(\xi)=\zeta^{+}\cdot\Omega.$$

We may choose (cf. [CP] (3.1)) ζ^+ and Ω with "constant coefficients":

$$\zeta^{+}=\Sigma a_{i,j}\partial/dz_i\wedge d\bar{z}_j \text{ mod. } \varepsilon^2$$

and

$$\Omega=\Sigma b_{i,j,k}dz_i\wedge dz_j\wedge d\bar{z}_k,$$

where $a_{i,j}$, $b_{i,j,k}$ are constant and the "dz" are invariant under the translations of J(C). Since ζ^+ and Ω have constant coefficients $\zeta^+\cdot\Omega$ is actually the zero form. Therefore we may take $\xi=0$.

We have just to compute the second term in (3.11). Since the j =-1 involution on J(C) is anti-invariant on the 3-forms with constant coefficient we get:

(3.12) $\delta\nu(\zeta\otimes\int_C)= -\int_Z i(\zeta^+)\Omega(\varepsilon)= -\int_{C^+} i(\zeta^+)\Omega(\varepsilon) +\int_{C^-} i(\zeta^+)\Omega(\varepsilon)=$

$$-\int_{C^+} i(\zeta^+)\Omega(\varepsilon) +\int_{C^+} i(\zeta^+)j^*\Omega(\varepsilon)= -2\int_{C^+} i(\zeta^+)\Omega(\varepsilon)$$

To contract $\Omega(\varepsilon)$ against the vector field $i(\zeta^+)$ we first restrict $\Omega(\varepsilon)$ to the surface $\mathcal{C}(\varepsilon)$. Since $\sigma'(\varepsilon)\wedge\gamma'(\varepsilon)_{/\mathcal{C}(\varepsilon)}=\sigma(\varepsilon)\wedge\gamma(\varepsilon)$ is of type 2.0 on $\mathcal{C}(\varepsilon)$ we have:

$$\Omega(\varepsilon)_{/\mathcal{C}(\varepsilon)}=\sigma(\varepsilon)\wedge\gamma(\varepsilon)\wedge\zeta\cdot\omega.$$

and since ζ^+ is a holomorphic field:

$$i(\zeta^+)\Omega(\varepsilon)=\{i(\zeta^+)((\sigma(\varepsilon)\wedge\gamma(\varepsilon))\}\wedge\zeta\cdot\omega.$$

An easy adjunction formula (see section 1 of [CP]) finally gives:

(3.13) $$i(\zeta^+)(\sigma(\varepsilon)\wedge\gamma(\varepsilon))=\alpha(\sigma(\varepsilon)\wedge\gamma(\varepsilon)).$$

and then $\int_{C^+} i(\zeta^+)\Omega(\varepsilon)=\int_C \alpha(\sigma(\varepsilon)\wedge\gamma(\varepsilon))\wedge\zeta\cdot\omega$ which proves (3.4).

References

[Ba] F. Bardelli, *Curves of genus three on a generic Abelian threefold and the non finite generation of the Griffiths group.* in **Arithmetic of complex manifolds**, Lectures notes in Math 1399 Springer Berlin Heidelberg New York (1989), 10-26.

[B] S.Bloch, *Algebraic cycles and values of L-funcions* . J. Reine.Angew.Math 350 (1984) 94-108.

[CGGH] J.Carlson, M.Green, P. Griffiths, J Harris, *Infinitesimal variations of Hodge structures I*, Compositio Math. 50 (1983) 109-205.

[Ce] G.Ceresa, *C is not algebraically equivalent to C⁻ in its Jacobian* Ann. of Math.117 (1983) 285-291.
[Cl 1] H. Clemens, *Homological equivalence, modulo algebraic equivalence, is not finitely generated* IHES publ. Math. 58 (1983) 19-38.
[Cl 2] H. Clemens *Some result about Abel-Jacobi mappings* in **Topics in Transcendental Algebraic geometry.** Annals of Math. Studies, N. 106 Princeton, New Jersey (1984) 289-304.
[CP] A. Collino, G. Pirola, *Griffiths' Infinitesimal Invariant for a curve in its Jacobian.* (preprint).
[G] M. Green, *Griffiths' Infinitesimal Invariant and the Abel-Jacobi Map.* J. Diff. Geometry 29 (1989) 545-555.
[Gr 1] P. Griffiths, *Infinitesimal variations of Hodge structures III: Determinantal varieties and the infinitesimal invariant of normal functions.* Compositio Math. 50 (1983) 267-324.
[Gr 2] P. Griffiths, *Periods of integrals I and II.* Amer. J. of Math 90, (1968) 568-626, 805-865.
[Hain 1] R. Hain, *"The Geometry of the fundamental group"* in Proceedings of Summer Research Institute of Algebraic Geometry (ed. S Bloch) Bowdoin College 1985. Ams Symp.Proc. Pure Math 46 part 2 (1987), 247-282.
[Hain 2] R. Hain, *Completions of mapping class groups and the cycle C- C⁻* (preprint).
[H 1] B. Harris, *Harmonic volumes* Acta Math 150 (1983) 91-123.
[H 2] B. Harris, *Homological versus algebraic equivalence in a Jacobian*, Proc. Nat. Acad. Sci. U.S.A. 80 (1983) 1157-1158.
[J 1] D. Johnson, *A Survey of the Torelli Group.* Contemporary Math. 20 (1983), 165-179.
[J 1] D. Johnson, *An Abelian quotient of the mapping class group* J_g. Math. Ann. 249 (1980) 225-242.[N 1] M. Nori, *Cycles on the generic Abelian threefold.* Proc. Indian Acad. Sci. 99 (1989) 191-196.
[N 2] M. Nori, *Algebraic Cycles and Hodge theoretic Connectivity* . Invent. Math. 111, (1993) 349-373.
[P1] G. Pirola, *On the Abel-Jacobi Image of Cycles for a generic Abelian Fourfold* to appear in Boll. Umi.
[Pu] M.Pulte, *The fundamental group of a Riemann surface: Mixed hodge structures and algebraic cycles* , Duke Math J. 57 No 3 (1988) 721-760.
[Sc] C. Schoen (personal communication)
[Top 1] J. Top. *Hecke L series related with algebraic cycles or with Siegel modular forms.* Ph. D Thesis, Dept of Math. Univ. of Utrecht (1989).
[Top 2] J. Top. *Detecting Cycles by reducing mod p* . (progress report)
[V1] C. Voisin, *Une remarque sur l'invariant infinitésimal des functions normal.* C. R.Acad. sci. Paris Sér. I 307 (1988) 157-160.
[V2] C. Voisin, *Un Approach infinitésimal du théorèm de H. Clemens sur le cycles d'un quintique générale de* P^4. J. of Alg. Geom. 1 (1992) 157-174.
[Z] D. Zelinski, *Some Abelian threfold with Nontrivial Griffiths Group*, Ph. D Thesis, Dept of Math. Univ. of Chicago (1987).

Gian Pietro Pirola: Dipartimento di Matematica Università di Pavia, Via Abbiategrasso 209,27100 Pavia Italy.

An introduction to the Hodge conjecture for abelian varieties

Bert van Geemen

1 Introduction

1.1 In this lecture we give a brief introduction to the Hodge conjecture for abelian varieties. We describe in some detail the abelian varieties of Weil-type. These are examples due to A. Weil of abelian varieties for which the Hodge conjecture is still open in general.

The Mumford-Tate groups are a very usefull tool for finding the Hodge classes in the cohomology of an abelian variety. We recall their main properties and illustrate it with an example.

Finally we discuss recent results on the Hodge conjecture for abelian fourfolds. Most of this material is well known, and we just hope to provide an easy going introduction.

I am indebted to F. Bardelli for his invitation to give this talk and for stimulating discussions on the Hodge conjecture and Mumford-Tate groups.

2 The Hodge (p,p)-conjecture for abelian varieties

2.1 Let X be a smooth, projective variety over the complex numbers. We denote by

$$Z^p(X)_{\mathbf{Q}} := Z^p(X) \otimes_{\mathbf{Z}} \mathbf{Q}$$

the group of codimension p cycles on X with rational coefficients. The group of Hodge classes (of codimension p) is:

$$B^p(X) := H^{2p}(X,\mathbf{Q}) \cap H^{p,p}(X) \quad (\subset H^{2p}(X,\mathbf{C})).$$

The cycle class map $Z^p(X)_{\mathbf{Q}} \to H^{2p}(X,\mathbf{Q})$ factors over $B^p(X)$ and defines:

$$\Psi : Z^p(X)_{\mathbf{Q}} \longrightarrow B^p(X),$$

we will usually write $[Z]$, the cohomology class of the cycle Z, for $\Psi(Z)$.

2.2 Hodge (p,p)-conjecture: The map Ψ is surjective.

2.3 The Hodge (p,p)-conjecture is known in case $p = 0,1$ and thus also in case $p = d-1, d = \dim X$. In fact, $B^0(X)$ is spanned by $[X]$ and for $p = 1$ the conjecture is proven using the exponential sequence (cf. [V]). However, for the other p's very little is known, [Mu]. It is not easy to determine the groups $B^p(X)$ for a given X. If X varies in a family, the dimension of $B^p(X)$ may change for example.

M. Green et al.: LNM 1594, A. Albano and F. Bardelli (Eds.), pp. 233–252, 1994.

2.4 One can exploit the ring structure (cup-product) on

$$B^{\cdot} := \oplus_p B^p(X)$$

to obtain information on the Hodge conjecture as follows. Let $D^{\cdot} \subset B^{\cdot}$ be the subring generated by $B^0(X)$ and $B^1(X)$. Then D^p is spanned by:

$$[D_1] \cup [D_2] \cup \ldots \cup [D_p] = [D_1 \cdot D_2 \cdot \ldots \cdot D_p], \qquad D_i \in Z^1(X),$$

with $D_i \cdot D_j$ the intersection product of cycles. Therefore we have the inclusions:

$$D^p \subset Im(\Psi) \subset B^p.$$

In particular, if $D^p = B^p$, then the Hodge (p,p)-conjecture is true for X.

2.5 Definition. An exceptional Hodge class (of codimension p) is an element of B^p which is not in D^p.

2.6 Example. Let $Q \subset \mathbf{P}^{2n+1}$ be a smooth quadric. Then one has $H^{p,q}(Q) = 0$ for $p \neq q$ and:

$$B^p(Q) = H^{2p}(Q, X) = \begin{cases} \mathbf{Q} & p \neq n, \\ \mathbf{Q}^2 & p = n. \end{cases}$$

Thus if $n > 1$ we have $B^1 = D^1 = \mathbf{Q}$ and thus $D^p = \mathbf{Q}$ for all p. Therefore Q has exceptional Hodge classes in codimension n. It is well known however that the Hodge conjecture is true for Q, in fact such a Q has two rulings (= families of $\mathbf{P}^n$'s on it), the cohomology classes of the $\mathbf{P}^n$'s span B^n.

3 Abelian varieties

3.1 Let X be an abelian variety over the complex numbers, that is $X \cong \mathbf{C}^g/\Lambda$ for some lattice Λ, and X is a projective variety. Note that $\Lambda = \pi_1(X) = H_1(X, \mathbf{Z})$ and $\mathbf{C}^g$ is the universal cover of X; also $\mathbf{C}^g = T_0X$, the tangent space to X at the origin. Thus $H_1(X, \mathbf{R}) = H_1(X, \mathbf{Z}) \otimes_{\mathbf{Z}} \mathbf{R} \cong T_0X$ and multiplication by $i \in \mathbf{C}$ on T_0X corresponds to an $\mathbf{R}$-linear map:

$$J : H_1(X, \mathbf{R}) \to H_1(X, \mathbf{R}) \qquad \text{with} \quad J^2 = -I.$$

The map J allows us to recover this structure of complex vector space on $H_1(X, \mathbf{R})$; multiplication by $a + bi \in \mathbf{C}$, with $a, b \in \mathbf{R}$, is given by the linear map $aI + bJ$: $H_1(X, \mathbf{R}) \to H_1(X, \mathbf{R})$. Thus

$$T_0X \cong (H_1(X, \mathbf{R}), J).$$

3.2 Any embedding $\theta : X \hookrightarrow \mathbf{P}^n$ defines a polarization $E := c_1(\theta^*\mathcal{O}(1)) \in B^1(X) \subset H^2(X, \mathbf{Q})$. By the duality, E defines a map, denoted by the same name:

$$E : \wedge^2 H_1(X, \mathbf{Q}) \longrightarrow \mathbf{Q}$$

and this map satisfies the Riemann Relations (here we extend E $\mathbf{R}$-linearly):

$$E(Jx, Jy) = E(x, y), \qquad E(x, Jx) > 0$$

for all $x, y \in H_1(X, \mathbf{R})$, with $x \neq 0$ for the last condition. That condition also implies that E is non-degenerate.

Conversely, $\mathbf{C}^g/\Lambda$ is an abelian variety iff there exists an $E : \wedge^2\Lambda \to \mathbf{Q}$ satisfying the Riemann Relations.

3.3 The cohomology of X and the Hodge structure on it is completely determined by $H^1(X, \mathbf{Q})$ and its Hodge structure:

$$H^p(X, \mathbf{Q}) = \wedge^p H^1(X, \mathbf{Q}), \qquad H^{p,q}(X) = (\wedge^p H^{1,0}(X)) \otimes (\wedge^q H^{0,1}(X)).$$

This remarkable fact can be exploited to determine the B^p's, see section 6.

3.4 One can (almost) recover X from the Hodge structure on $H^1(X, \mathbf{Q})$ and the polarization E. In fact, using the Hodge decomposition:

$$H^1(X, \mathbf{R}) \hookrightarrow H^1(X, \mathbf{C}) = H^{1,0}(X) \oplus H^{0,1}(X)$$

one defines a $\mathbf{C}$-linear map $H^1(X, \mathbf{C}) \to H^1(X, \mathbf{C})$ by defining it to be multiplying by $+i$ on $H^{1,0}$ and by $-i$ on $H^{0,1}$. This map restricts to an $\mathbf{R}$-linear map:

$$J' : H^1(X, \mathbf{R}) \longrightarrow H^1(X, \mathbf{R}), \qquad \text{with} \quad (J')^2 = -I.$$

(see [G]). Using the duality: $H^1(X, \mathbf{R}) \xrightarrow{\cong} H_1(X, \mathbf{R})^*$ the map J' defines a dual map:

$$J := (J')^* : H_1(X, \mathbf{R}) \to H_1(X, \mathbf{R}) \qquad \text{with} \quad J^2 = -I.$$

To obtain X, we have to take the quotient of the complex vector space $(H_1(X, \mathbf{R}), J)$ by a lattice $\Lambda \subset H_1(X, \mathbf{Q}) \subset H_1(X, \mathbf{R})$. Note that $H_1(X, \mathbf{Q})$ is just the dual of $H^1(X, \mathbf{Q})$, but since we are not given $H^1(X, \mathbf{Z})$ we cannot reconstruct $H_1(X, \mathbf{Z})$, that is, we don't know which lattice $\Lambda \subset H_1(X, \mathbf{Q})$ to choose. This leads to the following definitions.

3.5 Definition. Abelian varieties X and Y are said to be isogeneous, $X \approx_{isog} Y$, if there is a finite, surjective map (an isogeny) $\phi : Y \to X$. (If $X \approx_{isog} Y$ then there is actually also a finite, surjective map $X \to Y$.)

An abelian variety is said to be simple if X is not isogeneous to a product of abelian varieties (of dimension > 0).

3.6 Given an isogeny $\phi : Y \to X$, the group $\phi_*(\pi_1(Y))$ is a subgroup of finite index of $\pi_1(X)$, and thus $N\pi_1(X) \cong \pi_1(X) \subset \phi_*\pi_1(Y)$ for some integer N. Therefore one has a finite, surjective map $X \to Y$. The inclusion $\phi_* : H_1(Y, \mathbf{Z}) \to H_1(X, \mathbf{Z})$ extended $\mathbf{Q}$-linearly and dualized gives an isomorphism:

$$\phi^* : H^1(X, \mathbf{Q}) \xrightarrow{\cong} H^1(Y, \mathbf{Q}), \qquad \text{and} \quad \phi^*_{\mathbf{C}}(H^{1,0}(Y)) \subseteq H^{1,0}(X).$$

(The line above is equivalent to saying that ϕ is an isomorphism of Hodge structures.) Conversely, the existence of such a map ϕ^* implies that the abelian varieties X and Y are isogeneous.

An isogeny $\phi : Y \to X$ thus induces isomorphisms $B^p(X) \to B^p(Y)$. Moreover, using pull-back and push-forward of cycles we have the following consequence.

3.7 Lemma. Let $X \approx_{isog} Y$. Then the Hodge (p,p)-conjecture for X is true if and only the Hodge (p,p)-conjecture is true for Y.

4 Overview of results

4.1 With the definitons of the previous sections we can now state some of the results on the Hodge (p,p)-conjecture for abelian varieties. In the later sections we will discuss aspects of the proofs.

4.2 Theorem. (Mattuck, [Ma]) For a general abelian variety one has:

$$B^p(X) = D^p(X) = \mathbf{Q} \qquad \text{for all } p,$$

and thus the Hodge (p,p)-conjecture is true for X and all p.

4.3 Theorem. (Tate, [Tat]) For an abelian variety X which is isogeneous to a product of elliptic curves (one dimensional abelian varieties), one has:

$$B^p(X) = D^p(X) \qquad \text{for all } p,$$

and thus the Hodge (p,p)-conjecture is true for X and all p.

4.4 These two theorems deal with rather extreme cases and one could wonder whether one has $B^{\cdot} = D^{\cdot}$ for any abelian variety. This is not the case, but it still happens quite often. Theorem 4.6 is proven using Mumford-Tate groups.

4.5 Theorem. (Mumford, [Po]) There exist simple four dimensional abelian varieties with $B^2 \neq D^2$.

4.6 Theorem. (Tankeev, [Tan], [R]) For a simple abelian variety X whose dimension is a prime number one has:

$$B^p(X) = D^p(X) \qquad \text{for all } p,$$

and thus the Hodge (p,p)-conjecture is true for X and all p.

4.7 The example of Mumford concerned abelian varieties with a large endomorphism algebra (in fact a CM field L with $[L:\mathbf{Q}] = 2\dim X$). Weil observed that the field was a composite of a totally real field and an (arbitrary) imaginary quadratic field K. He found that the imaginary quadratic field was 'responsible' for the exceptional Hodge cycles.

4.8 An endomorphism $f: X \to X$ maps the origin $0 \in X$ to itself, and therefore induces a linear map $df_0: T_0X \to T_0X$. This extends $\mathbf{Q}$-linearly to a ringhomomorphism:

$$t: \mathrm{End}(X)_{\mathbf{Q}} \longrightarrow \mathrm{End}(T_0X), \qquad f \otimes 1 \mapsto t(f \otimes 1) := df_0,$$

here $\mathrm{End}(X)_{\mathbf{Q}} := \mathrm{End}(X) \otimes_{\mathbf{Z}} \mathbf{Q}$ is the endomorphism algebra of X.

Similarly, using the maps f^* and f_*, the algebra $\mathrm{End}(X)_{\mathbf{Q}}$ acts on $H^1(X,\mathbf{Q})$ and $H_1(X,\mathbf{Q})$ respectively.

4.9 Definition. An abelian variety of Weil-type of dimension $2n$ is a pair (X, K) with X a $2n$ dimensional abelian variety and $K \hookrightarrow \mathrm{End}(X) \otimes \mathbf{Q}$ is an imaginary quadratic field such that for all $x \in K$ the endomorphism $t(x)$ has n eigenvalues x and n eigenvalues $\overline{x}$:

$$t(x) \sim diag(x, \ldots, x, \overline{x}, \ldots, \overline{x})$$

(here we fix an embedding $K \subset \mathbf{C}$).

The space of Weil-Hodge cycles of (X, K) is defined to be the two dimensional $\mathbf{Q}$-vector space

$$\wedge_K^{2n} H^1(X, \mathbf{Q}) \hookrightarrow B^n(X) \subset H^{2n}(X, \mathbf{Q}) = \wedge_{\mathbf{Q}}^{2n} H^1(X, \mathbf{Q}),$$

where the K-vector space structure on $H^1(X, \mathbf{Q})$ is obtained via f^*, $f \in K \subset \mathrm{End}(X)_{\mathbf{Q}}$. (Note $\dim_{\mathbf{Q}} H^1(X, \mathbf{Q}) = 4n$, $\dim_K H^1(X, \mathbf{Q}) = 2n$).

A polarized abelian variety of Weil-type is a triple (X, K, E) where (X, K) is an abelian variety of Weil-type and where E is a polarization on X with $(\sqrt{-d})^* E = dE$ and $K = \mathbf{Q}(\sqrt{-d})$.

4.10 That $\wedge_K^{2n} H^1(X, \mathbf{Q}) \hookrightarrow B^n(X)$ follows from the condition 4.9 on $t(x)$, see 5.2.6. Any abelian variety of Weil type (X, K) with field $K = \mathbf{Q}(\sqrt{-d})$ has a polarization E with $(\sqrt{-d})^* E = dE$, see 5.2.1.

In the next section we recall that any $2n$-dimensional polarized abelian variety of Weil-type is a member of a n^2-dimensional family of polarized abelian varieties of Weil-type. The 'general' in the next Theorems refers to the general member of such a family.

Theorem 4.12 shows that in dimension four, the abelian varieties of Weil-type are the only simple ones for which the Hodge conjecture needs to be verified. The proof depends on a detailed study of endomorphism algebras of abelian fourfolds and their Mumford-Tate groups.

4.11 Theorem. (Weil, [W]; cf. Thm 6.12) For a general $2n$-dimensional abelian variety X of Weil-type (with $n > 1$) one has $B^1(X) = \mathbf{Q}$ (and thus $D^p(X) = \mathbf{Q}$ for all p) but $B^n(X) = \mathbf{Q}^3$, the direct sum of $D^n(X)$ and the space of Weil-Hodge cycles. Therefore:

$$B^n(X) \neq D^n(X).$$

4.12 Theorem. (Moonen-Zarhin, [MoZ]) Let X be a simple abelian variety of dimension 4 with $B^2(X) \neq D^2(X)$. Then X is of Weil-type.

4.13 Other examples of abelian varieties with $B^p \neq D^p$ are the abelan varieties with an endomorphism algebra of type III, see [Mur], [H]. In these papers also non-simple abelian varieties are considered. A simple abelian variety of dimension 4 of type III actually has $\dim B^1 = 1$, $\dim B^2 = 6$, however, its endomorphism algebra contains infinitely many imaginary quadratic fields and B^2 is spanned by D^2 and the spaces of Weil-Hodge cycles for these fields ([MoZ], 7.2). In higher dimensions there exist abelian varieties X with $End(X) = \mathbf{Z}$ but $B^p \neq D^p$ (cf. [MoZ], 7.5).

4.14 The following theorem provides some examples of four dimensional abelian varieties of Weil-type for which the Hodge conjecture has been verified. In 5.2.3 we will define a discrete invariant

$$det\, H \in \mathbb{Q}^*/Nm(K^*)$$

associated to (the isogeny class of) a polarized abelian variety of Weil-type (X, K, E) (here $Nm : K^* \mapsto \mathbb{Q}_{>0}$, $a \mapsto a\bar{a}$ is the norm map).

The group on the right is an infinite 2-torsion group, and for any $x \in \mathbb{Q}^*/Nm(K^*)$ with $(-1)^n x > 0$ we will construct an n^2-dimensional family of abelian varieties of Weil-type with $det\, H = x$.

4.15 Theorem. (Schoen, [S]) The Hodge (2,2)-conjecture is true for the general four dimensional abelian varieties of Weil-type $(X, \mathbb{Q}(\sqrt{-3}))$, $(X, \mathbb{Q}(i))$ with $det\, H = 1$.

4.16 The method of Schoen uses the theory of Prym varieties, and can also be used to find cycles on certain abelian varieties of Weil-type also in higher dimensions. In [vG] we use theta functions to give another proof of Theorem 4.15 for the field $\mathbb{Q}(i)$.

We will sketch the proof of the last theorem in section 7. The theorem implies easily that for any four dimensional abelian variety (X, K) of Weil-type with field $K = \mathbb{Q}(\sqrt{-3})$ or $\mathbb{Q}(i)$ and $det\, H = 1$, the space of Weil-Hodge cycles of (X, K) is spanned by cohomology classes of algebraic cycles.

5 Abelian varieties of Weil-type

5.1 As we saw in the previous section, the abelian varieties of Weil type provide an interesting test case for the Hodge conjecture. In this section we first study these abelian varieties (Lemma 5.2) and then we will construct families of such abelian varieties.

Let (X, K) be an abelian variety of Weil type. The action of K on $H_1(X, \mathbb{Q})$ gives $H_1(X, \mathbb{Q})$ the structure of a K-vector space. The space $H_1(X, \mathbb{R})$ has the structure of complex vector space via its identification as $T_0 X = (H_1(X, \mathbb{R}), J)$ (see 3.1).

5.2 Lemma. Let (X, K), with $K = \mathbb{Q}(\sqrt{-d}) \subset \mathbb{C}$ be an abelian variety of Weil-type of dimension $2n$.

1. There exists a polarization E on X such that (X, K, E) is a polarized abelian variety of Weil-type.
2. Let (X, K, E) be a polarized abelian variety of Weil-type. Then the map:

$$H : H_1(X, \mathbb{Q}) \times H_1(X, \mathbb{Q}) \longrightarrow K, \qquad H(x, y) := E(x, (\sqrt{-d})_* y) + \sqrt{-d}E(x, y)$$

is a non-degenerate Hermitian form on the K-vectorspace $H_1(X, \mathbb{Q})$. (Hermitian means that H is K-linear in the second factor and $H(y, x) = \overline{H(x, y)}$).

3. Let $\Psi \in M_{2n}(K)$ be the Hermitian matrix which defines H w.r.t. some K-basis of $H_1(X, \mathbf{Q})$. Then
$$det\,\Psi \in \mathbf{Q}^*/Nm(K^*)$$
does not depend on the choice of the K-basis, nor the lattice defining X. Thus $det\,\Psi$ is an isogeny invariant of (X, K, E) and will be denoted by $det\,H$ (cf. 4.14).

4. The signature of the Hermitian form H is (n, n).

5. Let $W \subset T_0(X)$ be the n-dimensional complex subspace on which K acts via scalar multiplication by $x \in K \subset \mathbf{C}$. Then
$$H_{|W} > 0,$$
where H is extended $\mathbf{R}$-linearly to $T_0X = (H_1(X, \mathbf{R}), J)$.

6.
$$\wedge_K^{2n} H^1(X, \mathbf{Q}) \hookrightarrow B^n(X) = H^{2n}(X, \mathbf{Q}) \cap H^{n,n}(X).$$

Proof. For the first statement we observe that $(\sqrt{-d})^*$ acts on $B^1(X)$, with eigenvalues in $\{-d,\, d\}$ (use that $B^1(X) \subset H^{1,1}(X) = H^{1,0}(X) \otimes H^{0,1}(X)$). Thus any polarization can be written as $E = E_+ + E_-$ with $E_\pm \in B^1(X)$ and $(\sqrt{-d})^* E_\pm = \pm d E_\pm$. We claim that E_+ is in fact a polarization. Since $E_+ \in B^1(X)$, the first Riemann condition is satisfied. The second one follows from adding the inequalities $dE(x, Jx) > 0$ and $E((\sqrt{-d})_* x, J(\sqrt{-d})_* x) = dE_+(x, Jx) - dE_-(x, Jx) > 0$ shows that $2dE_+(x, Jx) > 0$ for $x \neq 0$ (note $(\sqrt{-d})_*$ commutes with J since it is an endomorphism of X). Note that for the general abelian variety of Weil-type X one has $B^1(X) = \mathbf{Q}$ (see Theorem 4.11), so the polarization is unique and must be of Weil-type.

That H is Hermitian is an easy computation, using $E((\sqrt{-d})_* x, (\sqrt{-d})_* y) = dE(x, y)$.

The form H is thus given by $H(x, y) = {}^t\bar{x}\Psi y$ and ${}^t\Psi = \overline{\Psi}$. Changing the K-basis by linear map A changes Ψ to ${}^t\overline{A}\Psi A$ and thus $det\,\Psi$ changes to $Nm(a) \cdot det\,\Psi$ where $a = det\,A$. Let $\phi : Y \to X$ with (X, K, E) of Weil-type be an isogeny. Since isogenies are isomorphisms on $(H_1)_{\mathbf{Q}}$ which preserve $\mathrm{End}_{\mathbf{Q}}$, also $(Y, K, \phi^* E)$ must be of Weil-type, and the map $\phi^* : H^1(X, \mathbf{Q}) \to H^1(Y, \mathbf{Q})$ is an isomorphism of K-vector spaces. Thus $det\,H_E = det\,H_{\phi^* E}$.

The map $(\sqrt{-d})_*$ is a $\mathbf{C}$-linear map on $T_0(X) = (H_1(X, \mathbf{R}), J)$ (since it commutes with J), and has two eigenspaces $W_\pm$, each of dimension n, on which it acts as $\pm\sqrt{-d} = \pm\sqrt{d}J$. Thus restricted to $W_\pm$ we have $H(x, x) := E(x, \sqrt{-d}x) = \pm\sqrt{d}E(x, Jx)$. Since $E(x, Jx) > 0$ for $x \neq 0$ by the second Riemann condition, H is positive definite on W_+ (and negative definite on W_-). Note that W_+ and W_- are perpendicular w.r.t. H since
$$dH(x_+, x_-) = H((\sqrt{-d})_* x_+, (\sqrt{-d})_* x_-) = H(\sqrt{d}Jx_+, -\sqrt{d}Jx_-) = -dH(x_+, x_-)$$
with $x_\pm \in W_\pm$ (we use that $H(Jx, Jy) = H(x, y)$ which follows from the fact that J and $(\sqrt{-d})_*$ commute and the first Riemann condition).

First we show there is an inclusion $\wedge_K^{2n} H^1(X,\mathbf{Q}) \subset \wedge_{\mathbf{Q}}^{2n} H^1(X,\mathbf{Q}) = H^{2n}(X,\mathbf{Q})$. Let V be a finite dimensional K-vector space and let $W = V^* := \mathrm{Hom}_K(V,K)$ be its dual, note $V = W^*$. Let $Tr : K \to \mathbf{Q}$, $z \mapsto z + \bar{z}$ be the trace map, then:

$$W^* = \mathrm{Hom}_K(W,K) \xrightarrow{\cong} W^{*\mathbf{Q}} := \mathrm{Hom}_{\mathbf{Q}}(W,\mathbf{Q}), \qquad f \mapsto Tr \circ f$$

is an isomorphism of $\mathbf{Q}$-vector spaces (reduce to $W \cong K^n$ and then $W = K$ (and $f(z) = az$), now use that for $a \in K$ one has: $Tr(az) = 0$ for all $z \in K$ iff $z = 0$, which proves injectivity, surjectivity follows by comparing dimensions).

Using that $\wedge_K^n W^* = (\wedge_K^n W)^*$ and that K-linear maps are in particular $\mathbf{Q}$-linear we obtain:

$$\wedge_K^n V = \mathrm{Hom}_K(\wedge_K^n W, K) \longrightarrow \mathrm{Hom}_{\mathbf{Q}}(\wedge_{\mathbf{Q}}^n W, K) \xrightarrow{Tr\circ} \mathrm{Hom}_{\mathbf{Q}}(\wedge_{\mathbf{Q}}^n W, \mathbf{Q}),$$

where we compose with the trace in the last map. The space on the right is $(\wedge_{\mathbf{Q}}^n W)^{*\mathbf{Q}} \cong \wedge_{\mathbf{Q}}^n (W^{*\mathbf{Q}})$ and is thus isomorphic to $\wedge_{\mathbf{Q}}^n W^*$, so

$$\wedge_K^n V \longrightarrow \mathrm{Hom}_{\mathbf{Q}}(\wedge_{\mathbf{Q}}^n W, \mathbf{Q}) \cong \wedge_{\mathbf{Q}}^n W^* = \wedge_{\mathbf{Q}}^n V.$$

Actually $\wedge_K^n V$ is naturally a direct summand of $\wedge_{\mathbf{Q}}^n V$, since the $\mathbf{Q}$-linear, alternating map $V^n \to \wedge_K^n V$, $(v_1, \ldots, v_n) \mapsto v_1 \wedge_K \ldots \wedge_K v_n$ factors over a map $\wedge_{\mathbf{Q}}^n V \to \wedge_K^n V$.

To get the Hodge type, we tensor by $\mathbf{R}$ (note $K \otimes_{\mathbf{Q}} \mathbf{R} = \mathbf{C}$) and consider the space $\wedge_{\mathbf{C}}^{2n} H^1(X,\mathbf{R}) \subset \wedge_{\mathbf{R}}^{2n} H^1(X,\mathbf{R})$. The eigenspaces $W'_{\pm} \subset H_1(X,\mathbf{R})$ of $(\sqrt{-d})^*$ (the duals of the $W_{\pm}$'s) are $K \otimes_{\mathbf{Q}} \mathbf{R}$ stable, thus $\wedge_{\mathbf{C}}^{2n} H^1(X,\mathbf{R}) = \wedge_{\mathbf{C}}^{2n} W'_+ \oplus \wedge_{\mathbf{C}}^{2n} W'_-$. Since $W'_{\pm}$ are also stable under J', and the eigenspaces of J' in $H^1(X,\mathbf{C})$ are $H^{1,0}$ and $H^{0,1}$ (with eigenvalues i and $-i$), we have $\dim W'_{\pm} \otimes_{\mathbf{R}} \mathbf{C} \cap H^{1,0} = n$ and thus $(\wedge_{\mathbf{C}}^{2n} W'_{\pm}) \otimes_{\mathbf{R}} \mathbf{C} \subset H^{n,n}(X)$.

□

5.3 In the remainder of this section we will construct and investigate n^2 dimensional families of polarized abelian varieties of Weil type.

Such a family is constructed from data

$$(V, K, H, \Lambda).$$

Here V is a vector space over an imaginary quadratic field $K = \mathbf{Q}(\sqrt{-d}) \subset \mathbf{C}$ with $\dim_K V = 2n$. Furthermore $\Lambda \subset V$ is a lattice in $V_{\mathbf{R}} := V \otimes_{\mathbf{Q}} \mathbf{R}$. Finally H is a Hermitian form on V, with signature (n,n).

Note that a polarized abelian variety of Weil-type of dimension $2n$ (X, K, E) provides such data. In fact one takes $V = H_1(X,\mathbf{Q})$, $K = K$, H as in Lemma 5.2 and $\Lambda = H_1(X,\mathbf{Z})$. If, in the construction below, one puts $V_+ := W$, with W as in Lemma 5.2.5, one obtains again (X, K). In particular, any (X, K, E) is a member of an n^2 dimensional family of polarized abelian varieties of Weil-type.

5.4 Given the Hermitian form

$$H : V \times V \longrightarrow K$$

of signature (n,n), there exists a K-basis of V, on which H is given by:

$$H(z,w) \;=\; a\bar{z}_1 w_1 + \ldots + \bar{z}_n w_n - (\bar{z}_{n+1} w_{n+1} + \ldots + \bar{z}_{2n} w_{2n}), \qquad (5.4.1)$$

with $a \in \mathbf{Q}_{>0}$ (see [L]). Conversely, taking $V = K^{2n}$ and defining H by this formula with $a \in \mathbf{Q}_{>0}$ we obtain a Hermitian form on V of signature (n,n) with $\det H = (-1)^n a$.

5.5 Let $K = \mathbf{Q}(\sqrt{-d}) \subset \mathbf{C}$, so $K \otimes_{\mathbf{Q}} \mathbf{R}$ is naturally identified with $\mathbf{C}$ and thus $V_{\mathbf{R}} := V \otimes_{\mathbf{Q}} \mathbf{R} \cong \mathbf{C}^{2n}$. More precisely, multiplication by:

$$i := \sqrt{-d} \otimes (1/\sqrt{d}) : V_{\mathbf{R}} \longrightarrow V_{\mathbf{R}}$$

defines structure of complex vector space $(V_{\mathbf{R}}, i)$ on $V_{\mathbf{R}}$.

The abelian varieties which we construct are all obtained as $V_{\mathbf{R}}/\Lambda$. What changes will be the complex structure on $V_{\mathbf{R}}$. Such a complex structure is just an $\mathbf{R}$-linear map $J : V_{\mathbf{R}} \longrightarrow V_{\mathbf{R}}$. The polarization E will also be fixed, it will be given by the imaginary part of H:

$$E := Im\, H : V \times V \longrightarrow \mathbf{Q}$$

which is an alternating map since H is Hermitian.

To define the complex structures, we choose a complex subspace $V_+ \subset V_{\mathbf{R}}$ (with its complex structure $(V_{\mathbf{R}}, i)$) with $\dim_{\mathbf{C}} V_+ = n$ such that:

$$H_{|V_+} > 0, \qquad \text{define} \quad V_- := V_+^{\perp},$$

here we extend H $\mathbf{R}$-linearly from V to $V_{\mathbf{R}}$. Thus H is positive definite on V_+ and therefore negative definite on V_-, the perpendicular w.r.t. H of V_+. Then we have:

$$V_{\mathbf{R}} = V_+ \oplus V_-.$$

Now we define the complex structure on $V_{\mathbf{R}}$ corresponding to V_+ as follows:

$$J = J_{V_+} : V_{\mathbf{R}} \longrightarrow V_{\mathbf{R}}, \qquad Jv_+ = iv_+, \quad Jv_- = -iv_-$$

for all $v_\pm \in V_\pm$. Clearly $J^2 = -I$ and thus we obtain a complex vector space $(V_{\mathbf{R}}, J)$. We note that J is in fact $\mathbf{C}$-linear on $(V_{\mathbf{R}}, i)$, since J and i commute. Therefore J also commutes with the action of K on $V_{\mathbf{R}}$ (via multiplication on the left on $V_{\mathbf{R}} = V \otimes_{\mathbf{Q}} \mathbf{R}$).

5.6 With this choice of complex structure on $V_{\mathbf{R}}$, it remains to show that E satisfies the Riemann Relations, i.e. that

$$E(Jx, Jy) = E(x, y), \qquad E(x, Jx) > 0 \quad \text{for } x \neq 0.$$

The first condition is clear when we write $x = x_+ + x_-$ etc. with $x_\pm \in V_\pm$, use that $Jx_\pm = \pm iv_\pm$, and that $E = Im\, H$ for a Hermitian form H for which $V_+ \perp V_-$. For the second we recall that given $E = Im\, H$ one recovers (the $\mathbf{R}$-linear extension of) H as $H(x, y) = E(x, iy) + iE(x, y)$ (just write $H(x, y) = A(x, y) + iE(x, y)$ with $\mathbf{R}$-valued forms A, E and expand both sides of $iH(x, y) = H(x, iy)$). Writing x as before we get

$$E(x, Jx) = E(x_+, ix_+) - E(x_-, ix_-) = H(x_+, x_+) - H(x_-, x_-) > 0$$

for $x \neq 0$ since H is positive definite on V_+ and negative definite on V_-.

5.7 Thus from the data (V, K, H, Λ) and a $V_+ \subset V_{\mathbf{R}}$ we have constructed a polarized abelian variety $(X := V/\Lambda, E)$, with complex structure J on V.

Moreover, since the complex structure J commutes with the action of K, we have $K \subset \mathrm{End}(X)_{\mathbf{Q}}$. The complex vector space $(V_{\mathbf{R}}, J)$ is the tangent space at the origin of X. The subspaces $V_\pm$ are stable under the action of J and are thus also complex subspaces of (V, J). Since $i = \pm J$ on $V_\pm$ and $i := \sqrt{-d} \otimes (1/\sqrt{d})$, the action of $x \in K$ on V_+ is scalar multiplication by x whereas on V_- it is scalar multiplication by $\overline{x}$. Thus (X, K, E) is a polarized abelian variety of Weil-type.

5.8 The triple (X, K, E) we constructed is thus determined by the choice of V_+ in the Hermitian vectorspace $((V_{\mathbf{R}}, i), H)$. The only condition on V_+ is that H is positive definite on V_+, which is an open condition (in the analytic topology on the Grassmanian of n-dimensional subspaces in the $2n$ dimensional space $(V_{\mathbf{R}}, i)$). Thus X is a member of an $n^2 = \dim Grass(n, 2n)$ dimensional family of abelian varieties of Weil-type.

5.9 The global structure of this family can also be derived easily. Let

$$SU(n,n) := \{A \in GL((V_{\mathbf{R}}, i)) \cong GL(2n, \mathbf{C}) : \; H(x,y) = H(Ax, Ay), \; det(A) = 1\}$$

for all $x, y \in V_{\mathbf{R}}$ (we supress H from the notation, a more accurate notation would be $SU_H(\mathbf{R})$, see the next section).

5.10 Lemma. The group $SU(n,n)$ acts transitively on the set

$$\mathbf{H}_n := \{W \subset (V_{\mathbf{R}}, i) \cong \mathbf{C}^{2n} : \; W \cong \mathbf{C}^n, \; H_{|W} > 0\}.$$

The stabilizer of a $W \in \mathbf{H}_n$ is isomorphic to the group $S(U(n) \times U(n))$ (pairs of unitairy matrices with product of the determinants equal to one). Thus:

$$\mathbf{H}_n \cong SU(n,n)/S(U(n) \times U(n)).$$

Proof. Given any two n-dimensional subspaces V_+, W_+ on which H is positive definite, we can choose orthonormal bases (w.r.t. to H) $e_1, \ldots, e_n$ of V_+ and $f_1, \ldots, f_n$ of W_+ which can be extended with orthonormal bases $e_{n+1}, \ldots, e_{2n}$ of W_- and $f_{n+1}, \ldots,$ f_{2n} of W_- to $\mathbf{C}$-bases of $(V_{\mathbf{R}}, i)$. On each of these bases of $(V_{\mathbf{R}}, i)$, the form H is then given by the formula 5.4.1 with $a = 1$. Thus the matrix relating the e_i and the f_i preserves H and, after multiplying say f_1 by a suitable $\lambda \in \mathbf{C}$, $|\lambda| = 1$, the matrix will be in $SU(n,n)$. Thus the group $SU(n,n)$ acts transitively on the set of n-dimensional subspaces on which H is positive definite.

The stabilizer of V_+ in $SU(n,n)$ consists of maps mapping V_+, and thus also $V_- = V_+^\perp$, into itself and preserving the restriction of H on these subspaces, which is definite on each of them. Thus the stabilizer is isomorphic to $S(U(n) \times U(n))$. □

5.11 The set $\mathbf{H}_n$ actually has the structure of complex manifold on which the group $SU(n,n)$ acts by holomorphic maps, in fact it is a bounded Hermitian domain. The construction of Weil-type abelian varieties shows that there exist embeddings $\mathbf{H}_n \hookrightarrow \mathbf{S}_{2n}$, the Siegel space of positive definite $2n \times 2n$ period matrices, see [Sh] and [Mum].

Using the lattice $\Lambda \subset V \subset V_{\mathbf{R}}$ we define a group by:

$$\Gamma = \Gamma_\Lambda = \{A \in SU(n,n) : \; A\Lambda \subset \Lambda\}.$$

Then $\mathcal{H} := \mathbf{H}_n/\Gamma$ is a quasi-projective variety and it parametrizes abelian varietes of Weil-type (n,n); it is an example of a Shimura variety (although that name is nowadays in fact reserved for a more sophisticated but related object).

In the case that E is a principal polarization, one has $\mathbf{H}_n \subset \mathbf{S}_{2n}$ and $\Gamma :=$ $Sp(2g, \mathbf{Z}) \cap SU(n,n)$, and one obtains an algebraic subvariety $\mathcal{H} \subset \mathcal{A}_{2n}$, the moduli space of principally polarized, $2n$-dimensional, abelian varieties. For a general polarization one has to make the obvious changes.

5.12 Example. Usually one describes (principally polarized) abelian varieties via their period matrix. Here is an example of principally polarized abelian varieties of Weil-type with $K = \mathbf{Q}(i)$ and $det\, H = 1$.

Any principally polarized abelian variety of even dimension can be obtained as $\mathbf{C}^{2n}/\Lambda$ with the inclusion $\Lambda \cong \mathbf{Z}^{4n} \hookrightarrow \mathbf{C}^{2n}$ given by the 'period matrix'

$$\Omega := (I\,\tau) = \begin{pmatrix} I & 0 & \tau_{11} & \tau_{21} \\ 0 & I & \tau_{21} & \tau_{22} \end{pmatrix},$$

where each block on the right is an $n \times n$ matrix. The polarization $\Lambda \times \Lambda \to \mathbf{Z}$ is given by the alternating matrix (with $2n \times 2n$ blocks):

$$E := \begin{pmatrix} 0 & I \\ -I & 0 \end{pmatrix}.$$

The Riemann conditions on E are equivalent to $\tau \in \mathbf{S}_{2n}$ (the Siegel space), that is, τ satisfies ${}^t\tau = \tau$ and $Im\,\tau > 0$.

To give an endomorphism of an abelian variety $X \cong \mathbf{C}^{2n}/\Lambda$ we must give a $\mathbf{C}$-linear map $A : \mathbf{C}^{2n} \to \mathbf{C}^{2n}$ which satisfies $A(\Lambda) \subset \Lambda$. Define

$$A := \begin{pmatrix} 0 & -I \\ I & 0 \end{pmatrix}, \quad \text{then}: \quad A^2 = -I.$$

In particular, A has n eigenvalues i and n eigenvalues $-i$. Define

$$\mathsf{H}_A := \{\tau \in \mathbf{S}_{2n} : \tau_{11} = \tau_{22},\ \tau_{21} = -\tau_{12}\}.$$

Note that we have $(1/2)n(n+1)$ parameters for τ_{11} and $(1/2)n(n-1)$ parameters for τ_{12} since $\tau_{12} = -\tau_{21} = -{}^t\tau_{12}$, thus $\dim \mathsf{H}_A = n^2$. For $\tau \in \mathsf{H}_A$ one has:

$$A\Omega = \Omega B \qquad \text{with} \quad B := \begin{pmatrix} A & 0 \\ 0 & A \end{pmatrix},$$

which shows that A preserves the lattice. The automorphism B it induces on the lattice preserves E, so A defines an automorphism of X/Λ which preserves the polarization.

Thus we have an n^2 dimensional family of polarized abelian varieties of Weil type, with $K = \mathbf{Q}(i)$, over H_A. An easy computation shows $det\, H = 1$.

6 Mumford-Tate groups

6.1 The Mumford-Tate groups were introduced in [Mum]. In general, they are associated to $\mathbf{Q}$-Hodge structures, cf. [DMOS]. Here we restrict ourselves to the case of abelian varieties and do not discuss nor give general definitions. We use Mumford-Tate groups mainly to find the (dimensions of the) spaces of Hodge cycles B^p.

6.2 Let X be an abelian variety. In this section we write:

$$V = V_{\mathbf{Q}} = H^1(X, \mathbf{Q}), \qquad V_L := V \otimes_{\mathbf{Q}} L = H^1(X, L),$$

for any field $L \supset \mathbf{Q}$. In 3.4 we defined the structure of a complex vectorspace on $V_{\mathbf{R}}$ using a map J' with $(J')^2 = -I$. We define a group:

$$S^1 := \{z \in \mathbf{C}^* : |z| = 1\}$$

and a representation of this group on $V_{\mathbf{R}}$:

$$h = h_X : S^1 \longrightarrow SL(V_{\mathbf{R}}), \qquad h(a+bi)v := (aI + bJ')v,$$

with $a, b \in \mathbf{R}$ and $v \in V_{\mathbf{R}}$. So h just gives the scalar multiplication by complex numbers of length one on the complex vector space $(V_{\mathbf{R}}, J')$.

That $det(h(z)) = 1$ follows from the fact that J' acts as i on $H^{1,0}$, so $h(z)$ is scalar multiplication by z and $\bar{z}$ on $H^{1,0}$ and $H^{0,1}$ respectively. We also have representations

$$\wedge^n h : S^1 \longrightarrow GL(\wedge^n H^1(X, \mathbf{R})) = GL(H^n(X, \mathbf{R}))$$

and, by $\mathbf{C}$-linear extension, on $H^n(X, \mathbf{C})$. The Hodge decomposition on $H^n(X, \mathbf{C})$ can be recovered from $\wedge^n h$ since the action of $(\wedge^n h)(z)$ on $H^{p,q}(X)$, with $p + q = n$, is scalar multiplication by $z^p \bar{z}^q$.

6.3 Recall that an algebraic group defined over a field L is a quasi-projective variety over L whose group laws are given by morphisms defined over L. For example, the group $GL(n)_L$, with a field L, is an algebraic group (as a variety it is defined by $t \cdot det(A) - 1 = 0$, which is a polynomial equation in $L[\ldots, a_{ij}, \ldots, t]$, and the group laws are given by polynomials in the a_{ij} and $t\,(= (detA)^{-1})$ with coefficients in L).

Let G be an algebraic subgroup of $SL(n)_{\mathbf{Q}}$, defined over $\mathbf{Q}$ (so its ideal $I(G)$ is generated by polynomials in $\mathbf{Q}[\ldots, a_{ij}, \ldots]$). The ring of polynomial functions on G is, as usual,

$$\mathbf{Q}[G] := \mathbf{Q}[\ldots, a_{ij}, \ldots]/I(G).$$

For a $\mathbf{Q}$-algebra L we define

$$G(L) := \mathrm{Hom}_{\mathbf{Q}-algebra}(\mathbf{Q}[G], L),$$

the set (in fact group) of L-valued points of G. (If $\phi \in G(L)$ maps $a_{ij} \in \mathbf{Q}[G]$ to $l_{ij} \in L$, then ϕ defines a matrix with coefficients l_{ij} which satisfies the defining equations of G; conversely, such a matrix defines a $\phi \in G(L)$. Thus $G(L)$ is just the set of matrices with determinant one with coefficients in L which satisfy the equations defining G. The group law on $G(L)$ is just the matrix product of matrices with coefficients in L).

6.4 Definition. The Special Mumford-Tate group G (sometimes also called Hodge group) of the abelian variety X is the smallest algebraic subgroup $G \subset SL(V_{\mathbf{Q}})$, which is defined over $\mathbf{Q}$, such that:

$$h(S^1) \subset G(\mathbf{R}).$$

(The Mumford Tate group itself is $\mathbf{G}_m \cdot G$, i.e. one also allows scalar multiples of the identity.)

6.5 In this definition, note that $h(S^1) \subset SL(V_{\mathbf{R}})$ (thus $G \subset SL(V_{\mathbf{Q}})$ and that the intersection of two algebraic subgroups is again an algebraic subgroup, so the definition makes sense. Since G acts on $V_{\mathbf{Q}}$ it also acts on $\wedge^k V_{\mathbf{Q}} = H^k(X, \mathbf{Q})$ for all k.

Let $Sp(E)$ be the algebraic subgroup of $SL(V_{\mathbf{Q}})$ which fixes a polarization $E \in \wedge^2 V_{\mathbf{Q}}$ of X. Note that $Sp(E)$ is defined over $\mathbf{Q}$ and that $h(S^1) \subset Sp(E)(\mathbf{R})$, in fact, this is equivalent to $E(h^*(z)x, h^*(z)y) = E(x, y)$ for all $x, y \in H_1(X, \mathbf{R})$ (with $h^*(z) = aI + bJ$ the dual representation of h), which follows from $J^2 = -I$ and the Riemann Relation $E(x, y) = E(Jx, Jy) = E(x, y)$.

Therefore:

$$G \subset Sp(E).$$

A generalization of this argument gives the main result:

6.6 Theorem. For all p, the space of Hodge cycles is the subspace of G-invariants in $H^{2p}(X, \mathbf{Q})$:

$$B^p(X) = H^{2p}(X, \mathbf{Q})^G.$$

Proof. (Sketch.) First we show $B^p(X) \subset H^{2p}(X, \mathbf{Q})^G$. Let $\rho_k : GL(V) \to GL(\wedge^k V)$ be the kth-exterior power of the standard representation ρ_1 of $GL(V)$. Since $B^p(X)$ is a $\mathbf{Q}$-subspace of $H^{2p}(X, \mathbf{Q})$, the subgroup $P = P_p \subset SL(V_{\mathbf{Q}})$ which via ρ_{2p} acts as the identity on $B^p(X)$, is an algebraic subgroup defined over $\mathbf{Q}$. Because $(\wedge^{2p} h)(z)$ acts as $z^p \bar{z}^p$, that is, as the identity on $H^{p,p} \supset B^p$, we have $(\wedge^{2p} h)(S^1) = \rho_{2p}(h(S^1)) \subset \rho_{2p}(P(\mathbf{R}))$ and thus $h(S^1) \subset P(\mathbf{R})$. Therefore $G \subset P$ and thus $B^p \subset H^{2p}(X, \mathbf{Q})^P \subset H^{2p}(X, \mathbf{Q})^G$.

To show $B^p(X) \supset H^{2p}(X, \mathbf{Q})^G$ we must show $H^{p,p} \supset H^{2p}(X, \mathbf{Q})^G$ and $H^{2p}(X, \mathbf{Q}) \supset H^{2p}(X, \mathbf{Q})^G$. The last is trivial, the first follows from the fact that $h(S^1) \subset G(\mathbf{R})$, so $G(\mathbf{R})$-invariants must be $h(S^1)$-invariants. Since $h(S^1)$ acts as $z^a \bar{z}^b$ on $H^{a,b}$, the space of $h(S^1)$-invariants in $H^{2p}(X, \mathbf{C})$ is just $H^{p,p}(X)$. □

6.7 The Special Mumford-Tate group itself can be somewhat subtle, however one has

$$B^p(X) \otimes_{\mathbf{Q}} \mathbf{C} = (H^{2p}(X, \mathbf{Q})^G) \otimes \mathbf{C} = H^{2p}(X, \mathbf{C})^{G(\mathbf{C})}.$$

The group $G(\mathbf{C})$, a complex Lie group, is known to be a connected reductive Lie group [DMOS] (here one uses the polarization in an essential way) and thus it, and especially its Lie algebra, is well understood. To compute the space of Hodge cycles one determines the representation of $G(\mathbf{C})$ (or its Lie algebra) on $V_{\mathbf{C}}$ and then, using representation theory, one tries to find the invariants in $\wedge^{2p} V_{\mathbf{C}}$.

6.8 Example. For a general abelian variety of dimension g one has $G = Sp(E)$, the proof is similar to the one sketched in 6.11. Then $G(\mathbf{C}) \cong Sp(2g, \mathbf{C})$ and the representation of $Sp(2g, \mathbf{C})$ on $V_{\mathbf{C}}$ is just the standard representation. It is well known that the subspace of invariants of $Sp(2g, \mathbf{C})$ in $\wedge^{2p} V_{\mathbf{C}}$ is one dimensional and is spanned by $\wedge^p E$ (where $E \in H^2(X, \mathbf{Q})$ is the polarization), cf. Thm 17.5 from [Fu]. Thus we obtain another proof of Mattuck's result 4.2.

6.9 Example. Let now (X, K, E) be a $2n$-dimensional polarized abelian variety of Weil-type with $K = \mathbb{Q}(\sqrt{-d})$. Then $V = H^1(X, \mathbb{Q})$ has the structure of a Hermitian K-vector space, with H the Hermitian form associated to E. We will construct algebraic groups U_H and SU_H, defined over $\mathbb{Q}$. The group SU_H will be the Special Mumford Tate group of a general abelian variety of Weil-type.

For convenience, let $B_K := \{e_1, \dots, e_{2n}\}$ be a K-basis of V for which H is given by a diagonal Hermitian matrix Ψ, so:

$$\Psi = {}^t\Psi \in M_{2n}(\mathbb{Q}).$$

The condition that a $2n \times 2n$ matrix $A = (a_{ij})$ is unitary (that is ${}^t\overline{A}\Psi A = \Psi$) is *not* given by polynomials in $K[\dots, a_{ij}, \dots]$, since conjugation $x \mapsto \bar{x}$ is not given by a polynomial in $K[X]$. However, viewing the K-vector space V as a $\mathbb{Q}$-vectorspace we can define an algebraic group U_H over $\mathbb{Q}$ with the property that $U_H(\mathbb{Q})$ is isomorphic to the K-linear maps on V which preserve H.

A $\mathbb{Q}$-basis of V is given by $B_{\mathbb{Q}} := \{e_1, \dots, e_{4n}\}$, with $e_{2n+j} := (\sqrt{-d})^* e_j$. Since the K-linear maps are just the $\mathbb{Q}$-linear maps which commute with $(\sqrt{-d})^*$, we consider the algebraic group R of such (invertible) maps:

$$R := \left\{ r_{B,C} := \begin{pmatrix} B & -dC \\ C & B \end{pmatrix} \in GL(V_{\mathbb{Q}}) \right\},$$

where each block is an $2n \times 2n$ matrix (in fact $R = Res_{K/\mathbb{Q}}(GL(V_K)))$. Note that on the basis $B_{\mathbb{Q}}$, the map $(\sqrt{-d})^*$ is given by $r_{0,I} \in R(\mathbb{Q})$. One easily verifies, for any $\mathbb{Q}$-algebra L, that:

$$R(L) = \{A \in GL(4n, L) : Ar_{0,I} = r_{0,I}A\},$$

thus $R(L)$ consists of the invertible L-linear maps commuting with $(\sqrt{-d})^* \otimes 1$, i.e. $(\sqrt{-d})^* \otimes 1$-linear maps, on $V_{\mathbb{Q}} \otimes_{\mathbb{Q}} L$. In particular:

$$\kappa : R(\mathbb{Q}) \xrightarrow{\cong} GL(V_K)(K) \cong GL(2n, K) \qquad r_{B,C} \mapsto B + \sqrt{-d}C.$$

Next we observe that (in $\mathbb{Q}[\dots, b_{ij}, \dots, c_{ij}, \dots]$):

$${}^t(B - \sqrt{-d}C)\Psi(B + \sqrt{-d}C) = \Psi \iff \begin{cases} {}^tB\Psi B + d{}^tC\Psi C &= \Psi \\ {}^tB\Psi C - d{}^tC\Psi B &= 0. \end{cases}$$

This shows that the equations on the right define an algebraic subgroup of R, denoted by U_H, defined over $\mathbb{Q}$, with the property that $\kappa(U_H(\mathbb{Q}))$ are the K-linear maps preserving H.

The conditions above are moreover equivalent to:

$${}^tr_{B,C}E_{\Psi}r_{B,C} = E_{\Psi} \qquad \text{with} \quad E_{\Psi} = \begin{pmatrix} 0 & \Psi \\ -\Psi & 0 \end{pmatrix}.$$

This is not so surprising, since the Hermitian form H is determined by its imaginary part E which is given by the alternating matrix E_{Ψ} on the basis $B_{\mathbb{Q}}$. Thus we have:

$$U_H = R \cap Sp(E).$$

For any $\mathbf{Q}$-algebra L, $U_H(L) \subset GL(V_{\mathbf{Q}})(L) \cong GL(4n, L)$ is thus the subgroup of matrices which commute with the action $(\sqrt{-d})^* \otimes 1$ on $V_L := V \otimes_{\mathbf{Q}} L$, which preserve the L-linear extension of the $\mathbf{Q}$-bilinear form E.

Finally we define SU_H to be the subgroup of U_H defined by the two polynomial equations ('real' and 'imaginary part') in $\mathbf{Q}[\ldots, b_{ij}, \ldots, c_{ij}, \ldots]$ obtained from the condition:

$$det(B + \sqrt{-d}C) = 1.$$

Then one has $SU_H(\mathbf{R}) = SU(n, n)$, as in 5.8.

6.10 Lemma. With the notation from above, we have:

$$SU_H(\mathbf{C}) \cong SL(2n, \mathbf{C}).$$

Moreover, the representation of $SU_H(\mathbf{C})$ on $V \otimes_{\mathbf{Q}} \mathbf{C}$ is isomorphic to the direct sum of the standard representation of $SL(2n, \mathbf{C})$ and its dual representation.

Proof. The action of $(\sqrt{-d})^*$ on V can be diagonalized on $V \otimes_{\mathbf{Q}} \mathbf{C}$:

$$V_{\mathbf{C}} = W \oplus \overline{W}, \qquad \text{with} \quad W := \langle \ldots, e_i \otimes \sqrt{-d} + e_{i+2n} \otimes 1, \ldots \rangle_{i=1,\ldots,2n}$$

(use $e_{i+2n} := (\sqrt{-d})^* e_i$). Since the SU_H action on V commutes with $(\sqrt{-d})^*$, we see that both W and $\overline{W}$ are invariant subspaces of $SU_H(\mathbf{C})$. In terms of matrices one has, with $r_{B,C} \in SU_H \subset R$: $r_{B,C}S = SD$, with

$$D := \begin{pmatrix} B + \sqrt{-d}C & 0 \\ 0 & B - \sqrt{-d}C \end{pmatrix}, \qquad S := \begin{pmatrix} \sqrt{-d}I & -\sqrt{-d}I \\ I & I \end{pmatrix}.$$

Thus we have an injective homomorphism of groups:

$$\tau : SU_H(\mathbf{C}) \longrightarrow SL(W) \times SL(\overline{W}) \cong SL(2n, \mathbf{C})^2, \qquad r_{B,C} \mapsto (B + \sqrt{-d}C,\ B - \sqrt{-d}C).$$

The spaces W and $\overline{W}$ are isotropic subspaces w.r.t. the $\mathbf{C}$-linear extension of E. In fact, since (X, K, E) is of Weil-type we have $E((\sqrt{-d})^* x, (\sqrt{-d})^* y) = dE(x, y)$ and for $x \in W$ one has $(\sqrt{-d})^* x = \sqrt{-d}x$. Thus if $x, y \in W$ we have:

$$dE(x, y) = E((\sqrt{-d})^* x, (\sqrt{-d})^* y) = E(\sqrt{-d}x, \sqrt{-d}y) = -dE(x, y),$$

and so $H_{|W \times W} = 0$, the same with $\overline{W}$. Therefore E induces a duality between W and $\overline{W}$. Since E is invariant under $SU_H(\mathbf{C}) \subset Sp(E)(\mathbf{C})$, the representations induced on W and $\overline{W}$ are dual.

Since $E_{|W \times W} = 0$, any $\mathbf{C}$-linear map $Q : W \to W$ preserves the restriction of E to W. Using the duality of W and $\overline{W}$ defined by E, one gets a map $Q' : \overline{W} \to \overline{W}$ such that the pair $(Q, Q') \in GL(W) \times GL(\overline{W})$ is in (the image of) $U_H(\mathbf{C})$. Taking $Q \in SL(W)$ we get the isomorphism $SL(2n, \mathbf{C}) \cong SU_H(\mathbf{C})$.

(To show the map $SU_H(\mathbf{C}) \to SL(W)$ is surjective, we can use also a dimension argument. The dimension of $SL(2n, \mathbf{C})$ (as a complex manifold) is $(2n)^2 - 1$. The dimension of $SU(n, n)$ (as real manifold) is also $(2n)^2 - 1$, as is easily seen by a Lie algebra computation.)

Another way to prove the Theorem is to show that $su(n, n) \otimes_{\mathbf{R}} \mathbf{C}$, with $su(n, n)$ the Lie algebra of $SU(n, n)$, is isomorphic to the Lie algebra of $SL(2n, \mathbf{C})$, which is the vector space of matrices with trace zero. □

6.11 Theorem. (Weil, [W]) The Special Mumford-Tate group of a general polarized abelian variety (X, K, E) of Weil-type is SU_H.

Proof. (Sketch.) First we show $h(S^1) \subset SU_H(\mathbf{R})$. Recall that $U_H(\mathbf{R}) = R(\mathbf{R}) \cap Sp(E)(\mathbf{R})$. Since the complex structure J' commutes with $(\sqrt{-d})^* \otimes 1$, we have $J' \in R(\mathbf{R})$ and then also $h(S^1) \in R(\mathbf{R})$. That $h(z)$ fixes the polarization E we have already seen. Finally, the fact that $h(z)$ has n eigenvalues z and n eigenvalues $\bar{z}$ on the complex vectorspace $(V_\mathbf{R}, i)$ with $i = (\sqrt{-d})^* \otimes (1/\sqrt{d})$ shows that $h(z) \in SU_H(\mathbf{R})$.

Next we must show that in general SU_H is the smallest algebraic subgroup of $GL(V)$ defined over $\mathbf{Q}$ containing $h(S^1)$ for the general X of Weil type. With $z = \cos\phi + i\sin\phi \in S^1$, we have (since $(J')^2 = -I$):

$$h(z) = (\cos\phi)I + (\sin\phi)J' = exp(\phi J'), \qquad \text{with} \quad exp : \mathrm{End}(V_\mathbf{R}) \longrightarrow GL(V_\mathbf{R})$$

the exponential map ($exp(M) = \sum_{n=0}^{\infty} M^n/(n!)$). For an algebraic subgroup G' of $GL(V_\mathbf{Q})$, the group $G'(\mathbf{R})$ is a Lie group and we thus have:

$$h(S^1) \subset G'(\mathbf{R}) \quad \Longleftrightarrow J' \in Lie(G')_\mathbf{R} := Lie(G') \otimes_\mathbf{Q} \mathbf{R}.$$

The complex structure J' is determined by $V_+ \subset (V_\mathbf{R}, i)$. For $g \in SU(n,n) = SU_H(\mathbf{R})$ the complex structure $gJ'g^{-1}$ is then determined by the subspace $g(V_+)$. The manifold $\mathsf{H}_n = SU(n,n)/S(U(n) \times U(n))$ from 5.10, parametrizing $V_+ \subset (V_\mathbf{R}, i)$, may thus be identified with the submanifold of $Lie(SU_H)_\mathbf{R}$:

$$\mathsf{H}_n = \{gJ'g^{-1} : \ g \in SU_H(\mathbf{R})\} \quad \subset Lie(SU_H)_\mathbf{R}.$$

The Adjoint representation of $SU_H(\mathbf{R})$ on its Lie algebra $Lie(SU_H)_\mathbf{R} \subset \mathrm{End}(V_\mathbf{R})$ (via $g \cdot M := gMg^{-1}$) is irreducible (one may use for example that $SU_H(\mathbf{R}) \subset SU_H(\mathbf{C}) \cong SL(2n, \mathbf{C})$ is Zariski dense and that the Adjoint representation of $SL(2n, \mathbf{C})$ is irreducible). Therefore H_n does not lie in any (proper) linear subspace of $Lie(SU_H)_\mathbf{R}$.

Thus if $G' \subsetneqq SU_H$ (so $Lie(G')_\mathbf{R} \subsetneqq Lie(SU_H)_\mathbf{R}$), then $\mathsf{H}_n \cap Lie(G')_\mathbf{R}$ ($\subsetneqq \mathsf{H}_n$) is a real analytic submanifold of H_n. The algebraic group SU_H has only countably many (connected) algebraic subgroups G' defined over $\mathbf{Q}$ (because these are determined by their Lie algebra, which is a $\mathbf{Q}$-vector space in the finite dimensional Lie algebra of SU_H). Since H_n is not a countable union of lower dimensional submanifolds, the general $J' \in \mathsf{H}_n$ defines an abelian variety with Special Mumford-Tate group equal to SU_H. □

6.12 Theorem. (Weil [W]) Let (X, K) be an abelian variety of Weil-type of dimension $2n$. If the Special Mumford-Tate group of X is SU_H, then:

$$\dim B^p(X) = \begin{cases} 1 & p \neq n; \\ 3 & p = n, \end{cases} \qquad \text{and} \quad B^n(X) = D^n \oplus \wedge_K^{2n} H^1(X, \mathbf{Q}).$$

Proof. Since $B^p \cong B^{2n-p}$, it suffices to consider the $p \leq n$. In view of the previous results we have:

$$\dim_\mathbf{Q} B^p(X) = \dim_\mathbf{C}((\wedge^{2p} V) \otimes_\mathbf{Q} \mathbf{C})^{SL_{2n}(\mathbf{C})}.$$

Let W be the standard $2n$-dimensional representation of $SL_{2n}(\mathbb{C})$ and let W^* be its dual. Then:

$$V_{\mathbb{C}} := V \otimes_{\mathbb{Q}} \mathbb{C} = W \oplus W^* \cong W \oplus \wedge^{2n-1} W$$

(use the pairing $\wedge^k W \times \wedge^{2n-k} W \to \wedge^{2n} W \cong \mathbb{C}$ to identify $\wedge^k W^* \cong \wedge^{2n-k} W$). Thus

$$\begin{aligned}(\wedge^{2p} V) \otimes_{\mathbb{Q}} \mathbb{C} &\cong \wedge^{2p}(W \oplus W^*) \\ &= \oplus_{a=0}^{2p} (\wedge^{2p-a} W) \otimes (\wedge^{2n-a} W).\end{aligned}$$

Viewing W as (standard) $GL(2n,\mathbb{C})$ representation, the decomposition of $(\wedge^{2p-a}W)\otimes(\wedge^{2n-a}W)$ into irreducible $GL(2n,\mathbb{C})$ representations is given by formula (6.9) of [Fu]. The $SL(2n,\mathbb{C})$-invariants correspond to the one dimensional $GL(2n,\mathbb{C})$ representations. The formula (6.9) combined with Theorem 6.3.1 of [Fu] shows that:

$$\dim \left((\wedge^{2p-a} W) \otimes (\wedge^{2n-a} W) \right)^{SL(2n,\mathbb{C})} = 1 \qquad \text{iff} \quad n + p - a = kn \quad (k \in \mathbb{N}),$$

and in the other cases there no invariants.

In fact, the irreducible representations in $\wedge^a W \otimes \wedge^b W$, with, say $a \geq b$, correspond to partitions of $a+b$ of the form $\lambda = (2,\ldots,2,1,\ldots,1)$, with $\lambda_{a+1} \leq 1$ and with $\lambda_{2n+1} = 0$. The dimension of the corresponding representation is equal to one iff $\lambda_i = \lambda_j$ for all $1 \leq i,j \leq 2n$. The only such partitions are thus $\lambda = (1,\ldots,1)$ (and $a+b=2n$) or $\lambda = (2,\ldots,2)$ (and $a=b=2n$).

Thus, if $p < n$ we must have $a = p$, and (the dual of) the map $(\wedge^p W)\otimes(\wedge^{2n-p}W) \to \wedge^{2n} W \cong \mathbb{C}$ provides the invariant. In case $p = n$ we can take $a = 0, n, 2n$. The cases $a = 0, 2n$ give the invariant subspaces $\wedge^{2n} W$ and $\wedge^{2n} W^*$, which span $(\wedge_K^{2n} H^1(X,\mathbb{Q})) \otimes_{\mathbb{Q}} \mathbb{C}$.

Since we have $B^n(X) \supset \wedge_K^{2n} H^1(X,\mathbb{Q}) \oplus D^n(X)$ for any abelian variety of Weil-type the equality now follows for dimension reasons. $\square$

7 Sketch of proofs of Theorem 4.15

7.1 The method of Schoen to verify the Hodge conjecture is based on a geometrical construction. First of all, the abelian varieties are constructed in a geometrical way.

A curve C_3 of genus 3 and a subgroup G_3 of order three of $Jac(C_3)$ define an unramified cyclic 3:1 covering:

$$\pi : C_7 \longrightarrow C_3$$

with C_7 a genus seven curve. The map π induces a map π_* on divisors which again induces the norm map $Nm : J(C_7) \longrightarrow J(C_3)$. The connected component of $0 \in J(C_7)$ of $\ker(Nm)$ is called the Prym variety $P = P(C_7/C_3)$ of the covering. Then:

$$J(C_7) \approx_{isog} J(C_3) \times P.$$

The Prym variety P is an abelian variety of dimension 4. Let $\alpha \in Aut(C_7)$ be a generator of the covering group of π. Then α induces an automorphism α^* of order three on P, and thus $\mathbb{Q}(\sqrt{-3}) \subset \mathrm{End}(P)_{\mathbb{Q}}$. Using the holomorphic Lefschetz trace formula, one finds that P is Weil-type $(2,2)$. With the polarization E on P which is induced by the natural polarization on $J(C_7)$, $(P, \mathbb{Q}(\sqrt{-3}), E)$ is a polarized abelian variety of Weil-type.

An explicit computation, using for example the description of the action of α on the homology of C_7 given in [Fay], chap. IV, shows that $det\, H = 1$ (this remark was omitted in Thm 3.2 of [S]).

The construction $(C_3, G_3) \mapsto (P, E)$ extends to a morphism, the Prym map:

$$\mathcal{P} : \mathcal{M}_{3,\mathbf{Z}/3\mathbf{Z}} \longrightarrow \mathcal{A}_{4,E},$$

where $\mathcal{M}_{3,\mathbf{Z}/3\mathbf{Z}}$ is the (6 dimensional) moduli space of genus 3 curves with a subgroup of order three of $J(C_3)$ and $\mathcal{A}_{4,E}$ is moduli space of four dimensional abelian varieties with a polarization like the one on P.

From section 5.11 we know that the image of $\mathcal{P}$ lies in a $n^2 = 4$ dimensional subvariety $\mathcal{H}$ of $\mathcal{A}_{4,E}$, and Schoen proves that $\mathcal{P}$ has a Zariski dense image in $\mathcal{H}$. Thus the general abelian variety of Weil-type with $K = \mathbf{Q}(\sqrt{-3})$ and $det\, H = 1$ is isogeneous to such a Prym variety.

7.2 Schoen explictly constructs cycles on these Prym varieties with cycle classes that span the space of Weil-Hodge cycles. The construction is as follows.

$$\begin{array}{ccc} \cup \tilde{S}_i & \subset & S^4C_7 \\ \downarrow & & \downarrow \\ \cup S_i & \subset & T \\ \downarrow & & \tau \downarrow 3{:}1 \\ |K| & \subset & S^4C_3 \end{array}$$

Since C_7 is an unramified cyclic 3:1 covering of C_3, the map $\pi^{(4)} : S^4C_7 \to S^4C_3$ (with S^kC the k-fold symmetric product of the curve C) can be shown to factor over a fourfold T such that $\tau : T \to S^4C_3$ is an unramified cyclic 3:1 cover. In S^4C_3 lies a $\mathbf{P}^2 = |K|$, the linear system of effective, canonical divisors on C_3. Since $\mathbf{P}^2$ is simply connected, $\tau^{-1}(\mathbf{P}^2)$ must be reducible, say

$$\tau^*(\mathbf{P}^2) = S_1 + S_2 + S_3 \quad (\in Z^2(T)).$$

Thus $\pi^{(4)*}|K|$ has at least three irreducible components $\tilde{S}_i$ in S^4C_7. Using the composition

$$\phi : S^4C_7 \longrightarrow J(C_7) \approx_{isog} J(C_3) \times P \longrightarrow P,$$

one obtains cycles $\phi_*\tilde{S}_i$ in P. Schoen proves that linear combinations of the cycle classes of the $\phi_*\tilde{S}_i$ span the space of Weil-Hodge cycles for the general P, $[P] \in \mathcal{H}$. By specialization (over a one parameter family) one finds that the space of Weil-Hodge cycles on any P, $[P] \in \mathcal{H}$ is spanned by cycle classes.

7.3 Similarly, the general 6 dimensional abelian variety of Weil-type with $K = \mathbf{Q}(\sqrt{-3})$ and $det\, H = 1$ is obtained as the Prym of an unramified 3:1 cover $C_{10} \to C_4$ (cf. [F]). Schoen's results imply that the cycles obtained from $|K| \cong \mathbf{P}^3 \subset S^6C_4$ will again span the space of Weil-Hodge cycles, proving the Hodge (3,3) conjecture for such an abelian variety.

7.4 Schoen observed that one can also obtain abelian varieties of Weil-type with field $\mathbf{Q}(i)$ and $det\, H = 1$ via a Prym construction. Starting from a curve C_{h+1} of genus $h+1$ and a cyclic subgroup of order 4 of $J(C_{h+1})$, one has a tower of unramified 2:1 coverings:

$$C_{4h+1} \longrightarrow C_{2h+1} \longrightarrow C_{h+1}.$$

The Prym variety P of the 2:1 covering $C_{4h+1} \to C_{2h+1}$ is a (principally polarized) abelian variety of dimension $2h$, of Weil-type with field $\mathbf{Q}(i)$ (note that C_{4h+1} has an automorphism of order 4) and with $det\, H = 1$. Each P is in fact a member of the family constructed in Example 5.12 (see [vG]).

A variation on a proof in [S] shows that the general abelian fourfold of Weil-type with field $\mathbf{Q}(i)$ and $det\, H = 1$ is isogeneous to such a P (cf. [vG]) with $h = 2$. The cycle construction as before gives the proof of the Hodge conjecture for the general member of this family of abelian fourfolds.

7.5 We sketch the method used in [vG] to prove the Hodge conjecture for the general abelian varieties of Weil-type with $K = \mathbf{Q}(i)$ and $det\, H = 1$.

In Example 5.12 ('universal') families of such (principally polarized) abelian varieties were constructed. Using the easy description of any member X, one can actually obtain useful information on the multiplication maps $S^n H^0(X, L) \to H^0(X, L^{\otimes n})$ where L is an ample line bundle on the abelian variety X.

We will now restrict ourselves to the case $\dim X = 4$ and $L = \mathcal{O}(2\Theta)$, with Θ a symmetric divisor defining the principal polarization. The automorphism of order 4 of X acts as an automorphism of order 2 on $H^0(X, L)$, splitting it in a direct sum of a 10 dimensional even part H^0_+ and a 6 dimensional odd part H^0_-. Thus, by composing the natural map with the projection, we have a rational map

$$\Phi_L^- : X \longrightarrow \mathbf{P}^{15} = \mathbf{P}H^0(X, L) \longrightarrow \mathbf{P}^5 \cong \mathbf{P}H^0_-.$$

The (closure of) the image of X turns out to be a smooth quadric Q (for this one computes the kernel of the map $S^2 H^0(X, L) \to H^0(X, L^{\otimes 2})$ and shows that it contains a quadric which lies in the subspace $S^2 H^0_- \subset S^2 H^0(X, L)$).

Pulling back the rulings of Q to X along Φ_L^- produces cycles whose classes do not lie in $\wedge^2 B^1$. Using the action of $\mathbf{Q}(i)^*$ on $H^4(X, \mathbf{Q})$, one finds cycles which span B^2 for the general X in the family.

References

[DMOS] P. Deligne, *Hodge cycles on abelian varieties*, in: Hodge Cycles, Motives, and Shimura Varieties. LNM 900, Springer Verlag, pp. 9-100, (1982).

[F] C. Faber, *Prym varieties of triple cyclic covers*, Math. Z. **199**, 61-79 (1988).

[Fay] J. D. Fay, *Theta Functions on Riemann Surfaces*, LNM 352, Springer Verlag (1973).

[Fu] W. Fulton and J. Harris, *Representation Theory*, GTM 129, Springer-Verlag New York Inc. 1991.

[G] M. Green, lectures in this volume.

[vG] B. van Geemen, *Theta functions and cycles on some abelian fourfolds*, To appear in: Mathematische Zeitschrift.

[H] F. Hazama, *Algebraic cycles on certain abelian varieties and powers of special surfaces*, J. Fac. sci. Univ. Tokyo, Sect. IA, Math. **31**, 487-520 (1984).

[L] W. Landherr, *Äquivalenz Hermitescher Formen über einem beliebigen algebraischen Zahlkörper*, Abh. Math. Semin. Hamburg Univ. **11** 245-248 (1936).

[Ma] A. Mattuck, *Cycles on abelian varieties*, Proc. A.M.S. **9**, 88-98 (1958).

[MoZ] B. Moonen and Yu. Zarhin, *Hodge classes and Tate classes on simple abelian fourfolds*, Preprint Utrecht University (1993).

[Mum] D. Mumford, *Families of abelian varieties*, in: Algebraic Groups an Discontinuous Subgroups, Proc. Symp. Pure Math. **9**, A.M.S. Providence, R.I. 347-351 (1966).

[Mu] J. P. Murre, lectures in this volume.

[Mur] V. K. Murty, *Exceptional Hodge classes on certain abelian varities*, Math. Ann. **268**, 197-206 (1984).

[Po] H. Pohlman, *Algebraic cycles on abelian varieties of complex multiplication type*, Ann. of Math. **88**, 161-180 (1968).

[R] K. A. Ribet, *Hodge classes on certain types of abelian varieties*, Amer. J. Math. **105**, 523-538 (1983).

[S] C. Schoen, *Hodge classes on self-products of a variety with an automorphism*, Comp. Math. **65**, 3-32 (1988).

[Sh] G. Shimura, *On analytic families of polarized abelian varieties and automorphic functions*, Ann. Math. **78**, 149-192 (1963).

[Tat] J. Tate, *Algebraic cycles and poles of zeta functions*, in: Arithmetical Algebraic Geometry. Harper and Row, New York pp. 93-110 (1965).

[Tan] S. G. Tankeev, *On algebraic cycles on surfaces and abelian varieties*, Math. USSR Izv. **18**, 349-380 (1982).

[V] C. Voisin, lectures in this volume.

[W] A. Weil, *Abelian varieties and the Hodge ring*, in: Collected Papers, Vol. III, 421-429. Springer Verlag (1980).

Bert van Geemen
Department of Mathematics RUU
P.O.Box 80.010
3508TA Utrecht
The Netherlands

A REMARK ON HEIGHT PAIRINGS

Stefan Müller-Stach

Abstract: *We explain three definitions for the complex analytic height pairing of algebraic cycles on a complex algebraic manifold and give a refinement of these definitions in the case where one of the cycles has zero Abel-Jacobi invariant.*

§0. Introduction

These are extended notes of a lecture on the complex analytic part of the theory of height pairings. The notion of height pairing is of particular interest in Arakelov theory but is also useful for the study of algebraic cycles on a complex algebraic variety. Almost all the definitions and material presented here is due to several authors: we refer to [Arakelov74], [Beilinson87], [Bloch84], [Bloch89], [Bost90], [Gillet-Soulé87], [Hain90] for further study. Beilinson has formulated interesting conjectures about the height pairing for varieties defined over number fields which are related to the classical case of the Néron-Tate pairing. Our contribution consists only of the presentation of the material and some variations of it. Any incorrect statements should therefore be attributed to the author.

Let X be a compact Kähler manifold and $Z \in Z^p(X), W \in Z^q(X)$ two disjoint algebraic cycles of codimensions p and q such that $p + q = \dim X + 1$. Our goal is to associate to them a real number $\langle Z, W\rangle \in \mathbb{R}$ which is bilinear in its entries and satisfies some reasonable properties. The classical case is where Z and W are both the difference of two points on the projective line. Then if τ denotes the cross ratio of those 4 points, one defines $\langle Z, W\rangle = \log|\tau|$.

In §1 we present three definitions of $\langle Z, W\rangle$ that occur in the literature. Since it is not obvious from there that they coincide, we give a short indication of the proof of this in §2. Finally in §3 we refine this invariant in the case where the Abel-Jacobi invariant of - say - W is zero and obtain a $\mathbb{C}^*$-valued invariant. We also give some examples at the end of §3, including zero sections of vector bundles as one of the cycles.

It is a pleasure to thank Fabio Bardelli and Alberto Albano for organizing this fabulous C.I.M.E. workshop as well as Fondazione CIME and DFG for financial support. I would also like to thank several people for discussions about this topic, including H.Esnault, H. Gillet, M. Green, Ch. Soulé, C. Voisin and the participants of a seminar at Essen.

M. Green et al.: LNM 1594, A. Albano and F. Bardelli (Eds.), pp. 253–259, 1994.

§1. Definitions

Let (X,ω) be a compact Kähler manifold of dimension n and Z, W two disjoint cycles of codimensions p and q adding up to $p+q=n+1$. As usual we denote by $Z^k(X)$ the free group generated by integral cycles of codimension k.

(a) A current type definition

Let $\mathcal{D}_X^{r,s}$ be the sheaf of (r,s) currents on X. Via the choice of an orientation on X this is just the sheaf of (r,s)-forms with distribution coefficients. If $W \in Z^q(X)$ is a reduced and irreducible cycle of codimension q on X, then we denote by δ_W the current of integration over $W \setminus W_{\mathrm{sing}}$, defined by

$$\delta_W(\varphi) = \int_{W\setminus W_{\mathrm{sing}}} \varphi \,, \; \varphi \in \mathcal{A}_c^{n-q,n-q}(X).$$

as a functional on compactly supported smooth $(n-q,n-q)-$ forms. Then δ_W is a closed, positive current in $\mathcal{D}_X^{q,q}(X)$. This definition is extended to $Z^q(X)$ by linearity. Now consider the following real differential operators: $d = \partial + \overline{\partial}$ and $d^c = \frac{i}{2\pi}(\overline{\partial} - \partial)$. Then $dd^c = \frac{i}{\pi}\partial\overline{\partial}$. By the well known $\partial\overline{\partial}$-lemma on X, we can write

$$\delta_W = dd^c\eta_W + h_W$$

where h_W is the unique smooth harmonic form representing the cohomology class of W and $\eta_W \in \mathcal{D}_X^{q-1,q-1}$ a real current, which can be chosen smooth outside $|W|$. After a resolution of singularities

$$f : (Y, D) \longrightarrow (X, |W|)$$

where $D = f^{-1}(|W|)$ has simple normal crossings we may even assume that

$$\eta_W|_{X\setminus|W|} = f_*\varphi|_{Y\setminus D}$$

where φ is a smooth $(q-1,q-1)$-form on $Y \setminus D$ with logarithmic singularities along D. Hence locally on Y:

$$\varphi = \frac{1}{2}\sum_{i=1}^{k} \alpha_i \log|z_i|^2 + \beta$$

where α_i, β are smooth forms, the α_i are closed and $z_1 z_2 \ldots z_k = 0$ is the local equation for D. Note that η_W is unique up to expression of the form $\partial n + \overline{\partial} v$+*some harmonic form.* Hence if one assumes additionally that the projection of η_W onto the harmonic $(q-1,q-1)$ forms is zero, the following definition is unique:

Definition 1. *If $Z \in Z^p(X)$ is disjoint from W, let*

$$\langle Z, W\rangle = \int_Z \eta_W \in \mathbb{R}$$

(This is independent of all choices including the Kähler metric once Z and W are homologous to zero in X).

(b) A definition using Deligne-Beilinson cohomology

Here assume that X is projective and W is homologous to zero. For any smooth quasiprojective variety V and any subring $A \subset \mathbb{R}$ one has Deligne-Beilinson-groups

$$H^j_{\mathcal{D}}(V, A(k))$$

where $A(k) = (2\pi i)^k \cdot A \subset \mathbb{C}$. For a definition of these groups see Murre's lectures or [Esnault-Viehweg88]. They sit in long exact sequences

$$(*) \ldots \to H^{j-1}(V,\mathbb{C}) \to H^j_{\mathcal{D}}(V,A(k)) \to H^j(V,A(k)) \oplus F^k H^j(V,\mathbb{C}) \to \ldots$$

and satisfy various functorial properties. There is also the notion of Deligne-Beilinson cohomology with supports and a long exact sequence (assuming from now on that $V = X \setminus |W|$)

$$\ldots \to H^{2q-1}_{\mathcal{D}}(X \setminus |W|, A(q)) \to H^{2q}_{\mathcal{D},|W|}(X, A(q)) \to H^{2q}_{\mathcal{D}}(X, A(q)) \to \ldots$$

Now by $(*)$, taking $A = \mathbb{R}$,

$$H^{2q}_{\mathcal{D}}(X, \mathbb{R}(q)) \cong H^{2q}(X, \mathbb{R}(q)) \cap F^q H^{2q}(X, \mathbb{C})$$

and the cycle class γ_W of W in $H^{2q}_{\mathcal{D},|W|}(X, \mathbb{R}(q))$ maps to zero in $H^{2q}_{\mathcal{D}}(X, \mathbb{R}(q))$ since W is homologous to zero. Therefore it can be lifted non-canonically to a class $\hat{\gamma}_W \in H^{2q-1}_{\mathcal{D}}(X \setminus |W|, \mathbb{R}(q))$. The natural cup product structure on Deligne-Beilinson cohomology ([Esnault-Viehweg88]) defines a map

$$H^{2q-1}_{\mathcal{D}}(X \setminus |W|, \mathbb{R}(q)) \times H^{2p}_{\mathcal{D},|Z|}(X, \mathbb{R}(p)) \xrightarrow{\cup} H^{2n+1}_{\mathcal{D},|Z|}(X \setminus |W|, \mathbb{R}(n+1))$$

The latter group - since $Z \cap W$is empty - has a forgetful map to

$$H^{2n+1}_{\mathcal{D}}(X, \mathbb{R}(n+1))$$

which by $(*)$ is canonically isomorphic to $\mathbb{R}$. Now the fundamental class γ_Z of Z lies in $H^{2p}_{\mathcal{D},|Z|}(X, \mathbb{R}(p))$ and we can therefore take cup products:

Definition 2. *Assume X is projective and Z, W are homologous to zero. Then*

$$\langle Z, W \rangle := \hat{\gamma}_W \cup \gamma_Z \in \mathbb{R}$$

(This is independent of the lifting $\hat{\gamma}_W$ chosen).

(c) A definition via biextensions of mixed Hodge structures

For a general relation between cycles and biextensions we refer to [Bloch-Murre], [Bloch89] and [Hain90].
Let X be a projective manifold and Z, W be two disjoint homologically trivial algebraic cycles. For simplicity assume that $Z = Z_1 - Z_2, W = W_1 - W_2$ with Z_i, W_j irreducible and reduced. Let

$$H_0 = H^{2q-1}(X; \mathbb{Z}(q))/\text{torsion}$$

which is a pure Hodge structure of weight = -1. One has a commutative diagram (note that $q = \mathrm{codim} W = \dim Z + 1$)

$$\begin{array}{ccccc} H_0 & \longrightarrow & H^{2q-1}(X \setminus |W|; \mathbb{Z}(q))_0 & \longrightarrow & \mathbb{Z} \\ \uparrow & & \uparrow & & \| \\ H^{2q-1}(X, |Z|; \mathbb{Z}(q))_0 & \longrightarrow & B(Z,W) & \longrightarrow & \mathbb{Z} \\ \uparrow & & \uparrow & & \\ \mathbb{Z}(1) & = & \mathbb{Z}(1) & & \end{array}$$

where

$$\mathbb{Z} := \mathrm{Ker}(H^{2q}_{|W|}(X, \mathbb{Z}(q)) \to H^{2q}(X, \mathbb{Z}(q))),$$

$$\mathbb{Z}(1) := \mathrm{Coker}(H^{2q-2}(X, \mathbb{Z}(q)) \to H^{2q-2}(|Z|, \mathbb{Z}(q)))$$

and

$$B(Z,W) := H^{2q-1}(X \setminus |W|, (Z); \mathbb{Z}(q))_0.$$

$B(Z,W) \otimes \mathbb{C}$ carries a mixed Hodge structure, which by this diagram is a biextension of H_0 by $\mathbb{Z}$ and $\mathbb{Z}(1)$, and hence has natural weights -2, -1 and 0. Denote by $\mathcal{B}_{\mathbb{Z}}$ the set of all such biextensions modulo those isomorphisms of mixed Hodge structure that induce the identity on the graded quotients $\mathbb{Z}$ and $\mathbb{Z}(1)$. Let $0 \in \mathcal{B}_{\mathbb{Z}}$ be the trivial Hodge structure with lattice $V_{\mathbb{Z}} = \mathbb{Z} \oplus H_{\mathbb{Z}} \oplus \mathbb{Z}(1)$. By [Carlson85] we have an isomorphism

$$G_{\mathbb{Z}} \setminus G/F^0G \xrightarrow{\cong} \mathcal{B}_{\mathbb{Z}}, \quad \text{where } G = \begin{pmatrix} 1 & H_{\mathbb{C}} & \mathbb{C} \\ 0 & 1 & H^*_{\mathbb{C}} \\ 0 & 0 & 1 \end{pmatrix}$$

and the isomorphism is given by $g \mapsto g(V_{\mathbb{Z}})$. There is a forgetful map

$$\nu : \mathcal{B}_{\mathbb{Z}} \to \mathcal{B}_{\mathbb{R}} = G_{\mathbb{R}} \setminus G/F^0G \cong \mathbb{C}/\mathbb{R}(1) \cong \mathbb{R}$$

where $\mathcal{B}_{\mathbb{R}}$ classifies only the real biextension forgetting the $\mathbb{Z}$-lattice structure.

Definition 3. $\langle Z, W \rangle = \nu([B(Z,W)]) \in \mathbb{R}$

§2. The equivalence of the definitions

The equivalence of definition (1) and (3) is shown in [Hain90]. We give a slightly shorter version of his argument using the theory of logarithmic currents (residue forms; see also Green's lectures and [Bost-Gillet-Soulé93]). We also give a proof of the equivalence of (1) and (2) thus completing the circle of ideas. The most exciting correspondence then of course is the one between (2) and (3), revealing a connection between mixed Hodge structures and Deligne cohomology.

Throughout this chapter we assume that X is projective and Z, W are homologous to zero in X with $Z = Z_1 - Z_2$, $W = W_1 - W_2$ if necessary. Let η_W be a current with $dd^c\eta_W = \delta_W$. We choose η_W in such a way that it has only logarithmic poles along $|W|$ after a resolution of singularities i.e. locally

$$\eta_W = \frac{1}{2}\sum_{i=1}^{k} \alpha_i \log|z_i|^2 + \beta$$

and $f^{-1}(|W|) = \{z_1 \dots z_k = 0\}$. η_W defines a cohomology class in $H^{2q-2}(X \setminus |W|, \mathbb{C})$ which maps by $(*)$ to a class $\tilde{\eta}_W \in H^{2q-1}_{\mathcal{D}}(X \setminus |W|, \mathbb{R}(q))$. We have

Lemma 1. $\tilde{\eta}_W = \hat{\gamma}_W$, *i.e.* $\tilde{\eta}_W$ *is a natural candidate to lift the fundamental class of* W. *In particular definition (1) and (2) give the same result.*

Proof. One only has to check, that the composition of maps

$$H^{2q-2}(X \setminus |W|, \mathbb{C}) \to H^{2q-1}_{\mathcal{D}}(X \setminus |W|, \mathbb{R}(q)) \to H^{2q}_{\mathcal{D},|W|}(X, \mathbb{R}(q))$$

maps η_W to $dd^c\eta_W = \delta_W$. But this is obvious from the construction of these maps. □

Lemma 2. *With the same assumptions:*

$$\nu([B(Z,W)]) = \int_Z \eta_W$$

and therefore definitions (1) and (3) coincide.

Proof. Let s_F be the section of the map $B(Z,W) \otimes \mathbb{C} \to \mathbb{Z} \otimes \mathbb{C} \cong \mathbb{C}$ which is given by $s_F(1) := (2\pi i)^q(2\partial\eta_W)$. In the local description

$$s_F(1) = (2\pi i)^q(\sum_{i=1}^{k} \alpha_i \frac{dz_i}{z_i} + \partial\beta)$$

since the α_i are closed. s_F preserves the Hodge filtration and is a well defined element of $B(Z,W) \otimes \mathbb{C}$ by type reasons ($q > \dim Z$). Let $r_{\mathbb{Z}}$ be the integral retraction of $\mathbb{Z}(1) \hookrightarrow B(Z,W)$ which is given by

$$\alpha \mapsto (2\pi i)^{-q} \int_{\Gamma_Z} \alpha$$

with Γ_Z an integral chain, $\partial\Gamma_Z = Z_1 - Z_2$. Now the crucial computation is in [Hain90]: Given s_F and $r_{\mathbb{Z}}$ one computes in this special case

$$\nu([B(Z,W)]) = Re(r_{\mathbb{Z}} \circ s_F(1))$$

Hence we obtain:

$$\nu([B(Z,W)]) = Re \int_{\Gamma_Z} 2\partial\eta_W = \int_{\Gamma_Z} d\eta_W = \int_Z \eta_W$$

by Stokes' theorem. □

§3. A refinement and some examples

For the beginning of this chapter let us stay with the notations of §1(b). Assume that W is not only homologous to zero, but also that its Deligne class is zero in $H^{2q}_{\mathcal{D}}(X, \mathbb{Z}(q))$. We will show how to obtain a $\mathbb{C}^*-$ valued height pairing $\langle Z, W\rangle$ in this situation.

Let again $\hat{\gamma}_W \in H_{\mathcal{D}}^{2q-1}(X \setminus |W|, \mathbb{Z}(q))$ be a lifting. By assumption the lifting can be obtained by using $\mathbb{Z}$−coefficients, since γ_W maps to zero under

$$H^{2q}_{\mathcal{D},|W|}(X, \mathbb{Z}(q)) \to H^{2q}_{\mathcal{D}}(X, \mathbb{Z}(q))$$

Cup product with γ_Z lands in $H_{\mathcal{D}}^{2n+1}(X, \mathbb{Z}(n+1))$ which again by $(*)$ is isomorphic to

$$\frac{H^{2n}(X, \mathbb{C})}{F^{n+1} \oplus H^{2n}(X, \mathbb{Z}(n+1))} \cong \mathbb{C}/\mathbb{Z}(1) \cong \mathbb{C}^*$$

This is the desired $\mathbb{C}^*$−valued pairing. Using the current type definition, Mark Green has refined the height pairing also in his lectures. Of course by §2 all these definitions coincide.

Example 1. If s is a rational section in a hermitian line bundle L on X, then $W = \mathrm{div}(s)$ has the following current-type description:

$$\delta_W = dd^c \log||s|| + c_1(L)$$

Hence if $Z = \sum_i n_i P_i$ is a set of points on X:

$$\langle Z, W\rangle = \sum_i n_i \log||s(P_i)||.$$

In particular on $\mathbb{P}^1$ this is the classical pairing described in the introduction. We also see that $\langle Z, W\rangle$ does not respect rational equivalence at all!

Example 2. If one of the cycles - say - W is the zero section of a certain vector bundle E on X, there exist formulas for the current η_W in [Griffiths69] or [Bost-Gillet-Soulè93]. Thus we get $\langle Z, W\rangle$ by definition (1).

REFERENCES

[Arakelov 74] S.Arakelov: Intersection theory of divisors on an arithmetic surface; Izv. Akad. Nauk SSSR **38**, 1167-1180 (1974)

[Beilinson 87] A. Beilinson: Height pairings between algebraic cycles; Cont. Math. **67**, 1-24 (1987)

[Bloch 84] S. Bloch: Height pairings for algebraic cycles; J. Pure Appl. Algebra **34**, 119-145 (1984)

[Bloch 89] S. Bloch: Cycles and biextensions; Cont. Mathematics Vol. **83**, 19-30 (1989)

[Bloch,Murre] S. Bloch, J. Murre: unpublished notes;

[Bost 90] B. Bost: Green's Currents and height paring on complex tori; Duke Math. Journal **61**, 899-912 (1990)

[Bost-Gillet-Soulé 93] B. Bost, H. Gillet, C. Soulé: Heights of projective varieties and positive Green forms; preprint Publ. IHES (1993)

[Carlson 85] J. Carlson: The geometry of the extension class of a mixed Hodge structure; Proc. Bowdoin conference Vol. 2 of the AMS, 199-222 (1987)

[Esnault-Viehweg 88]: H. Esnault, E. Viehweg: Deligne-Beilinson Cohomology; Perspectives in mathematics Vol.4, Acad. Press, 43-91 (1988)

[Gillet-Soulé 84] H. Gillet, Ch. Soulé: Intersection sur les variétés de Arakelov; CRAS sér I. Math. **299**, 563-566 (1984)

[Griffiths 69] Ph. Griffiths: Algebraic cycles on algebraic manifolds; Amer. Journal of Math., 93-191 (1969)

[Hain 90] R. Hain: Biextensions and heights associated to curves of odd genus; Duke Math. Journal **61**, 859-898 (1990)

STEFAN MÜLLER-STACH, UNIVERSITÄT ESSEN
E-MAIL: MAT930 at VM.HRZ.UNI-ESSEN.DE

C.I.M.E. Session on "Algebraic Cycles and Hodge Theory"

List of Participants

D. AFFANY, UER de Mathématiques, Univ. de Nancy I, 54506 Vandoeuvre-lès-Nancy Cedex

A. ALZATI, Dip. di Matematica, Università, Via C. Saldini 50, 20133 Milano

E. AMERIK, Dept. of Math., Leiden Univ., P.O. Box 9512, 2300 RA Leiden

J. AMOROS, Univ. de Barcelona, Dep. d'Alg. i Geom., Gran Via 585, 08007 Barcelona

E. BALLICO, Dip. di Matematica, Univ. di Trento, 38050 Povo, Trento

L. BARBIERI-VIALE, Dip. di Matematica, Università, Via L.B. Alberti 4, 16132 Genova

B. BENDIFFALAH, CIRM-Luminy, Case 916, 13288 Marseille Cedex 9

M. BERTOLINI, Dip. di Matematica, Università, Via C. Saldini 50, 20133 Milano

J. BISWAS, School of Mathematics, Tata Inst. of Fundamental Research,
Homi Bhabha Road, Bombay 400 005

J.-P. BRASSELET, CIRM-Luminy, Case 916, 13288 Marseille Cedex 9

S. BRIVIO, Dip. di Matematica, Università, Via Carlo Alberto 10, 10123 Torino

A. COLLINO, Corso Re Umberto, 66, 12039 Verzuolo, Como

E. COLOMBO, Dip. di Matematica, Università, Via C. Saldini 50, 20133 Milano

A. CONTE, Dip. di Matematica, Università, Via Carlo Alberto 10, 10123 Torino

P. CRAGNOLINI, Dip. di Matematica, Università, Via F. Buonarroti 2, 56126 Pisa

A. D'AGNOLO, Dip. di Metodi e Modelli matematici, Via A. Scarpa 10, 00161 Roma

S. del BANO, Dept. de Mathem., Univ. Autonoma de Barcelona,
Bellaterra, Barcelona, E-08193

J. ESSER, Univ. GHS Essen, FB 6 Mathematik, Universitatsstr. 2, D-4300 Essen

B. FANTECHI, Dip. di Matematica, Univ. di Trento, 38050 Povo, Trento

A.M. FINO, Dip. di Matematica, Università, Via Carlo Alberto 10, 10123 Torino

D. FRANCO, SISSA, Via Beirut 2-4, 34014 Trieste

M. GOBBINO, Scuola Normale Superiore, Piazza dei Cavalieri 7, 56126 Pisa

M. KADDAR, UER de Mathématiques, Univ. de Nancy I, 54506 Vandoeuvre-lès-Nancy Cedex

M. KWIECINSKI, CIRM-Luminy, Case 016, 13288 Marseille Cedex 9

J. LI, MSRI, Berkeley, 1000 Centennial dr., Berkeley, CA 94720

A. LOPEZ, Dip. di Matematica, Università, Strada Nuova 65, 27100 Pavia

L. MIGLIORINI, Dip. di Matematica appl., Università, Via S. Marta 3, 50139 Firenze

A. MORO, Dip. di Matematica, Università, Via Carlo Alberto 10, 10123 Torino

S. MULLER-STACH, Fachbereich 6, Univ. GH Essen, Universitatsstr. 3, D-4300 Essen

J. NAGEL, Dep. of Math., Leiden Univ., P.O.Box 9512, 2300 RA Leiden

P.A. OLIVERIO, Dip. di Matematica, Università della Calabria,
87036 Arcavacata di Rende, Cosenza

R. PARDINI, Dip. di Matematica, Università, Via F. Buonarroti 2, 56127 Pisa

G. PARESCHI, Dip. di Matematica, Università, Via Machiavelli 35, 44100 Ferrara

C. PAULY, Univ. de Paris-Sud, Mathématiques, Bât. 435, 91-405 Orsay Cedex

L. PICCO BOTTA, Dip. di Matematica, Università, Via Carlo Alberto 10, 10123 Torino

G.P. PIROLA, Dip. di Matematica, Università, Strada Nuova 65, 27100 Pavia

D. PORTELLI, Dip. di Scienze Matematiche, Università, P.le Europa 1, 34127 Trieste

N. RAMACHANDRAN, Dept. of Math., Brown Univ., Box 1917, Providence, RI 02912

L. RAMERO, Dept. of Math., MIT, Room 2-230, Cambridge, MA 02139

E. ROGORA, Dip. di Matematica, Università, P.le Aldo Moro 7, 00185 Roma

M. ROSSI, Dip. di Matematica, Università, Via Carlo Alberto 10, 10123 Torino

G.K. SANKARAN, Dept. of Pure Math., Univ. of Cambridge, 16 Mill Lane,
Cambridge CB2 1SB

A. TORTORA, U.C.L.A., Dept. of Math., Los Angeles, CA 90024-1555

B. VAN GEEMEN, Dept. of Math., Univ. of Utrecht, Budapestlaan 6, Utrecht

M. VIOLO, Dip. di Matematica, Università, Via Carlo Alberto 10, 10123 Torino

G. WELTERS, Universidad de Barcelona, Gran Via 585, 08007 Barcelona

C. ZAAL, Plantage Muidergracht24, 1018 TV Amsterdam

P. ZAPPA, Dip. di Matematica, Università, Via Vanvitelli 1, 06100 Perugia

J. ZHOU, CIRM-Luminy, Case 916, 13288 Marseille Cedex 9

F. ZUCCONI, Dip. di Matematica, Università Via F. Buonarroti 2, 56126 Pisa

FONDAZIONE C.I.M.E.
CENTRO INTERNAZIONALE MATEMATICO ESTIVO
INTERNATIONAL MATHEMATICAL SUMMER CENTER

"Recent Mathematical Methods in Nonlinear Wave Propagation"

is the subject of the First 1994 C.I.M.E. Session.

The Session, sponsored by the Consiglio Nazionale delle Ricerche and the Ministero dell'Università e della Ricerca Scientifica e Tecnologica, will take place under the scientific direction of Prof. TOMMASO RUGGERI (Università di Bologna) at Villa "La Querceta", Montecatini Terme (Pistoia), **from May 23 to May 31, 1994.**

Courses

a) **Non Linear Fields and Waves. (6 lectures in English)**
Prof. Guy BOILLAT (Univ. Clermont Ferrand)

Contents

Non linear conservative systems - Involutive constraints - Symmetrization - Waves - Ordinary and characteristic shocks - Applications to mechanics and theoretical physics.

References

- S.K. Godunov, An interesting class of quasi-linear systems. Sov. math. 2, 947 (1961).
- K.O. Friedrichs and P.D. Lax, Systems of conservation equations with a convex extension. Proc. Nat. Acad. Sci. USA 68, 1686 (1971).
- A.M. Anile, Relativistic fluids and magneto-fluids. Cambridge Monographs on Mathematical Physics, Cambridge University Press (1989).
- I. Müller and T. Ruggeri, Extended thermodynamics. Springer Tracts on Natural Philosophy, 37, Springer Verlag, New York (1993).
- G. Boillat, Sur l'existence et la recherche d'équations de conservation supplémentaires pour les systèmes hyperboliques. C. R. Acad. Sci. Paris 278-A 909 (1974); Symétrisation des systèmes d'équations aux derivées partielles avec densité d'énergie convexe et contraintes. Ibid 295-I, 551 (1982); Involutions des systèmes conservatifs, Ibid 307-I, 891 (1988); Expression explicite des chocs caractéristiques de croisement, Ibid, 312-I, 653 (1991).

b) **The Theory of Hyperbolic Conservation Laws: Current Trends and Open Problems. (6 lectures in English)**
Prof. Constantin M. DAFERMOS (Brown University)

Abstract

The lectures will survey emerging trends in the study of hyperbolic systems of conservation laws. They will emphasize the deep influence of ideas from Physics in the development of the analysis, addressing in particular the following issues: functional analytic techniques and the quenching of oscillations; interplay of ideas from statistical and continuum Physics; relaxation phenomena; wave interactions and the geometric structure of BV solutions.

The course will also describe the challenge posed by hyperbolic systems of conservation laws in several space dimensions.

Reference

A basic general reference for background information is the book "Shock Waves and Reaction Diffusion Equation" by J. Smoller, Springer Verlag.

c) **Dispersive Systems.** (5 lectures in English)
Prof. Peter D. LAX (Courant Institute, NYU)

Outline

The first lecture will give an overview of the subject describing a number of examples such as KdV, non linear Schroedinger equation and Toda.

The second lecture will deal with Hamiltonian systems and the remaining lectures will be devoted to the zero dispersion limit and the oscillatory structures that emerge.

References

- P.D. Lax, C.D. Levermore, S. Venakides, The Generation and Propagation of Oscillations in Dispersive IVPs and their Limiting Behavior, Important Developments in Soliton Theory 1980-1990, T. Fokas and V.E. Zakharov eds., Springer Verlag, Berlin (1992).
- Peter D. Lax, Almost Periodic Solutions of the KdV Equation, SIAM Review, vol. 18, No. 3, July 1976, 351-375.

d) **Nonlinear Waves for Quasilinear Hyperbolic-Parabolic Partial Differential Equations.** (6 lectures in English)
Prof. Tai-Ping LIU (Stanford University)

Abstract

Many physical models in gas dynamics, fluid and mechanics are of the form of system of quasilinear hyperbolic-parabolic partial differential equations. For instance, the compressible Navier-Stokes, MHD, visco-elasticity, shallow water model, kinetic models, combustions models are among them.

We plan to describe recent progresses on the qualitative theory for these systems. We will start with the Burgers equation and then other physical models in gas dynamics, MHD and elasticity, multi-phase flows and non-equilibrium phenomena.

Analytical techniques, energy-characteristic method, pointwise estimates, approximate Green functions, time-asymptotic expansions, and nonlinear superpositions and interactions of waves will be explained for the physical models.

These techniques are introduced in the attempts to overcome analytical difficulties arised from the coupling of nonlinear hyperbolicity, which is coupled with the effects of dissipations, relaxations, reactions and damping.

References

1. Chen, G.Q., Levermore, D. and Liu, T.-P., Stiff relaxation and weakly nonlinear limits for hyperbolic conservation laws, Comm. Pure Appl. Math. (to appear).
2. Freistuhler, H. and Liu T.-P., Nonlinear stability of overcompressive shock waves in a rotationally invariant system of viscous conservation laws, Comm. Math. Phys., 153 (1993), 147-158
3. Hsiao, L. and Liu, T.-P., Convergence to nonlinear diffusion waves for solutions of a system of hyperbolic conservation laws with damping, Comm. Math. Phys., 143 (1992), 599-605.
4. Goodman, J., Stability of viscous shock fronts in several dimensions, Trans. Amer. Math. Soc., 311 (1989), 683-695.
5. Kawashima, S., Large-time behavior of solutions to hyperbolic-parabolic systems of conservation laws and applications, Proc. Roy. Soc. Edinburgh, Sect. A 105 (1987), 169-194.
6. Liu, T.-P., Nonlinear stability of shock waves for viscous conservation laws, Memoirs, Amer. Math. Soc., No. 328, 1985.
7. Liu, T.-P., and Zeng, Y., Large-time behavior of solutions of general systems of hyperbolic-parabolic equations in conservation form, (to appear).
8. Liu, T.-P., and Zumbrun, K., Nonlinear stability of general undercompressive shock waves for viscous conservation laws, (to appear).
9. Whitham, J., Linear and Nonlinear Waves, Wiley, 1974.

FONDAZIONE C.I.M.E.
CENTRO INTERNAZIONALE MATEMATICO ESTIVO
INTERNATIONAL MATHEMATICAL SUMMER CENTER

"Dynamical Systems"

is the subject of the Second 1994 C.I.M.E. Session.

The Session, sponsored by the Consiglio Nazionale delle Ricerche and the Ministero dell'Università e della Ricerca Scientifica e Tecnologica, will take place under the scientific direction of Prof. RUSSELL JOHNSON (Università di Firenze) at Villa "La Querceta", Montecatini Terme (Pistoia), from 13 to 22 June, 1994.

Courses

a) **Random Dynamical Systems. (6 lectures in English)**
Prof. Ludwig ARNOLD (Universität Bremen)

Contents

1) Metric, topological and smooth dynamics
2) Random dynamical systems: concepts, invariant measures
3) Generation of random dynamical systems by random and stochastic differential equations
4) The multiplicative ergodic theorem
5) Invariant manifolds, stochastic stability
6) Normal forms. The Hartman-Grobman theorem
7) Stochastic bifurcation theory
8) Dynamics and measurability

References

1) L. Arnold: Random dynamical systems. In preparation (preliminary version November 1992).
2) L. Arnold, H. Crauel, J.-P. Eckmann (eds.): Lyapunov exponents. Lecture Notes in Mathematics Vol. 1486 Springer, Berlin 1991.
3) Y. Kifer: Ergodic theory of random transformations. Birkhäuser, Boston 1986.
4) H. Kunita: Stochastic flows and stochastic differential equations. Cambridge University Press, Cambridge 1990.
5) R. Mane: Ergodic theory and differentiable dynamics. Springer, Berlin 1987.

b) **Geometric Singular Perturbation Theory. (6 lectures in English)**
Prof. Christopher K. R. T. JONES (Brown University, Universität Stuttgart)

Abstract

Complex systems of dynamical systems in dimension greater than 2 can support rich dynamical behavior but, in general, defy any complete analysis. If the systems incorporate differing time scales then a reduction can often exist to lower dimensional systems through singular perturbation theory. While making the systems thus susceptible to analysis, this procedure leaves much of the richness of the systems intact. This is possible through the matching of motion in separate phases of fast and slow time.

The basic theorems of geometric singular perturbation theory are due to Fenichel. These assert the existence of invariant manifolds and attendant stable and unstable manifolds, as well as certain important foliations. The basic theory will be presented, including motivation and proofs. Furthermore, a basic normal form will be derived for singularly perturbed problems. After the establishment of these basic results, the focus will be on the problem of constructing homoclinic, heteroclinic and periodic orbits that involve both fast and slow time scales. The recently developed technique known as the Exchange Lemma, due in its original form to Jones and Kopell, for tracking invariant manifolds in singularly perturbed systems will be explained in detail. Applications will be given in a variety of areas such as nonlinear optics, nerve impulse propagation and atmospheric sciences.

Reference

1) N. Fenichel, Persistence and smoothness of invariant manifolds for flows, Indiana U. Math. J., 21 (1971), 193 226.
2) N. Fenichel, Geometric singular perturbation theory for ordinary differential equations, JDE 31 (1979), 53 98.
3) G. Haller & S. Wiggins, Orbits homoclinic to resonances: the Hamiltonian case, Physica D 66 (1993), 298-346.
4) C. Jones & N. Kopell, Tracking invariant manifolds with differential forms in singularly perturbed systems, to appear JODE, winter 1993.
5) C. Jones, N. Kopell & R. Langer, Construction of the FirzHugh-Nagumo pulse using differential forms, in Patterns and Dynamics in Reactive Media, IMA volumes in Mathematics and its Applications, 37, H. Swinney, G. Aris and D. Aronson eds., Springer Verlag, New York, 1991, 101-116.
6) G. Kovacic & S. Wiggins, Orbits homoclinic to resonances with an application to chaos in the damped and forced sine-Gordon equation, Physica D 57 (1992) 185.
7) K. Sakamoto, Invariant manifolds in singular perturbation problems for ordinary differential equations. Proc. Roy. Soc. Ed. 116A (1990), 45-78.
8) P. Szmolyan, Heteroclinic and homoclinic orbits in singular perturbation problems, JDE 92 (1991), 255-281.
9) S. Wiggins, Global bifurcation and chaos, Springer-Verlag, New York, 1988.

c) **The Conley Index Theory**. (6 lectures in English)
Prof. Konstantin MISCHAIKOW (Georgia Inst. of Tech. Atlanta)

Abstract

The goal of this series of lectures is an up to date description of the Conley Index Theory and will be divided into three general topics.

1. Isolated invariant sets and the Conley index.

The Conley index is an index of isolated invariant sets which on the one hand generalizes the Morse index of gradient flows and on the other preserves the homotopy properties of degree theory. We shall define and discuss the index in the case of both continuous and discrete dynamical systems with considerable emphasis on its continuation properties. As in the case of Morse theory, the Conley index allows one to draw conclusions about global dynamics from local information. However, to do so requires an understanding of how to decompose isolated invariant sets, which will lead us to the topics of chain recurrent sets, Morse decompositions, and Lyapounov functions. The process of going from local to global can be done algebraically and is codified in the connection matrix which will be described. Finally, in order to measure changes in the dynamics along homotopies we shall discuss transition matrices.

2. Extracting Dynamics from the Conley index.

The most obvious question is what does the Conley index of an isolated invariant set say about the dynamics of the invariant set. The answer comes in two forms; existence results and semiconjugacies. Examples of the first, include the existence of heteroclinic orbits, fixed points, and periodic orbits. In the case of semi-conjugacies the goal is to describe a model dynamical system (i.e., one for which the dynamics is completely understood) and then use that index information to conclude that the isolated invariant set can be mapped continuously and surjectively onto the model system in such a way that the dynamics commute. Several recent theorems to this effect will be presented.

3. Methods for computing the Conley index.

Applying the results which are described in 2. presupposes knowledge of the index, thus it is important to be able to compute this index for specific systems. We shall briefly discuss two very recent developments along this line.
The first involves numerical approximations to rigorously compute the index.
The second is an attempt to develop a topological singular perturbation theory.

References

1) C.C. Conley, Isolated invariant sets and the Morse index, CBMS Lecture Notes 38, A.M.S., Providence R.I. 1978.
2) C.C. Conley, A qualitative singular perturbation theory, in "Global Theory of Dynamical Systems", eds. Z. Nitecki and C. Robinson, Lecture Notes in Mathematics 829, Springer-Verlag, 1980.
3) R. Franyosa, The connection matrix theory for Morse decompositions, Trans. A.M.S. 311 (1989), 781-803.
4) R. Franyosa and K. Mischaikow, The connection matrix theory for semiflows on (not necessarily locally compact) metric space, J.D.E. 71 (1988), 270-287
5) C.Mc Cord and K. Mischaikow, Connected simple systems, transition matrices, and heteroclinic bifurcations, Trans. A.M.S. 333 (1992), 397-422
6) M. Mrozek, Leray functor and cohomological index for discrete dynamical systems, Trans. A.M.S. 318 (1990),149-178.
7) R. Reineck, Connectivity orbits in one-parameter families of flows, Erg. Thy. Dyn. Sys. 8 (1988), 359-374.
8) K.P. Rybakowski, The homotopy index and partial differential equations, Universitext, Springer-Verlag, 1987.
9) D. Salamon, Connected simple systems and the Conley index of isolated invariant sets, Trans. A.M.S. 291 (1985), 1-41.

d) **Dynamics of partial differential equations on thin domains.** (6 lectures in English)
Prof. Geneviève RAUGEL (Université de Paris-Sud)

Abstract

In many applications, we encounter partial differential equations (PDE) defined on domains for which the size of the domain in some directions is much smaller than the size in others. Such problems arise in engineering, mechanics, fluid dynamics...

Let us suppose that we are given a partial differential equation with homogeneous Neumann boundary conditions, on the thin (n+1)-dimensional product domain W x $(0, \varepsilon)$, where W is a bounded n-dimensional domain. In such a situation it is natural to ignore the thin direction and to consider the PDE on the lower dimensional domain W. This type of reduction in the dimension of the domain is still valid for more general thin domains and often leads to a simpler problem for which we can give a more complete description of the dynamics. One purpose of this course is the comparison of the dynamics of the full equation $(P)_\varepsilon$ on the thin domain with the dynamics of the reduced equation $(P)_0$ on the lower dimensional domain. In particular, in the case of dissipative systems, we compare the attractors of $(P)_\varepsilon$ and $(P)_0$. For conservative systems on thin domains, one compares their solutions on long time intervals, the length of which goes to infinity with ε^{-1}. Topics in the course will include:

- Examples of thin domains, the limit equations (or reduced equations) and the influence of the shape of the domain on the limit equations.
- For parabolic equations and damped wave equations, upper and lower semicontinuity of attractors. More precise comparison of the dynamics in some particular case (Morse-Smale systems; case where the limit equation is given on an one-dimensional domain). Use of the Conley index theory.
- Convergence results if the reduced domain is one-dimensional (i.e., the ω- limit set of an orbit is a singleton, even without assumptions of hyperbolicity).
- Effect of domain shape on dynamics and bifurcation problems.
- Navier-Stokes equations on thin bounded three-dimensional domains (global existence of strong solutions for data which are bigger than those classically expected). Euler equation on thin three-dimensional domains.
- Generalizations to thin L-shaped or T-shaped domains.

References

1) Hale J.K., Asymptotic behavior of dissipative systems, Math. Surveys and Monographs 25, Am. Math. Soc., Providence, USA 1988.
2) Henry D., Geometric theory of semilinear parabolic equations, Lecture Notes in Math., Vol. 840, Springer-Verlag, Berlin 1981.
3) Temam R., Infinite dimensional dynamical systems in mechanics and physics, Springer-Verlag, Berlin 1988.

FONDAZIONE C.I.M.E.
CENTRO INTERNAZIONALE MATEMATICO ESTIVO
INTERNATIONAL MATHEMATICAL SUMMER CENTER

"Transcendental Methods in Algebraic Geometry"

is the subject of the Third 1994 C.I.M.E. Session.

The Session, sponsored by the Consiglio Nazionale delle Ricerche and the Ministero dell'Università e della Ricerca Scientifica e Tecnologica, will take place under the scientific direction of Prof. FABRIZIO CATANESE (Università di Pisa) and Prof. CIRO CILIBERTO (Università di Roma "Tor Vergata") at Grand Hotel San Michele, Cetraro (Cosenza), **from 4 to 12 July, 1994.**

Courses

a) **L^2 Vanishing Theorems for Positive Line Bundles and Adjunction Theory.** (8 lectures in English)
Prof. Jean-Pierre DEMAILLY (Institut Fourier, Université de Grenoble I)

Outline

1) Connections on holomorphic line bundles, curvature and positivity
2) Elliptic operators on compact manifolds. Hodge Theory
3) Hörmander's L^2 estimates for the $\bar{\partial}$-operators on complete Kähler manifolds
4) Basic vanishing theorems for ample or numerically effective line bundles
5) Positive currents, Lelong numbers and intersection theory
6) Construction of plurisubharmonic potentials through complex Monge-Ampère equations
7) Towards Fujita's conjecture: a numerical criterion for very ample line bundles
8) An effective version of the big Matsusaka theorem (after Y. T. Siu)

References

- A. Andreotti and E. Vesentini, Carleman estimates for the Laplace-Beltrami equation in complex manifolds, Publ. Math. I.H.E.S., 25 (1965), 81-130.
- T. Aubin, Equations du type Monge-Ampère sur les variétés kahlériennes compactes, C.R. Acad. Sci. Paris Ser. A. 285 (1976), 119-121; Bull. Sci. Math., 102 (1976), 63-95
- M. Beltrametti, P. Francia and A.J. Sommese, On Reider's method and higher order embeddings, Duke Math. J., 52 (1989), 425-439.
- M. Beltrametti and A.J. Sommese, On K-jet ample line bundles and Bogomokov's inequality, Manuscript in preparation, (February 1990).
- E. Bombieri, Algebraic values of meromorphic maps, Invent. Math., 10 (1970), 267-287 and Addendum, Invent. Math. 11 (1970), 163-166.
- E. Bombieri, Canonical models of surfaces of general type, Publ. Math. I.H.E.S., 42 (1973), 171-219.
- F. Catanese, Footnotes to a theorem of I. Reider, Proc. Intern. Conf. on Algebraic Geometry (L'Aquila, June 1988), Lecture Notes in Math., Vol. 1417, Spring - Verlag, Berlin, 1990, 67-74.
- J.P. Demailly, Nombres de Lelong généralisés, théorème d'intégralité et d'analyticité, Acta Math., 159 (1987), 153-169.
- J.-P. Demailly, Une généralisation du théorème d'annulation de Kawamata-Viehweg, C.R. Acad. Sci. Paris Sér. I Math., 309 (1989), 123-126.
- J.-P. Demailly, Singular hermitian matrics on positive line bundles, Proc. Conf. Complex algebraic varieties (Bayreuth, April 2-6, 1990), edited by K. Hulek,
- T. Peternell, M. Schneider, F. Schreyer, Lecture Notes in Math., Vol. 1507, Springer Verlag, Berlin, 1992.
- J.-P. Demailly, Regularization of closed positive currents and Intersection Theory, J. Alg. Geom., 1 (1992), 361-409.
- J.-P. Demailly, Monge-Ampère operators, Lelong numbers and intersection theory, Complex Analysis and Geometry, Univ. Series in Math., edited by V. Ancona and A. Silva, Plenum Press, New York, 1993.
- J.-P. Demailly, A. numerical criterion for very ample line bundles, J. Differential Geom., 37 (1993); 323-374.
- L. Ein and R. Lazarsfeld, Seshadri constants on smooth surfaces, Preprint 1992, to appear in Ann. Ec. Norm. Sup.
- L. Ein and R. Lazarsfeld, Global generation of pluricanonical and adjoint linear series on smooth projective threefolds, Preprint 1992.
- T. Fujita, Semipositive line bundles, J. Fac. Sci. Univ. of Tokyo, 30 (1983), 353-378.
- T. Fujita, On polarised manifolds whose adjoint bundles are not semipositive, Algebraic Geometry, Sendai, 1985, Adv. Stud. in Pure Math., vol. 10, North-Holland, Amsterdam, 1987, 167-178.
- T. Fujita, Problem list, Conference held at the Taniguchi Foundation, Kataia, Japan, August 1988.
- T. Fujita, Remarks on Ein-Lazarsfeld criterion of spannedness of adjoint bundles of polarized threefolds, Preprint 1993.
- R. Hartshorne, Ample vector bundles, Publ. Math. I.H.E.S., 29 (1966), 319-350.
- R. Hartshorne, Ample subvarieties of algebraic varieties, Lecture Notes in Math., Vol. 156, Springer-Verlag, Berlin, 1970.
- L. Hörmander, An introduction to Complex Analysis in several variables, 1966, 3rd edition, North-Holland Math. Libr., Vol. 7, Amsterdam, 1991.

- A.G. Hovanski, Geometry of convex bodies and algebraic geometry, Uspehi Mat. Nauk, 34 (4) (1979), 160-161.
- Y. Kawamata, A generalization of Kodaira-Ramanujan's vanishing theorem, Math. Ann., 261 (1982), 3-46.
- J. Kollar, Effective base point freeness, Preprint 1992.
- J. Kollar and T. Matsusaka, Riemann-Roch type inequalities, Amer. J. of Math., 105 (1983), 229-252.
- P. Lelong, Intégration sur un ensemble analytique complexe, Bull. Soc. Math. France, 85 (1957), 239-262.
- P. Lelong, Plurisubharmonic functions and positive differential forms, Gordon and Breach, New York, and Dunod, Paris, 1969.
- T. Matsusaka, Polarized varieties with a given Hilbert polynomial, Amer. J. of Math., 94 (1972), 1027-1077.
- A.M. Nadel, Multiplier ideal sheaves and Kiblar-Einstein metrics of positive scalar curvature, Proc. Nat. Acad. Sci. USA, 86 (1989), 7299-7300 and Annals of Math., 132 (1990), 549-596.
- T. Ohsawa, On the extension of L^2 holomorphic functions, II, Publ. RIMS, Kyoto Univ., 24 (1988), 265-275.
- Y.T. Siu, Analyticity of sets associated to Lelong numbers and the extension of closed positive currents, Invent. Math., 27 (1974), 53-156.
- Y.T. Siu, An effective version of the Matsusaka big theorem, Preprint 1993, to appear in Ann. Inst. Fourier
- H. Skoda, Sous ensembles analytiques d'ordre fini ou infini dans C^n, Bull. Soc. Math. France, 100 (1972), 353-408.
- H. Skoda, Applications des techniques L^2 à la théorie des idéaux d'une algèbre de fonctions holomorphes avec poids, Ann. Scient. Ec. Norm. Sup. 4e Série, 5 (1972), 545-579.
- H. Skoda, Estimation L^2 pour l'opérateur $\bar\partial$ applications arithmétiques, Séminaire P. Lelong (Analyse), années 1975/76, Lecture Notes in Math., Vol. 538, Springer-Verlag, Berlin, 1977, 314-323.
- H. Skoda, Prolongements des courants positifs fermés de masse finie, Invent. Math., 66 (1982), 361-
- B. Teissier, Du théorème de l'index de Hodge aux inégalités isopérimetriques, C.R. Acad. Sc. Paris, Sér, A, 288 (29 Jan. 1979), 287-289.
- B. Teissier, Bonnesen-type inequalities in algebraic geometry, Sem. on Diff. Geom. edited by S.T. Yau,1982, Princeton Univ. Press, 85-105.
- E. Viehweg, Vanishing Theorems, J. Reine Angew. Math., 335 (1982), 1-8.
- S.T. Yau, On the Ricci curvature of a complex Kähler manifold and the complex Monge-Ampère equation I, Comm. Pure and Appl. Math., 31 (1978), 339-411.

b) **Complex Varieties With Semi Positive Curvature.** (8 lectures in English)
Prof. Thomas PETERNELL (Universität Bayreuth)

Outline

The aim of the course is to study complex algebraic manifolds and compact Kähler manifolds via their tangent and (anti)-canonical bundles both from an algebraic and a transcendental point of view. More specifically the following topics will be discussed.

- Mori's theorem on ample tangent bundles (solutions of the Hartshorne-Frankel conjecture) with a proof based on the base point free theorem; relation to the methods of Siu-Yau.
- Uniformisation of compact Kähler manifolds with semi-positive bisectional curvature: Mok's theorem.
- Manifolds with nef tangent bundles; structure theorems (after Demailly-Peternell-Schneider), analytic versus algebraic methods.
- Manifolds with semipositive Ricci curvature and with nef anticanonical class: Albanese map, fundamental group, structure theorems.

Reference

1) S. Mori: Projective manifolds with ample tangent bundles. Ann. Math. 110, 593-606 (1979).
2) N. Mok: The uniformisation theorem for compact Kähler manifolds of non negative bisectional curvature. J. Diff. Geom. 27, 179-214 (1988)
3) J.P. Demailly, T. Peternell, M. Schneider: Compact complex manifolds with numerically effective Ricci class. To appear in J. Alg. Geometry (1993/94); Kähler manifolds with numerically effective Ricci class. To appear in Comp. Math. (1993/94).
4) Y.T. Siu, S.Y. Yau: Compact Kähler manifolds with positive bisectional curvature. Inv. Math. 59, 189-204, (1980).

c) **Kähler Metrics on Algebraic Manifolds.** (8 lectures in English)
Prof. Gang TIAN (Courant Institute, NY)

Outline

The lectures will focus on special Kähler metrics on algebraic manifolds and their applications. The following topics will be discussed.

- The connection between Donaldson functionals and Mumford stability of underlying manifolds.
- Existence and uniqueness of Kähler-Einstein metrics. The lower boundedness of Donaldson functionals on Kähler-Einstein manifolds.

. Chern number inequalities and uniformization theorems.

. Kähler-Einstein metrics on quasi-projective manifolds and singular varieties. Their applications to problems in algebraic geometry, for instance, stability of tangent bundles, moduli problems.

. Calabi-Yau manifolds and special Kähler geometry.

Prerequisites: 1. Some familiarity with complex manifolds and vector bundles; 2. Elementary differential geometry.

References

1) A.L. Besse, Einstein manifolds (Chapter 11), Ergebnisse der Math. und ihrer Grenzgebiete, 3. Folge, band 10, Springer Verlag.
2) S.T. Yau, Calabi's conjecture and some new results in algebraic geometry, Proc. Nat. Acad. Sci. USA 74 (1979).
3) Kähler metrics and moduli spaces, Advanced studies in pure math., vol. 18; part. II, edited by T. Ochiai, Academic Press.
4) G. Tian, Kähler-Einstein metrics on algebraic manifolds, Proc. of ICM, Kyoto, 1990.
5) G. Tian, The K-energy on hypersurfaces and stability, to appear in Comm. in Geom. and Analysis.
6) A. Strominger, Special geometry, Comm. Math. Phys., vol. 133 (1990).

d) **Smooth structure of algebraic surfaces. (6 lectures in English)**
Prof. Andrej N. TYURIN (Steklov Institute of Mathematics)

Lecture 1. Introduction and survey of the results.
P-r. (Pre-requisites): The topological classification of 4-manifolds and algebraic surfaces, obtained by conventional methods.
Ref.: 1. S. Donaldson, P. Kronheimer. The Geometry of Four-Manifolds, Clarendon Press, Oxford, 1990, Ch. 1.
2. W. Barth, C. Peters and Van de Ven. Compact complex surfaces, Springer Verlag, 1984, Ch. 1.

Lecture 2. Reducibility of standard representation of Diffs.
P-r.: Lefschetz-theory, monodromy representations.
Ref.: 1. P. Griffiths, J. Harris. Principles of algebraic geometry, v. 2.
2. W. Ebeling, C. Okonek. Donaldson invariants, monodromy groups and singularities. Preprint, Eindhoven, 1990.

Lecture 3. Jacobians, theta-loci and polynomials.
P-r.: Gieseker closure, deformation to the normal cones.
Ref.: 1. D. Gieseker. On the moduli of vector bundles on an algebraic surface. Ann. of Math., 106 (1977), 45-60.
2. W. Fulton. Intersection theory, Springer Verlag, Oxford, 1987.

Lecture 4. Relation to Sarkisov program.
P-r.: Newstead construction, Bertram - Thaddeus construction.
Ref.: 1. A. Bertram. Moduli of rank 2 vector bundles, theta divisors and the geometry of curves in projective space, J. Diff. Geom., 35 (1992), 429-469.
2. M. Thaddeus. Stable pairs, linear system and the Verlinde formula. Preprint, 1992.
3. A. Tyurin. The Spin polynomial invariants of the smooth structures of algebraic surfaces. Mathematica Gottingensis, Sonderforschungsbereich Geometry and Analysis, Heft 6, (1933), 1-48.

Lecture 5. GA-procedure.
P-r.: Elementary transformations.
Ref.: 1. S. Langton. Valuative criteria for families of vector bundles on algebraic varieties, Ann. of Math., 101 (1975), 88-110.
2. A. Tyurin. The Spin polynomial invariants of the smooth structures of algebraic surfaces. Mathematica Gottingensis, Sonderforschungsbereich Geometry and Analysis, Heft 6, (1933), 1-48.

Lecture 6. Riemann geometry. Smooth invariance of canonical class.
P-r.: Yang-Mills moduli spaces, the polynomial invariants.
Ref.: 1. S. Donaldson, P. Kronheimer. The Geometry of Four-Manifolds. Clarendon Press, Oxford, 1990, Ch. 4,5,6,9.
2. A. Tyurin. The Spin polynomial invariants of the smooth structures of algebraic surfacces. Mathematica Gottingensis, Sonderforschungsbereich Geometry and Analysis, Heft 6 (1933), 1-48.

LIST OF C.I.M.E. SEMINARS

		Publisher
1954 -	1. Analisi funzionale	C.I.M.E.
	2. Quadratura delle superficie e questioni connesse	"
	3. Equazioni differenziali non lineari	"
1955 -	4. Teorema di Riemann-Roch e questioni connesse	"
	5. Teoria dei numeri	"
	6. Topologia	"
	7. Teorie non linearizzate in elasticità, idrodinamica,aerodinamica	"
	8. Geometria proiettivo-differenziale	"
1956 -	9. Equazioni alle derivate parziali a caratteristiche reali	"
	10. Propagazione delle onde elettromagnetiche	"
	11. Teoria della funzioni di più variabili complesse e delle funzioni automorfe	"
1957 -	12. Geometria aritmetica e algebrica (2 vol.)	"
	13. Integrali singolari e questioni connesse	"
	14. Teoria della turbolenza (2 vol.)	"
1958 -	15. Vedute e problemi attuali in relatività generale	"
	16. Problemi di geometria differenziale in grande	"
	17. Il principio di minimo e le sue applicazioni alle equazioni funzionali	"
1959 -	18. Induzione e statistica	"
	19. Teoria algebrica dei meccanismi automatici (2 vol.)	"
	20. Gruppi, anelli di Lie e teoria della coomologia	"
1960 -	21. Sistemi dinamici e teoremi ergodici	"
	22. Forme differenziali e loro integrali	"
1961 -	23. Geometria del calcolo delle variazioni (2 vol.)	"
	24. Teoria delle distribuzioni	"
	25. Onde superficiali	"
1962 -	26. Topologia differenziale	"
	27. Autovalori e autosoluzioni	"
	28. Magnetofluidodinamica	"

1963 - 29. Equazioni differenziali astratte "
30. Funzioni e varietà complesse "
31. Proprietà di media e teoremi di confronto in Fisica Matematica "

1964 - 32. Relatività generale "
33. Dinamica dei gas rarefatti "
34. Alcune questioni di analisi numerica "
35. Equazioni differenziali non lineari "

1965 - 36. Non-linear continuum theories "
37. Some aspects of ring theory "
38. Mathematical optimization in economics "

1966 - 39. Calculus of variations Ed. Cremonese, Firenze
40. Economia matematica "
41. Classi caratteristiche e questioni connesse "
42. Some aspects of diffusion theory "

1967 - 43. Modern questions of celestial mechanics "
44. Numerical analysis of partial differential equations "
45. Geometry of homogeneous bounded domains "

1968 - 46. Controllability and observability "
47. Pseudo-differential operators "
48. Aspects of mathematical logic "

1969 - 49. Potential theory "
50. Non-linear continuum theories in mechanics and physics and their applications "
51. Questions of algebraic varieties "

1970 - 52. Relativistic fluid dynamics "
53. Theory of group representations and Fourier analysis "
54. Functional equations and inequalities "
55. Problems in non-linear analysis "

1971 - 56. Stereodynamics "
57. Constructive aspects of functional analysis (2 vol.) "
58. Categories and commutative algebra "

1972 -	59. Non-linear mechanics		"
	60. Finite geometric structures and their applications		"
	61. Geometric measure theory and minimal surfaces		"
1973 -	62. Complex analysis		"
	63. New variational techniques in mathematical physics		"
	64. Spectral analysis		"
1974 -	65. Stability problems		"
	66. Singularities of analytic spaces		"
	67. Eigenvalues of non linear problems		"
1975 -	68. Theoretical computer sciences		"
	69. Model theory and applications		"
	70. Differential operators and manifolds		"
1976 -	71. Statistical Mechanics		Ed Liguori, Napoli
	72. Hyperbolicity		"
	73. Differential topology		"
1977 -	74. Materials with memory		"
	75. Pseudodifferential operators with applications		"
	76. Algebraic surfaces		"
1978 -	77. Stochastic differential equations		"
	78. Dynamical systems		Ed Liguori, Napoli and Birhäuser Verlag
1979 -	79. Recursion theory and computational complexity		"
	80. Mathematics of biology		"
1980 -	81. Wave propagation		"
	82. Harmonic analysis and group representations		"
	83. Matroid theory and its applications		"
1981 -	84. Kinetic Theories and the Boltzmann Equation	(LNM 1048)	Springer-Verlag
	85. Algebraic Threefolds	(LNM 947)	"
	86. Nonlinear Filtering and Stochastic Control	(LNM 972)	"
1982 -	87. Invariant Theory	(LNM 996)	"
	88. Thermodynamics and Constitutive Equations	(LN Physics 228)	"
	89. Fluid Dynamics	(LNM 1047)	"

1983 - 90. Complete Intersections (LNM 1092) Springer-Verlag
91. Bifurcation Theory and Applications (LNM 1057) "
92. Numerical Methods in Fluid Dynamics (LNM 1127) "

1984 - 93. Harmonic Mappings and Minimal Immersions (LNM 1161) "
94. Schrödinger Operators (LNM 1159) "
95. Buildings and the Geometry of Diagrams (LNM 1181) "

1985 - 96. Probability and Analysis (LNM 1206) "
97. Some Problems in Nonlinear Diffusion (LNM 1224) "
98. Theory of Moduli (LNM 1337) "

1986 - 99. Inverse Problems (LNM 1225) "
100. Mathematical Economics (LNM 1330) "
101. Combinatorial Optimization (LNM 1403) "

1987 - 102. Relativistic Fluid Dynamics (LNM 1385) "
103. Topics in Calculus of Variations (LNM 1365) "

1988 - 104. Logic and Computer Science (LNM 1429) "
105. Global Geometry and Mathematical Physics (LNM 1451) "

1989 - 106. Methods of nonconvex analysis (LNM 1446) "
107. Microlocal Analysis and Applications (LNM 1495) "

1990 - 108. Geoemtric Topology: Recent Developments (LNM 1504) "
109. H_∞ Control Theory (LNM 1496) "
110. Mathematical Modelling of Industrial Processes (LNM 1521) "

1991 - 111. Topological Methods for Ordinary Differential Equations (LNM 1537) "
112. Arithmetic Algebraic Geometry (LNM 1553) "
113. Transition to Chaos in Classical and Quantum Mechanics (LNM 1589) "

1992 - 114. Dirichlet Forms (LNM 1563) "
115. D-Modules, Representation Theory, and Quantum Groups (LNM 1565) "
116. Nonequilibrium Problems in Many-Particle Systems (LNM 1551) "

1993 - 117.	Integrable Systems and Quantum Groups	to appear	Springer-Verlag
118.	Algebraic Cycles and Hodge Theory	(LNM 1594)	
119.	Phase Transitions and Hysteresis	(LNM 1584)	"
1994 - 120.	Recent Mathematical Methods in Nonlinear Wave Propagation	to appear	"
121.	Dynamical Systems	to appear	"
122.	Transcendental Methods in Algebraic Geometry	to appear	"

9 783540 586920